# Medical Statistics for Cancer Studies

# Chapman & Hall/CRC Biostatistics Series

*Series Editors*
**Shein-Chung Chow,** Duke University School of Medicine, USA
**Byron Jones,** Novartis Pharma AG, Switzerland
**Jen-pei Liu**, National Taiwan University, Taiwan
**Karl E. Peace**, Georgia Southern University, USA
**Bruce W. Turnbull**, Cornell University, USA

*Recently Published Titles*

**Methodologies in Biosimilar Product Development**
*Sang Joon Lee, Shein-Chung Chow (eds.)*

**Statistical Design and Analysis of Clinical Trials**
Principles and Methods, Second Edition
*Weichung Joe Shih, Joseph Aisner*

**Confidence Intervals for Discrete Data in Clinical Research**
*Vivek Pradhan, Ashis Gangopadhyay, Sandeep Menon, Cynthia Basu, and Tathagata Banerjee*

**Statistical Thinking in Clinical Trials**
*Michael A. Proschan*

**Simultaneous Global New Drug Development**
Multi-Regional Clinical Trials after ICH E17
*Edited by Gang Li, Bruce Binkowitz, William Wang, Hui Quan, and Josh Chen*

**Quantitative Methodologies and Process for Safety Monitoring and Ongoing Benefit Risk Evaluation**
*Edited by William Wang, Melvin Munsaka, James Buchanan and Judy Li*

**Statistical Methods for Mediation, Confounding and Moderation Analysis Using R and SAS**
*Qingzhao Yu and Bin Li*

**Hybrid Frequentist/Bayesian Power and Bayesian Power in Planning Clinical Trials**
*Andrew P. Grieve*

**Advanced Statistics in Regulatory Critical Clinical Initiatives**
*Edited by Wei Zhang, Fangrong Yan, Feng Chen, Shein-Chung Chow*

**Medical Statistics for Cancer Studies**
*Trevor F. Cox*

For more information about this series, please visit: https://www.routledge.com/
Chapman--Hall-CRC-Biostatistics-Series/book-series/CHBIOSTATIS

# Medical Statistics for Cancer Studies

Trevor F. Cox

## CRC Press
Taylor & Francis Group
Boca Raton London New York

CRC Press is an imprint of the
Taylor & Francis Group, an **informa** business
A CHAPMAN & HALL BOOK

First edition published 2022
by CRC Press
6000 Broken Sound Parkway NW, Suite 300, Boca Raton, FL 33487-2742

and by CRC Press
4 Park Square, Milton Park, Abingdon, Oxon, OX14 4RN

**Library of Congress Cataloging-in-Publication Data**

Names: Cox, Trevor F., author.
Title: Medical statistics for cancer studies / authored by Trevor F. Cox.
Description: First edition. | Boca Raton : Chapman & Hall, CRC Press, 2022.
| Series: Chapman & hall/crc biostatistics series | Includes
bibliographical references and index.
Identifiers: LCCN 2021061413 (print) | LCCN 2021061414 (ebook) | ISBN
9780367486150 (hardback) | ISBN 9781032285870 (paperback) | ISBN
9781003041931 (ebook)
Subjects: LCSH: Cancer--Research--Statistical methods. | Medical
statistics.
Classification: LCC RC267 .C69 2022 (print) | LCC RC267 (ebook) | DDC
616.99/400727--dc23/eng/20211222
LC record available at https://lccn.loc.gov/2021061413
LC ebook record available at https://lccn.loc.gov/2021061414

ISBN: 978-0-367-48615-0 (hbk)
ISBN: 978-1-032-28587-0 (pbk)
ISBN: 978-1-003-04193-1 (ebk)

DOI: 10.1201/9781003041931

Typeset in CMR10
by KnowledgeWorks Global Ltd.

*Publisher's note:* This book has been prepared from camera-ready copy provided by the authors.

Access the Support Material: www.Routledge.com/9780367486150

*To my sister Jean, who has spent a lifetime caring for others, my wife Jill, for her love, support and patience, and my grandchildren Matthew, Eve, Emily and Avery, may they be healthy, wealthy and wise.*

# Contents

# *Preface*

I worked for seven years at the Liverpool Cancer Trials Unit (LCTU), based at the University of Liverpool. This was in complete contrast to my previous employments at Newcastle University, teaching and researching statistical theory, and at Unilever's Research and Development Laboratory at Port Sunlight. At the LCTU, I learnt about cancer, working with cancer surgeons, oncologists, cancer research scientists, palliative care medics, medical statisticians, cancer bioinformaticians and all the clinical trial staff of the LCTU. This text has arisen from those enjoyable and enlightening seven years. The only downside was the fact that we were dealing with a nasty disease, but on the positive side, due to all the cancer researchers in the world, cancer diagnosing and treatments are continually improving.

This text is aimed at final year undergraduate and postgraduate students studying medical statistics. It will also be accessible to others who have some experience in statistics. In the text, I have tried to explain in detail some of the statistical theory and how the algebra associated with it works. But, much of the algebra can be passed over without detracting from the main results. The Appendix, Statistics Revision, might be useful for reminding the reader of various aspects of statistical theory, but it is not detailed enough to act as a text for learning the theory. The only section that covers theory in detail is the one on Markov Chain Monte Carlo (MCMC) Bayesian statistical inference, as this is less well known than the usual frequentist inference.

As the title says, this text is about medical statistics used in cancer studies, but the statistical techniques described can be used in any area of medicine, as appropriate. In fact, you could argue that "medical statistics" is not a branch of "statistics" at all. It is just statistics applied to medical data! The difference here from texts on medical statistics in general is that all the examples are about cancer, and the statistical techniques described are those heavily used in cancer studies.

Cancer studies are of a diverse type: there are clinical trials, meta-analyses, biomarker studies, laboratory studies, genetic studies and epidemiological studies. This text covers all of these, hopefully giving an oversight of cancer studies in general. If each was treated with a lot of depth, you would probably be reading a text of 1,000 pages.

The text is written not only to describe statistical theory, but to help the reader understand cancer, what causes it and how it is measured in the patient and in the population as a whole. Some basic cell biology and genetics are covered in one chapter, but note, this has been written from a statistician's understanding of it and is intended for those, like myself, who only studied biology to a basic level. I am sure many readers can pass this over. Also of a descriptive nature, the setting up of a clinical trial and the statistician's role in such, is described in two chapters.

The main statistical technique covered is *survival analysis*, as that is the main one used in cancer clinical trials and many other cancer studies. About 80 pages are devoted to it. The other statistical techniques cover those used in: epidemiology (*cohort studies* and *case-control studies*), meta-analyses (*fixed effects, random effects and Bayesian models, network meta-analysis models*), biomarker studies (*ROC curve theory* and *survival analysis*) and cancer informatics (*multivariate analysis* and *survival analysis*)

Lastly, I would like to thank my colleagues from the LCTU and the Department of Molecular and Clinical Cancer Medicine for making this possible, in particular, Dr. Sarah Berhane, Dr. Seema Chuahan, Prof. Paula Ghaneh, Prof. Bill Greenhalf, Dr. Richard Jackson, Prof. Philip Johnson, Prof. John Neoptolemos, Prof. Daniel Palmer, Prof. Andrew Pettitt, Ms. Eftychia Psarelli, Ms. Charlotte Rawcliffe and Dr. Paul Silcocks. I would like to thank the LCTU for permission to use ESPAC clinical trial data for the illustration of statistical techniques.

# 1

## *Introduction*

Cancer is a dreaded disease. One in two people will be diagnosed with cancer within their lifetime. There are over 200 different types of cancer, however, and they are often grouped together under one name. For example, kidney cancer, otherwise known as renal cell cancer, can be of three main types: clear cell, papillary and chromophobe, but there are also some rare types of renal cell cancers. After grouping some of the cancers together, there are probably about 100 cancers.

Table 1.1 shows the twenty cancers with the most recorded cases per year averaged over the years 2016–2018 in the UK, as well as the twenty cancers with the most recorded deaths, for the year 2018. The data were taken from the world's largest independent cancer research charity, Cancer Research UK (CRUK) (website: https://www.cancerresearchuk.org/health-professional/cancer-statistics-for-the-uk, and other pages, accessed November, 2021).

The total number of cases recorded for all cancers was 375,400, averaged over the years 2016–2018 and the total number of deaths recorded for all cancers was 166,533, averaged over the years 2015–2017. Two more statistics from the CRUK website are that 50% of patients survive cancer for ten years or more and 38% of cancer cases are preventable. Other data shown in Table 1.1, are the risk factors associated with the various cancers, indicated by "1" for an association, however large or small, and a blank as no evidence of a risk. A lot of risk factors are not shown, for example "being inactive" produces a small increase in risk of breast cancer, and many risk factors will only increase cancer risk slightly. The "−1" for the contraceptive pill shows it lowers the risk.

We see that breast and prostate cancers lead the table with the most cases per year, but lung and bowel cancer have the most deaths per year. Tobacco, obesity and certain occupations are risk factors for several cancers.

This textbook shows how cancer data can be analysed in a variety of ways, covering cancer clinical trial data, epidemiological data, biological data and genetic data.

DOI: 10.1201/9781003041931-1

## 1.1   About cancer

Where does the word "cancer" come from? Cancer has been with mankind, probably forever. Fossilised bone tumours have been found in Egyptian mummies, and there is a description of cancer from around 3000 BC. The Greek physician, Hippocrates (460–375 BC), named the disease as "carcinos", the Greek word for crab, as he thought tumours resembled a crab, the legs possibly representing swollen blood vessels. Celsus, a Roman physician (28–50 BC) changed the word to "cancer". The Greek physician Galen described tumours as "oncos", the Greek word for swelling and hence we have the word "oncology", the study of tumours and cancer in general.

There are four main types of cancer:

- *Carcinomas* form in *epithelial tissue*, which is skin tissue and the tissue that lines organs, such as the liver, lung, kidney and pancreas. When there is abnormal cell division, a *neoplasm* forms (a new growth), which might be an ulcer say, or it may be a tumour – a carcinoma. Carcinomas account for about 85% of cancers. They are common because, epithelial cells divide often and the epithelium is where the body meets the outside world. It gets bombarded with food, smoke, alcohol, UV radiation, etc. An *adenocarcinoma* originates in an organ or gland.

- *Sarcomas* form in connective tissue, such as bone, muscles, fat, tendons and cartilage. There is not a direct connection to the outside world. Sarcomas probably make up about 1% of cancers.

- *Leukaemias* are cancers of the blood. The four main types are *chronic myeloid leukaemia* (CML), *acute myeloid leukaemia* (AML), *chronic lymphocytic leukaemia* (CLL) and *acute lymphocytic leukaemia* (CLL). Leukaemias are not solid tumours.

- *Lymphomas* occur in the lymphatic system (the system that removes waste products from cells and produces white blood cells and other immune cells). There are two main types: Hodgkin lymphoma and non-Hodgkin lymphoma.

Most of the data we encounter in the upcoming chapters relate to carcinomas.

What actually causes cancer? My answer is, "Cancer is caused when the biology of a cell "goes wrong". Chapter 2 covers some basic cell biology and genetics, intended for those readers, like myself, who did not study biology very much. Also included in Chapter 2 is a section on how cancer is measured within a patient.

The main treatments for cancer are surgery, chemotherapy and radiotherapy. Surgery – cut it out; chemotherapy – hit it with drugs, radiotherapy – hit it with X-rays or electron beams. These are described in Section 2.4.

## 1.2   Cancer studies

Cancers are studied using cancer clinical trials and other studies, in laboratories and using cancer databases. We outline some of the various types of study.

**Clinical trials**

Many cancer clinical trials are run throughout the world, under strict regulations. The trials are usually split into four groups: Phase I trials check on the safety of a new cancer treatment, using only a few participants; Phase II trials look at safety, efficacy and checks dose levels of a new treatment, using up to about 100 patients; Phase III trials confirm whether or not one treatment is superior to another, and can use hundreds of patients; Phase IV trials check the side effects and efficacy of a new treatment over the long term. Chapter 4 describes the setting up and the running of a clinical trial, and Chapter 6 describes the statistician's role in clinical trials.

We itemise a few points about cancer clinical trials.

- Cancer trials are generally run in cancer trials units or centres and hospitals. Cancer treatments are tested and so an intervention is made to the patient, e.g. the patient is given a new chemotherapy drug. Note, a state of *equipoise* is needed prior to a trial being planned or run. This means that with current knowledge, it is not known whether the proposed new treatment is any better or worse for patients that the current treatments available. Trials are tightly regulated.

- Patients are recruited to *arms* of the trial, with one arm per treatment being used in the trial, e.g. a two-arm trial might test a new drug against a placebo. Patients are *randomised* to the treatments. A large trial may use several, or many *centres* or *sites* (usually hospitals) to recruit the required number of patients. Recruitment occurs over an interval of time (*recruitment period*), and patients are followed up for a set time after treatment stops (*follow-up time*).

- Randomisation of patients to the arms of a trial often involves *stratification*, so that there is equal allocation of subgroups of patients to the arms. For instance, it may be desirable to have a balance of males and females across the arms. Or a balance of the number of patients with small size tumours to the number with large size tumours.

- To help reduce bias, if possible, patients and clinicians are *blinded* to which treatment a patient is receiving, giving a *double-blinded* trial. If only one of the clinician or patient can be blinded to the treatment, that is a *single-blinded* trial. If neither can be blinded, it is an *open label* trial.

- All sorts of data are collected from patients during the course of a trial: measurements made on blood samples, measurements on tumours and, as it is cancer, time to death or time to when the cancer progresses. Side effects of the treatments are also recorded. Section 2.5 discusses these measurements. Not every measurement is used in statistical analyses. The important ones that are analysed are called *endpoints*, there usually being one main endpoint, the *primary endpoint*, and the others being *secondary endpoints*.

- Often, samples of tumour and blood are collected from the patient and stored in freezers for future research.

- The most common statistical analysis for cancer trial data is *survival analysis*. This and *survival regression analysis* are covered in Chapter 3 and Chapter 5.

- Results of clinical trials are usually published in scientific journals and presented at cancer scientific meetings.

As an example, the results of the ESPAC3 trial were published in the Journal of the American Medical Association, entitled,

"Adjuvant chemotherapy with Fluorouracil plus Folinic Acid vs Gemcitabine following pancreatic cancer resection: a randomized controlled trial".

This was an open-label, randomised controlled trial, conducted in 159 pancreatic cancer centres in Europe, Australasia, Japan and Canada, testing two chemotherapy treatments, Fluorouracil plus Folinic Acid (5FU) against Gemcitabine (GEM). There were 1008 patients, who had undergone cancer *resection*. (surgery to remove the tumour), randomised to the two arms. Recruitment was over seven years and the follow-up time was two years.

**Systematic reviews and meta-analyses**
For a particular cancer, there might be several clinical trials or studies of one particular treatment against another particular treatment, for example, several like the pancreatic study mentioned above. A systematic review does how it reads – it systematically reviews the set of trials and draws overall conclusions. In the process of doing this, meta-analyses are carried out which statistically combine the data from the individual studies. This is covered in Chapter 8.

**Cancer epidemiology**
We started with some cancer statistics for the UK, part of *cancer epidemiology*. Epidemiology studies the distribution of diseases in populations, the factors which cause or are associated with them, and how they change in the population over time. Chapter 7 gives an introduction to cancer epidemiology.

**Laboratory studies**

Much cancer research is carried out in laboratories and is part of *translational medicine*, or *translational research* – "from bench to bedside", where research in the laboratory helps develop new treatments for patients. Biological, chemical and pharmacological research is needed in the process. A promising new treatment then moves into the clinical trial stage. Other laboratory research involves measurements (biomarkers) made using samples of tissue and blood, e.g. amounts of particular proteins, and genetic factors. Biomarker studies look for biomarkers that can help with diagnosis of cancers and a patient's prognosis. Chapter 9 gives some examples. Tailoring treatments to the patient, based on biomarkers, is in the realms of *personalised medicine*.

**Cancer informatics**

Cancer informatics is the science of obtaining cancer data, storing it, analysing it and summarising it for dissemination to the wider community. Often, but not always, the data collected are genetic data, and involves storing and analysing huge datasets. Chapter 10 gives a brief introduction.

## 1.3  R code

The only programming language used for the analyses and graphs in this text is R. R is freely available from https://www.r-project.org/. Some of the R code used can be downloaded from the publisher's website (https://www.routledge.com/Medical-Statistics-for-Cancer-Studies/Cox/p/book/9780-367486150, or https://www.routledge.com/9780367486150), but note, only the functions that do the statistical analyses, not that for manipulation of data prior to analysis. However, for the cancer bioinformatics chapter, whole programs are given that download data from data repositories, prepare it, carry out the analyses and displays the results.

I hope you find this textbook useful and you enjoy reading it as much as I have writing it. In places the mathematics gets rather involved and lengthy, but it can be passed over without detracting from the main themes of the statistical analyses and discussions. The Appendix can be an aide-memoire for statistical techniques. The section on Bayesian inference using Markov Chain Monte Carlo (MCMC) statistical inference is quite substantial, as this is less well understood than classical frequentist statistical inference.

Finally, good luck in your studies or research, and quoting CRUK's slogan, "Together we will beat cancer".

**TABLE 1.1**
Twenty most common cancers in the UK, together with risk factors

| Site | Cases | Rank | Deaths | Rank | Tob | Alc | Obes | UV | Occ | Diet | EBV | H.py | HIV | Hep | Para | Apol | HRT | Pil | Gen |
|---|---|---|---|---|---|---|---|---|---|---|---|---|---|---|---|---|---|---|---|
| Breast | 55,920 | 1 | 11,605 | 4 | | 1 | 1 | | | | | | | | | | 1 | 1 | 1 |
| Prostate | 52,254 | 2 | 11,890 | 3 | | | | | | | | | | | | | | | 1 |
| Lung | 48,549 | 3 | 34,594 | 1 | 1 | | | | | | | | | | | 1 | | | |
| Bowel | 42,885 | 4 | 16,659 | 2 | 1 | 1 | 1 | | | 1 | | | | | | | | | 1 |
| Melanoma Skin Cancer | 16,744 | 5 | 2,353 | 19 | | | | 1 | | | | | | | | | | | |
| Non-Hodgkin Lymphoma | 14,176 | 6 | 4,978 | 11 | | | | | | | 1 | 1 | 1 | | | | | | |
| Kidney | 13,323 | 7 | 4,677 | 13 | 1 | | 1 | | | | | | | | | | | | 1 |
| Head and Neck | 12,423 | 8 | 4,078 | 16 | 1 | 1 | | | | | 1 | | 1 | | | | | | |
| Brain | 12,287 | 9 | 5,545 | 9 | | | 1 | | | | | | | | | | | | 1 |
| Pancreas | 10,452 | 10 | 9,573 | 6 | 1 | | 1 | | | 1 | | | | | | | | | 1 |
| Bladder | 10,292 | 11 | 5,457 | 10 | 1 | | | | 1 | | | | | | 1 | | | | |
| Leukaemia | 9,907 | 12 | 4,693 | 12 | 1 | | | | 1 | | | | | | | | | | |
| Uterus | 9,703 | 13 | 2,496 | 18 | | | 1 | | | | | | | | | | 1 | -1 | 1 |
| Oesophagus | 9,272 | 14 | 8,061 | 7 | 1 | 1 | 1 | | 1 | | | 1 | | | | | | | |
| Cancer of Unknown Primary | 8,589 | 15 | 9,775 | 5 | 1 | | 1 | | | | | | | | | | | | |
| Ovary | 7,495 | 16 | 4,219 | 15 | 1 | | 1 | | 1 | | | | | | | | 1 | -1 | 1 |
| Stomach | 6,453 | 17 | 4,222 | 14 | 1 | | 1 | | | 1 | | 1 | | | | | | | |
| Liver | 6,214 | 18 | 5,771 | 8 | 1 | 1 | 1 | | 1 | | | | | 1 | | | | | |
| Myeloma | 5,951 | 19 | 3,084 | 17 | | | 1 | | | | | | | | | | | | |
| Thyroid | 3,865 | 20 | 400 | >20 | | | 1 | | | | | | | | | | | | 1 |
| Mesothelioma | 2,718 | >20 | 2,345 | 20 | | | | | 1 | | | | | | | | | | |

*Note:* Tob - tobacco use; Alc - alcohol use; Obes - overweight and obesity; UV - exposure to UV radiation; Occ - occupational exposure to radiation, chemicals,etc.; Diet - red meat, nitrates,etc.; EBV - Epstein-Barr virus; H.py - Helicobacter pylori bacteria; HIV - human immunodeficiency virus; Hep - hepatitis; Para - parasites; Air - air polution; Pil - contraceptive pill; Gen - family history and inherited genes

# 2

## Cancer Biology and Genetics for Non-Biologists

In the first part of this chapter, we cover basic cell biology and genetics which is intended for those readers unfamiliar with the areas. When some of this biology "goes wrong" in our bodies, we can end up with cancer. This is covered in the latter part of the chapter as well as how we can measure a patient's cancer.

## 2.1  Cells

All living things are made up of cells, from the simple unicellular amoeba to the complex human composed of about 37 trillion ($37 \times 10^{12}$) cells. Cells that contain a *nucleus* are called *eukaryotic* cells; cells without a nucleus are called *prokaryotic* cells. Bacteria are examples of prokaryotic cells. Humans are *eukaryotes* consisting of eukaryotic cells, such as bone, nerve and stem cells. In fact, there are about 200 types of cells in our bodies. Figure 2.1 shows a typical eukaryotic cell, illustrating its structure. Cells come in different shapes and sizes; neurons in the brain and nervous system are long and thin, blood cells are roughly spherical, some bone cells are cuboidal and columnar while others have many branches. The size of a red blood cell is 9 μm, the size of a skin cell is 30 μm, an ovum 100 μm, whilst the length of some nerve cells can be over 1m.

Cells take in raw ingredients, process them and output products. Oxygen and nutrients from food have to be inputted, the nutrients are metabolised with the help of oxygen and energy is released to be used by the cell. Each type of cell has a function, so for example, acini cells in a saliva gland produce saliva as output. Cells do not act alone and they combine together to form body tissue. Cells die and are replaced by new cells generated by cell division (*mitosis*). About 60 billion of your cells die each day and luckily about 60 billion new cells are created each day. That's 0.16% of your cells each day. Different types of cells have different lifetimes; red blood cells last about 120 days, pancreas cells about one year, cells in your eye lens last as long as you do. The process of cell death is termed *apoptosis*. The human body is made

DOI: 10.1201/9781003041931-2

up almost entirely of proteins – apart from the water and fat. Proteins are abundant in cells. The set of human proteins is called the *proteome*, which consists of tens of thousands of proteins.

Figure 2.1 shows the structure and component parts of a typical cell.

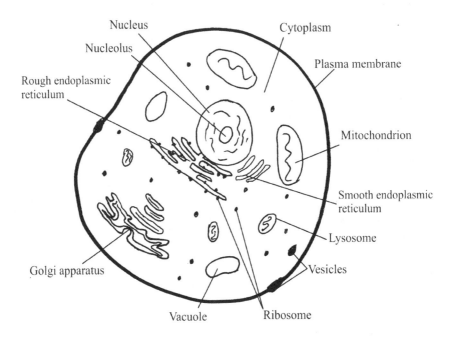

**FIGURE 2.1**
Structure of a typical eukaryotic cell

There is a lot going on in a cell. These are the jobs of the component parts:

**Plasma membrane** The plasma membrane is a complex structure at the boundary of the cell. It is selectively permeable so that specific molecules can enter or leave the cell. Passive transport is by diffusion across the plasma membrane and does not need energy; e.g. oxygen enters the cell, and carbon dioxide leaves the cell. Active transport needs energy where transport proteins help molecules across the membrane; e.g. nutrients, sodium ions leave the cell and potassium ions enter the cell. Receptors are proteins that detect signals from outside the cell.

**Cytoplasm** The cytoplasm is the fluid within the cell where much activity takes place, chemical reactions, protein construction, metabolism, transportation of molecules. The various components within the cell are called **organelles**, and they are all within the cytoplasm except for the nucleus.

**Nucleus** The nucleus holds most of the cell's genetic material. It has its own membrane separating it from the cytoplasm but which allows molecules to pass in and out of the nucleus. The **nucleolus** within the nucleus is where ribosomes are produced and assembled, and then transported into the cytoplasm.

**Ribosome** Ribosomes build chains of amino acids that go on to form proteins. They are in the cytoplasm or bound to the rough endoplasmic reticulum.

**Endoplasmic reticulum** The endoplasmic reticulum is connected to the nucleus and it builds proteins. The rough endoplasmic reticulum takes the amino acid chains of the ribosomes and turns them into proteins. Proteins are then collected in the membrane of the rough endoplasmic reticulum, which pinches around and seals to form a **vesicle**. The vesicle enters the cytoplasm and takes the protein to the Golgi apparatus. The smooth endoplasmic reticulum creates and stores lipids and steroids.

**Golgi apparatus** The Golgi apparatus modifies the protein from the rough endoplasmic reticulum and then forms another vesicle which transports the protein to the plasma membrane, whereupon it is released from the cell.

**Mitochondria** Mitochondria generate energy for the cell. They use nutrients to generate energy in a process called **cellular respiration**.

**Vacuole** Vacuoles are storage areas where nutrients and some waste products are stored.

**Lysosome** Lysosomes contain digestive enzymes that can breakdown molecules. They can breakdown the cell when it dies.

### Cell division
Some cells divide to produce sex cells (sperm, eggs), the process being referred to as *meiosis*, whilst others divide to replicate cells *mitosis*. Here we are only interested in the latter.

When a cell divides into two daughter cells, it passes through two gap phases (G1 and G2), a synthesis phase (S) and a mitosis phase (M). A brief description of the phases is:

- G1 phase: the cell enlarges, organelles are copied, preparation for cell division
- S phase: DNA replicates
- G2 phase: the cell further enlarges, proteins and organelles are made
- M phase: the splitting of the cell, first mitosis consisting of the phases: prophase, metaphase, anaphase and telophase, followed by cytokinesis where the cytoplasm of the cell is split in two, resulting in the two new cells

That is a brief description of the workings of a eukaryotic cell. Look at your hand and think of what is happening in each of the cells and then realise this is happening in most of the trillions of cells in your body. Unbelievable!

## 2.2    DNA, genes, RNA and proteins

Next, we look at DNA, genes, RNA and proteins.

### 2.2.1    DNA

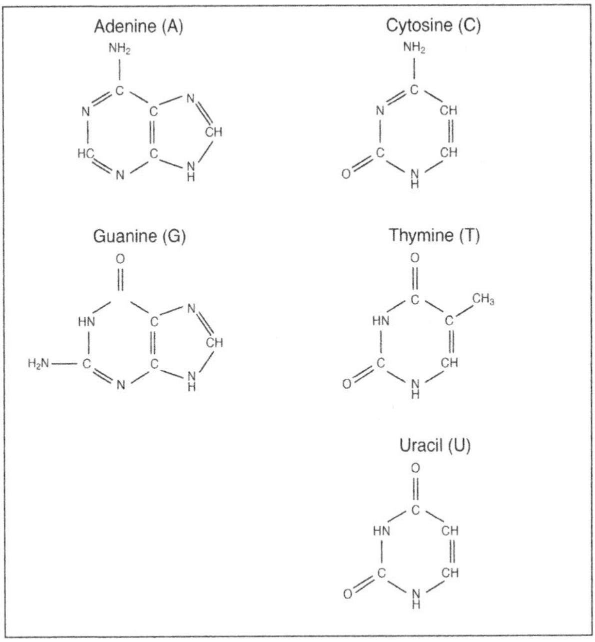

**FIGURE 2.2**
The four DNA bases: Adenine, Guanine, Cytosine and Thymine, plus Uracil

Your DNA is the blueprint that makes you. It dictates your eye colour, hair colour, sex and everything else that is physically you. DNA stands for *deoxyribonucleic acid*, a huge molecule, over two metres long contained in the nucleus of nearly every cell in your body. It is made up of a sequence of thousands of the four bases, *Adenine, Guanine, Cytosine* and *Thymine*, which are referred to by the letters **A, G, C** and **T**. Their chemical structure is shown in Figure 2.2. They are flat molecules with Adenine and Guanine being *Purines* and Cytosine and Thymine *Pyrimidines*. The other base shown in the figure, *Uracil*, which plays a part in RNA, as we will see later.

A deoxyribose sugar molecule and a phosphate molecule are attached to each base, forming a *nucleotide* (structure not shown). The nucleotides join together in pairs (base-pairs) so that Adenine always joins with Thymine (**A-T**) and Guanine always with Cytosine (**G-C**). DNA is a very long sequence (32 billion) of these pairs of nucleotides, joined by the deoxyribose sugar/phosphate molecules, as shown in Figure 2.3. The two strands of nucleotides run anti-parallel with the phosphate end of each strand denoted by 5′ and the deoxyribose sugar end denoted by 3′. By convention, the base sequence is read from the 5′ end to the 3′ end.

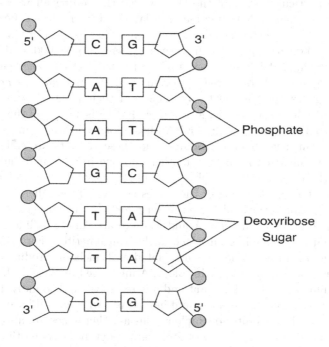

**FIGURE 2.3**
Seven pairs of nucleotides forming part of a DNA molecule

In practice, DNA does not look like the ladder in the figure but forms the famous double helix so often seen in illustrations; a vague representation is shown here in Figure 2.4. Most of the time, DNA is "bundled-up" and packed into the nucleus of the cell. Also, your DNA does not exist as a single, very long molecule, but is divided up into *chromosomes*. In humans, there are 22 *autosomal* (non-sex) pairs of chromosomes and one pair of sex chromosomes, each pair having one chromosome inherited from the father and one from the mother.

**FIGURE 2.4**
DNA double helix

## 2.2.2 Genes

Genes are particular areas of the DNA where the coding allows for protein production or some RNA function (see later). Humans have about 20,000 genes located on the chromosomes, each containing between a few thousand to over 2 million base pairs, with an average length of 27,000, and if you do the calculation, that is only about 1.5% of the total DNA. The gene regions of the DNA are said to be *coding* regions and the rest *non-coding* regions. Particular genes are referred to by name and their position (*locus*) on the chromosome where they reside, for example *KRAS* is a gene on the short arm of chromosome 12, position 12.1. KRAS has a lot to do with cancer.

A gene can have different codings, and these variants are called *alleles*. One allele is inherited from your mother and one from your father for each gene, and these determine your physical traits (*phenotype*), such as hair colour, height, etc. The combination of the alleles that you have inherited make up your *genotype*, but we can also use this term for a particular gene. If there are two possible alleles for a gene, A and a, there are three possible genotypes, AA, Aa and aa, depending on which alleles you inherited. For example, the OCA2 gene on chromosome 15 is associated with melanin production, which is a pigment for hair, eye and skin colour. A might be the allele for brown eyes and a for blue eyes. A is dominant and a is recessive, and so individuals with AA and Aa will have brown eyes, and individuals with aa will have blue eyes. But eye colour is not quite so simple as this as other genes are also involved. The allele that codes for the most common phenotype is called the *wild type allele*.

### 2.2.2.1 RNA and proteins

Within a gene's DNA coding region, there are sub-regions that code for protein, called *exons*, and regions that do not code for protein, called *introns*. The making of a protein from the gene is called *gene expression*. It happens via RNA (*ribonucleic acid*) using two processes, *transcription* and then *translation*. This is the *central dogma of life*,

$$\text{DNA} \xrightarrow{transcription} \text{RNA} \xrightarrow{translation} \text{Protein}$$

First, in transcription a copy of a single strand of a gene's DNA is made using RNA polymerase enzymes, but the thymine (T) bases are replaced by uracil (U) bases. This is messenger RNA (mRNA) which then goes to the ribosome. Then, translation occurs, where the mRNA codes for a series of *amino acids,* based on *codons* which are triples of bases read along the RNA. For example, for the portion of mRNA,

....ACGCGUA(AUG)(GUG)(CGC)............(CCC)(AAU)(UAA)AGACCA....

the triples in brackets are codons that together will code for a particular protein. The AUG is a start codon for starting the translation, the GUG, codes for an amino acid called valine, CGC for arginine, ...., CCC for proline, AAU for asparagine and UAA is a stop codon. Table 2.1 shows the 64 possible codons and their amino acids. Transfer RNA (tRNA) brings in the amino acids to match the mRNA codon coding. The chain of amino acids is a *polypeptide,* which is then folded into a three-dimensional structure, which is the protein. Cells do not make proteins all the time, only when they are signalled to do so.

**TABLE 2.1**

Codons coding for amino acids

| UUU | phenylalanine | CUU | leucine | AUU | isoleucine | GUU | valine |
|-----|---------------|-----|---------|-----|------------|-----|--------|
| UUC | phenylalanine | CUC | leucine | AUC | isoleucine | GUC | valine |
| UUA | leucine | CUA | leucine | AUA | isoleucine | GUA | valine |
| UUG | leucine | CUG | leucine | AUG | methionine and START | GUG | valine |
| UCU | serine | CCU | proline | ACU | threonine | GCU | alanine |
| UCC | serine | CCC | proline | ACC | threonine | GCC | alanine |
| UCA | serine | CCA | proline | ACA | threonine | GCA | alanine |
| UCG | serine | CCG | proline | ACG | threonine | GCG | alanine |
| UAU | tyrosine | CAU | histidine | AAU | asparagine | GAU | aspartate |
| UAC | tyrosine | CAC | histidine | AAC | asparagine | GAC | aspartate |
| UAA | STOP | CAA | glutamine | AAA | lysine | GAA | glutamate |
| UAG | STOP | CAG | glutamine | AAG | lysine | GAG | glutamate |
| UGU | cysteine | CGU | arginine | AGU | serine | GGU | glycine |
| UGC | cysteine | CGC | arginine | AGC | serine | GGC | glycine |
| UGA | STOP | CGA | arginine | AGA | arginine | GGA | glycine |
| UGG | tryptophan | CGG | arginine | AGG | arginine | GGG | glycine |

**Cell signalling**

Your cells have to communicate, and they do not all live in isolation. Cells send and receive messages via signalling molecules. The signals can tell a cell to grow, produce particular proteins, produce a metabolic response or to die (*apoptosis*). The signals sent and received are usually called *ligands* and can be molecules, proteins or ions. Signals can be received from other nearby cells or from afar, the four types being,

- *Paracrine signalling*: Signalling by nearby cells allowing similar cells to act similarly

- *Synaptic signalling*: Signalling by electrical impulses through nerve cells
- *Autocrine signalling*: Signalling of a cell to itself
- *Endocrine signalling*: Signalling from distant endocrine cells, e.g. the pancreas

Cells are tightly controlled, when to divide, when to produce proteins and when to die.

**Cell signalling pathways**

Cell signalling is very complex. The sending of an initial signal to the final outcome induced by the signal usually involves many proteins and hence genes. The route taken to achieve the outcome is via a cell signalling pathway. For example, the MAPK/ERK pathway, also known as the Ras/Raf/MEK/ERK pathway, plays a significant role in cell growth and differentiation (the process where immature cells develop to reach their specialised form and function). The MAPK family is a set of mitogen-activated protein kinases (a kinase is a particular type of enzyme) produced by particular genes such as MAPK1, MAPK3, ... MAPK15. These have other names, for example, MAPK1 is also known as ERK. A very simplified version of the MAPK/ERK pathway is,

Growth factor → Ras → Raf → MEK → ERK → Cell proliferation

First, receptors at the cell membrane receive a signal from growth factors, which activates the protein Ras. Ras causes the activation of the protein Raf, which activates MEK, which activates ERK. Lastly, ERK activates transcription factors, leading to genes encoding proteins and resulting in cell proliferation. All this is unimaginable!

## 2.3   Cancer – DNA gone wrong

Cancer is not one disease but a collection of diseases. It occurs when the DNA of a cell is damaged, and as there are about 200 different types of cells in the body, there can be 200 types of cancer. DNA damage is caused by mutations (changes) of the genetic code. The mutation can affect a single gene or, more widely, a chromosome.

For a gene, the types of mutation are,

- Accidental substitution of one nucleotide with another in the process of cell division. Sometimes this does not matter. For example, suppose at a particular point of the DNA coding, the coding should be CAA, which during transcription and translation, will code for glutamine in the mRNA if the CAA is a codon. Now suppose the last A mutates to a G during cell division, giving CAG, but CAG also codes for glutamine, and so the resulting protein is not changed. If this is not the case, the resulting protein will change.

- Insertion of one or more nucleotides into the DNA coding, again causing the structure of the protein to be altered.

- Deletion of one or more nucleotides from the DNA coding.

- Frame-shift where the codons are shifted along by one nucleotide, for example, ....(AAA)(GGT)(CTT)(A... is read as ....A(AAG)(GTC)(TTA)....
This will severely change the protein and may be non-functional.

For a chromosome, the types of mutation are,

- Deletion of part of the chromosome.

- Duplication of a segment of the chromosome.

- Inversion where a segment of the chromosome is reversed.

- Translocation where a segment of a chromosome attaches to another chromosome.

Environmental factors can also cause mutations in the DNA. These are called *mutagens* and include radiation, chemicals and some viruses. For example, UV radiation, smoke, nitrites and the human papillomavirus.

Most damaged cells will not be viable and will die (apoptosis), or a cell signal will tell them to die, or they will be killed off by other cells. A damaged cell that survives can go on to have more DNA damage, which can then ultimately lead to cancer.

There are three special types of gene, *proto-oncogenes*, *tumour suppressor genes* and *DNA repair genes*. Their functions are:

- DNA repair genes do the job of repairing genes.

- Proto-oncogenes make proteins that stimulate cell division, inhibit cell differentiation (stop a cell changing) and preventing apoptosis. If a proto-oncogene mutates, it can become an *oncogene* that does the opposite to what it should be doing, and it can make a cell divide uncontrollably.

- Tumour suppressor genes slow down cell division, repair DNA and send signals to cells to die. When these mutate, they no longer suppress the tumour.

So, once a cell becomes a cancer cell, it can tell itself to continually multiply and not to die. It can establish its own blood supply, it can ignore cell signals, and it can spread to other parts of the body. When a cancer spreads to other parts of the body, we say it has *metastasized*. Note, if you have lung cancer and it spreads to the pancreas, you do not have pancreatic cancer as well as lung cancer; it is still lung cancer but in a different organ. The cancer in the lung is the *primary* cancer, and the cancer in the pancreas is a *secondary* cancer. Each patient's cancer is different in its structure and mutations from other patients' cancers, and it keeps mutating, possibly becoming immune to treatment.

## 2.4    Cancer treatments

These are the main treatments for cancer.

### Surgery

Surgery has been used to remove tumours for at least two thousand years. It is usually the first-line treatment for many cancers and possibly in combination with other treatments, such as radiotherapy or chemotherapy. Surgery may remove all the cancer, but sometimes not. Often surgery is not possible. There was a time when radical surgery (extensive surgery) was thought to be the best option, but it was found that this was not always true. For breast cancer, radical mastectomy is generally not carried out, and most women have a lumpectomy where only the tumour is removed, followed by radiotherapy.

### Radiography or radiation therapy

Roentgen discovered X-rays in 1895. The fact that X-rays could burn skin, soon led to radiotherapy using radium to destroy tumour cells. Radiotherapy has improved immensely since the early days, and now we have different types, such as:

- *3D conformal radiotherapy*: CT scans help create a three-dimensional computer model of the area to be treated. This helps to reduce the amount of healthy tissue that also gets radiated.

- *Intensity-modulated radiotherapy (IMRT)*: Photon beams are applied from several directions so that the strength of the beams can be adjusted.

- *Electron beam therapy (EBT)*: Electrons are used instead of photons.

- *Proton beam therapy*: Protons are used instead of photons.

- *Brachytherapy*: A radioactive source is planted in or near the tumour.

Radiation is not given to a patient all in one dose. Radiation is measured in *Grays*. The total amount of radiation to be given is divided into *fractions* and a daily dose given. A typical treatment regime would be a daily dose, five days a week for six weeks. There are usually side-effects with radiotherapy, such as hair loss and fatigue.

### Chemotherapy

Chemotherapy has its roots in World Wars I and II, when mustard gas was used as a weapon. It was noticed that those not killed had immune cells destroyed by the gas. That lead to a nitrogen mustard as the first chemotherapy agent that could kill tumour cells. Since then more and more chemotherapy drugs have been discovered, and now over 100 are in current use. The drugs kill cancer cells that are dividing, but at the same time, they also kill normal cells. However, as cancer cells divide much more often than normal cells, more cancer cells than normal cells are destroyed. It is a balance between too little of the drug, and the cancer is not treated enough, or too much of the drug that will kill the cancer completely, but also the patient.

There are various types of chemotherapy drugs: alkylating agents (e.g. Cyclophosphamide); antimetabolites (e.g. Gemcitabine); anti-tumour antibiotics (e.g. Doxorubicin); topoisomerase inhibitors (e.g. Topotecan); mitotic inhibitors (e.g. Paclitaxel); platinum compounds (e.g. Cisplatin). They act on specific parts of the cell cycle, for instance mitotic inhibitors block mitosis. Combination chemotherapy involves using two or more types of drugs so that multiple parts of the cell cycle can be targeted (e.g. CVP which is Cyclophosphamide, Vincristine and Prednisolone).

When chemotherapy is given after surgery to remove a tumour or part of the tumour, it is called *adjuvant chemotherapy*. When chemotherapy is given before surgery, in order to shrink the tumour first, it is called *neoadjuvant chemotherapy*.

Typically, chemotherapy treatment is given over a number of *cycles*. For example, the drug is given for six days and then the patient rests for 15 days, giving one cycle of three weeks. The patient is given a set number of cycles, but treatment may stop if the patient cannot tolerate any more of the treatment. There are side-effects from chemotherapy, such as low white cell count (leucopoenia), anaemia, nausea, hair loss, fatigue and kidney damage.

## Immunotherapy

Immunotherapy is a more recent treatment for cancer, not widely used yet. It helps the immune system to recognise cancer cells and then kill them. There are various types of immunotherapy, for instance, drugs called *immune checkpoint inhibitors* block immune checkpoints, which stop the immune system from being too aggressive and *T-cell transfer therapy* where immune cells are taken from the tumour and modified, and then injected back into the body, where they become more effective. Other treatments include *monoclonal antibody therapy drugs*, called immune system modulators, and treatment by vaccines that strengthen the immune system.

## Palliative treatment

Palliative treatment of cancer is designed to relieve symptoms and improve the quality of life for patients. It can involve treatments as above, but these are not expected to particularly extend life.

## 2.5   Measuring cancer in the patient

A patient's tumour has to be described and measured in various ways. Apart from the type of cancer and its location, there are various biological and descriptive measurements that are made that help determine the patient's treatment and prognosis. We give an outline of some of these.

- *Tumour extent*: measures the size of the primary tumour. There are six categories, TX, T0, T1–T4, where TX means the size cannot be assessed and T0 means there is no evidence of a tumour. The categories T1–T4 represent increasing size of the tumour, the criteria for which depend on the type of cancer.

- *Lymph node involvement*: is categorised similarly to tumour extent, with NX, N0, N1–N4, this time assessing the extent to which tumour cells have infiltrated the lymph nodes.

- *Metastases*: measures whether the tumour has spread, with MX – cannot be assessed, M0 – no evidence of metastases, M1 – evidence of metastases.

- *Tumour stage*: is a combined measure of other measurements. TNM staging uses tumour extent, lymph node involvement and metastases to determine stage, with categories, Stage 1 to Stage 4, with the possibility of subcategories, e.g. 2a and 2b. We give an example, the staging of pancreatic adenocarcinoma.

  - *Tumour* T
    * T1: Tumour contained in pancreas, $T \leq 2$cm
      T1a: $T \leq 0.5$cm, T1b: $0.5 < T \leq 1$cm, T1c: $1 < T \leq 2$cm
    * T2: $2 < T \leq 4$cm
    * T3: $T > 4$cm
    * T4: Tumour grown outside the pancreas
  - *Node* N
    * N0: No cancer cells in the nearby lymph nodes, $N = 0$
    * N1: $1 \leq N \leq 3$ lymph nodes contain cancer cells
    * N2: $N \geq 4$ lymph nodes contain cancer cells
  - *Metastasis* M
    * M0: The cancer has not spread to other areas of the body
    * M1: The cancer has spread to other areas of the body
  - TNM staging of pancreatic adenocarcinoma
    * Stage 1a: T1, N0, M0
    * Stage 1b: T2, N0, M0

    * Stage 2a: T3, N0, M0
    * Stage 2b: T1, T2 or T3, N1, M0
    * Stage 3:  T1, T2 or T3, N2, M0.
    * Stage 4:  Any T, any N, M1

- *Tumour grade*: measures how similar the cancer cells look compared to normal cells; this is called degree of differentiation. The scale is: Grade 1 – well differentiated – cancer cells look mostly like normal cells; Grade 2 – moderately differentiated – cancer cells look more abnormal; Grade 3 – poorly differentiated – cancer cells look very abnormal; Grade 4 – undifferentiated.

- *Biomarkers*: are various biological measurements made from blood and other samples, for example amounts of particular proteins and hormones. For breast cells, high levels of the HER2 growth-promoting protein are associated with cancer, as is the prostate-specific antigen (PSA) associated with prostate cancer.

- *Performance status*: is a measure of how a cancer is affecting a patient's general ability. There are various performance indices, for example the *ECOG/WHO performance status* (Eastern Cooperative Oncology Group/World Health Organisation), which ranges from 0 to 5, with 0 – fully active, and 5 – completely disabled.

- *Quality of Life (QOL)*: is measured in clinical trials because cancer and cancer treatments can be very debilitating. A QOL questionnaire might be administered to a patient before, during and after treatment. There are several tried and tested questionnaires that can be used. A popular one for cancer is the EORTC (European Organisation for Research and Treatment of Cancer) QLQ-C30 questionnaire, consisting of 30 questions. Based on the responses to the questions, scores are given to the following: global health status/QoL; physical functioning; role functioning; emotional functioning; cognitive functioning; social functioning; fatigue; nausea and vomiting; pain; dyspnoea; insomnia; appetite loss; constipation; diarrhoea and financial difficulties. Analyses of QOL data can range from simple summaries of the data, to longitudinal analyses with sophisticated allowances for missing data.

Next, we need a measure of a tumour's response to treatment. The *RE-CIST* (Response Evaluation Criteria In Solid Tumours) guidelines suggest how the response should be measured. Table 2.2 outlines the possible responses, but these will depend on the cancer type and tumours and the number of lesions (areas of abnormal tissue).

Lastly, we look at the most important measurement, *survival*. To measure survival, a starting point needs to be chosen, for example, when the patient first had the cancer diagnosed. When a patient is recruited to a clinical trial,

**TABLE 2.2**

RECIST responses

| | | |
|---|---|---|
| CR | Complete response | The tumour has gone. |
| PR | Partial response | The tumour has shrunk by a pre-specified amount. |
| SD | Stable disease | The tumour has not changed. |
| PD | Progressive disease | The tumour has increased in size by a pre-specified amount |
| NE | Not Evaluable | |

the starting point is usually the date of their randomisation to treatment but can be some other time point.

For clinical trials, *overall survival* (OS), i.e. time until death is the most reliable, easily measured endpoint and is the preferred endpoint. However, for cancers where survival is good, then it may not be possible to run a long enough study that captures enough deaths. To overcome this problem, *surrogate endpoints* can be used so that a shorter trial can be designed. *Progression free survival* (PFS) using CT, PET and MRI scans is often used, where the event time is the time to when the cancer progresses or when death occurs. Table 2.3 shows some of the endpoints that have been approved by the United States Food and Drug Administration (FDA) for cancer drugs.

**TABLE 2.3**

FDA approved endpoints for cancer studies

| Endpoint | Abbr. | Definition |
|---|---|---|
| Overall Survival | OS | Time from randomisation until death from any cause |
| Progression-Free Survival | PFS | Time from randomisation until objective tumour progression or death, whichever occurs first |
| Time to Progression | TTP | Time from randomisation until objective tumour progression (death not included) |
| Disease-Free Survival | DFS | Time from randomisation until disease recurrence or death from any cause |
| Event-Free Survival | EFS | Time from randomisation until disease recurrence or death from any cause |
| Complete Response | CR | No detectable evidence of tumour |
| Objective Response Rate | ORR | Proportion of patients with tumour size reduction of a predefined amount and for a minimum time period. |

In measuring survival time for a group of patients, there are usually some *censored observations*, where survival times cannot be measured. For instance, when a trial ends before a patient dies, or when a patient leaves the trial because they have moved abroad, then their survival time cannot be measured. We say the survival time is censored and we can only measure the time until the trial ended, or when the patient moved abroad. We know the actual survival time is longer than the censored survival time, but not how much longer.

These cancer measurements that we have briefly discussed, will play a role in our analyses of cancer clinical trials, and other data.

# 3

## Survival Analysis

This is the place to start considering some statistical theory. The most often used statistical method in the analysis of data from cancer clinical trials and other studies is survival analysis, and so we start there. Survival analysis is also known as is "time to event analysis". The event of interest is often "death", but there are other time to event variables/outcomes that are of interest, such as time of tumour progression, as discussed in Chapter 2.

We cover the basics of survival analysis theory. Collett [6] covers it in more detail and is a very good introductory text. Two other books on survival analysis are: O'Quigley [46] and Moore [42].

## 3.1 The amazing survival equations

Most mathematicians and statisticians will have favourite topics and equations. Which of Einstein's equations would be his most precious? Possibly, "$E = mc^2$", but it might be his field equations of general relativity. For me, the four key survival equations hold a particular fascination. These four survival equations defined at the start of any study of survival analysis are simple and elegant, but as you delve deeper and into the counting theory approach to survival analysis, theory becomes much harder and understanding more difficult, but elegance remains. Martingale theory is used – an interesting name for a mathematical theory. Dictionary definitions say "martingale" relates to either horses or gambling.

### 3.1.1 The survival function

The starting point of survival analysis theory is the definition of the *survival function*. This is the probability that a subject (patient, individual) survives beyond time $t$ and usually written as $S(t)$. Let $T$ be the time variable, and so

$$S(t) = \Pr(T > t),$$

and this is just $1 - \Pr(T \le t) = 1 - F(t)$, where $F(t)$ is the cumulative distribution function. Generally, in statistics, we deal with probability density functions or probability mass functions, looking at the shape of histograms

DOI: 10.1201/9781003041931-3

and bar charts constructed from data, helping us decide upon an appropriate distribution. In survival analysis, we are more interested in the survival function. The other point to note is that we only have $T \geq 0$, there clearly being no negative survival times. Figure 3.1 shows four survival curves, examples from an exponential distribution, a Weibull distribution, a log-normal distribution and a log-logistic distribution.

Table 3.1 shows the formulae for these and other survival functions, together with their probability densities, means, standard deviations and median values. The piecewise exponential survival distribution is a series of exponential distributions, each covering an interval of the time axis, so an $\text{Exp}(\lambda_1)$ distribution covers $(0, t_1]$, an $\text{Exp}(\lambda_2)$ distribution covers $(t_1, t_2]$, etc.

Generally, survival functions start at the value one when $t = 0$ and decrease to zero as time progresses. However, the decrease to zero does not always occur; examples being (i) the time interval of interest is up to 10 years from treatment and those surviving beyond 10 years are said to be "cured" and (ii) for time to progression of the disease, the progression may not occur in some patients and the survival curve will level off at the value of the probability that a patient's disease does not progress.

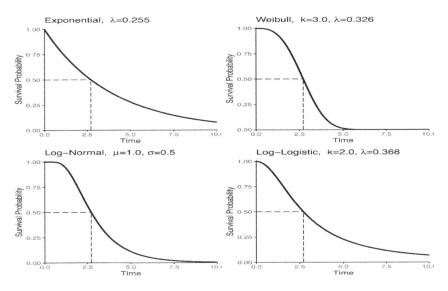

**FIGURE 3.1**
Survival curves for the exponential, Weibull, log-normal and log-logistic distributions

For the four survival functions in Figure 3.1, the horizontal dashed line at survival probability 0.5 and the corresponding vertical dashed line, indicate the median survival time for each curve. The parameters of each distribution have been chosen so that the median survival time is 2.718 for each of these four distributions, making it easier to compare the overall shapes among the

## TABLE 3.1
Some survival distributions

| | | Distribution | |
|---|---|---|---|
| | **Exponential** | **Gamma** | **Weibull** |
| Notation | $\text{Exp}(\lambda)$ | $\Gamma(\lambda, k)$ | $W(\lambda, k)$ |
| $S(t)$ | $e^{-\lambda t}$ | $1 - \frac{\gamma(k, \lambda t)}{\Gamma(k)}$ | $e^{-(\lambda t)^k}$ |
| $f(t)$ | $\lambda e^{-\lambda t}$ | $\frac{\lambda(\lambda t)^{k-1} e^{-\lambda t}}{\Gamma(k)}$ | $\lambda k(\lambda t)^{k-1} e^{-(\lambda t)^k}$ |
| $h(t)$ | $\lambda$ | $\frac{\lambda(\lambda t)^{k-1} \exp(-\lambda t)}{\Gamma(k) - \gamma(k, \lambda t)}$ | $\lambda k(\lambda t)^{k-1}$ |
| mean | $\lambda^{-1}$ | $k\lambda^{-1}$ | $\Gamma(1 + 1/k)/\lambda$ |
| variance | $\lambda^{-2}$ | $k\lambda^{-2}$ | $\{\Gamma(1 + 2/k) - [\Gamma(1 + 1/k]^2\}/\lambda^2$ |
| median | $\log(2)\lambda^{-1}$ | no closed form | $\log(2)^{1/k}\lambda^{-1}$ |

| | **Log Normal** | **Log Logistic** | **Piecewise Exponential** |
|---|---|---|---|
| Notation | $LN(\mu, \sigma)$ | $LL(\lambda, k)$ | $PE(\lambda, \theta)$ |
| $S(t)$ | $1 - \Phi\left(\frac{\ln t - \mu}{\sigma}\right)$ | $\frac{1}{(1+(\lambda t)^k)}$ | |
| $f(t)$ | $\frac{1}{\sqrt{2\pi}\sigma t} e^{\left\{\frac{(\ln t - \mu)^2}{\sigma^2}\right\}}$ | $\frac{\lambda k(\lambda t)^{k-1}}{(1+(\lambda t)^k)^2}$ | |
| $h(t)$ | $\frac{\phi(\ln t - \mu/\sigma)}{\Phi(-\ln t + \mu)/\sigma}$ | $\frac{\lambda k(\lambda t)^{k-1}}{(1+(\lambda t)^k)}$ | $h(t) = \lambda_i$ for $i$ for $i$th interval |
| mean | $e^{(\mu + \frac{1}{2}\sigma^2)}$ | $\frac{2\pi}{k\lambda \sin(\pi/k)}$ | |
| variance | $e^{\{2(\mu+\sigma^2) - \exp(2\mu+\sigma^2)\}}$ | $\frac{\pi}{k\lambda^2}\left\{\frac{2}{\sin(2\pi/k)} - \frac{\pi}{k\sin^2(\pi/k)}\right\}$ | |
| median | $e^\mu$ | $1/\lambda$ | |

Note: The mean and variance for the log-logistic distribution are only defined for $k > 1$ and $k > 2$ respectively; $\Gamma(k, z) = \int_0^z x^{k-1} e^{-x} dx$, $\gamma = \Gamma(k, \infty)$.

four curves. The exponential distribution is a special case, it being about the simplest survival distribution available. The curve falls quickly away from the value 1 at $t = 0$, as $t$ increases. How quickly it falls away will depend on the value of $\lambda$. The other three curves hover around the value 1 for a short time after the start, but then fall away, this being typical of survival curves in cancer studies. These three curves look similar, but look for the differences. Percentiles other than the 50th (median), can be read from the graphs, or calculated. For example the 90th percentiles for the four survival curves are: 9.030, 4.067, 5.159 and 8.155 respectively.

So far we have not discussed the time axis for the survival curves. A scale has to be chosen and this will depend on context. Here, the time axes are all marked 0 to 10 and this could represent days, months, years, etc. And of course the unit of measurement should be marked on the plot. For advanced pancreatic cancer, the unit of measurement might be months; for chronic lymphocytic leukaemia (CLL), the unit of measurement might be years.

The probability density function (pdf) associated with the survival function, $S(t)$, is easily derived as,

$$\frac{dS(t)}{dt} = \lim_{\delta t \to 0} \frac{S(t + \delta t) - S(t)}{\delta t} = \lim_{\delta t \to 0} \frac{F(t) - F(t + \delta t)}{\delta t} = -f(t).$$

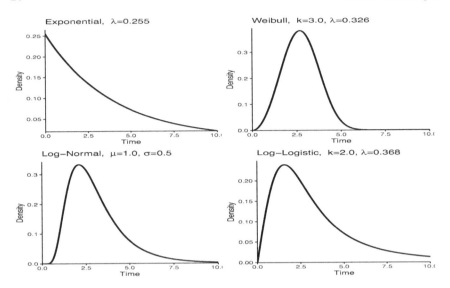

**FIGURE 3.2**
Probability density functions for the exponential, Weibull, log-normal and log-logistic distributions

Figure 3.2 shows the probability density functions for the corresponding survival curves in Figure 3.1. Look at the similarities in the pdfs for the Weibull, log-normal and log-logistic distributions and how different they are from that for the exponential distribution.

## 3.1.2   The hazard function

The second most important function in survival analysis is the *hazard function*. What is the chance that you will die in the next few seconds. Hopefully, this will not happen. The event of interest does not have to be death, but could be a seizure, discharge from hospital, or the failure of a heart pacemaker. We have to take into account the fact that you have survived up to current time. Here, we will take the event of interest to be death.

Loosely defined, the hazard function, $h(t)$, is the probability that a subject dies in the next moment in time, given the subject has survived from time zero up to the present time. Let $t$ be the present time point. The probability of dying in the time interval $(t, t + \delta t]$ is $S(t) - S(t + \delta t)$. The conditional probability of dying in the time interval $(t, t + \delta t]$, given survival up to time $t$ is $\{S(t) - S(t + \delta t)\}/S(t)$. We need to divide this probability by the length of this tiny interval, $\delta t$, in order to get the hazard rate, and then let $\delta t$ tend to zero. Thus

$$h(t) = \lim_{\delta t \to 0} \frac{\{S(t) - S(t + \delta t)\}}{S(t)\delta t} = \frac{f(t)}{S(t)},$$

and is just the pdf divided by the survival function.

Figure 3.3 shows the hazard functions corresponding to the survival curves in Figure 3.1. The hazard function for the exponential survival curve is constant over time, i.e. the chance of dying at any moment in time, given survival up to that moment, is the same for any chosen moment. For the Weibull survival curve, the hazard function increases with time. For a cancer following this survival curve, the longer a patient lives, the higher the risk of dying in the next few days/weeks/months. For the log-normal and log-logistic distributions, the hazard is low at the start, then steadily increases, reaches a maximum and then declines steadily. An example of when a constant hazard might be appropriate, is the time to the next diagnosis of a basal cell carcinoma nodule (a type of skin cancer) for a patient, given the patient has already had one nodule diagnosed. An example of the rising hazard function of the Weibull distribution, might be the hazard of developing breast cancer above the age of 30. An example of the hazard functions of the log-normal and log-logistic, is the hazard of dying after surgery for a particular cancer, where complications can set in within the first few months, but after surviving those, the risk of dying reduces.

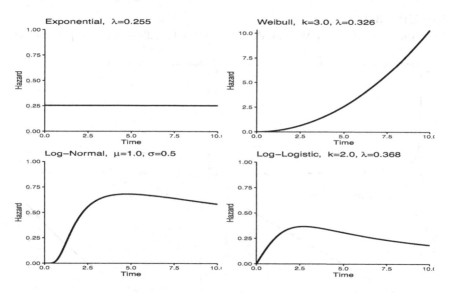

**FIGURE 3.3**
Hazard functions for the exponential, Weibull, log-normal and log-logistic distributions

### 3.1.2.1   Shapes of Weibull distributions

The shape of the hazard function for a Weibull distribution depends on the shape parameter $k$. Figure 3.4 shows the survival curves for four Weibull distributions with different values of the shape parameter. Figure 3.5 shows the

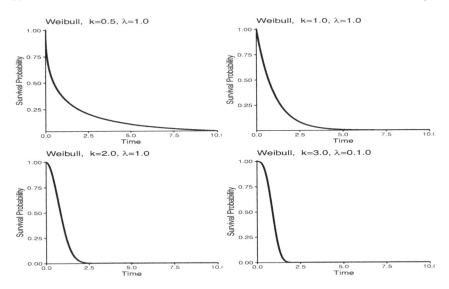

**FIGURE 3.4**
Survival curves for four Weibull distributions with different shape parameters

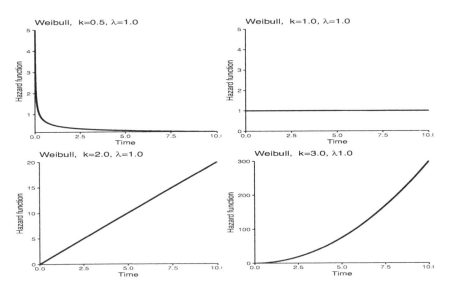

**FIGURE 3.5**
Hazard functions curves for the four Weibull distributions

corresponding hazard functions. As $k$ increases the survival functions approach
zero faster as $t$ increases. For $k < 1$, the hazard function is decreasing with
time. When $k = 1$, we have the exponential distribution with a constant

hazard function. When $k > 1$ the hazard function is increasing with time. An interesting case is $k = 2$, where the hazard function is a straight line.

### 3.1.2.2 Proportional hazards and the hazard ratio

It is worth mentioning here, the special case of *proportional hazards*, as it is an assumption made many times in our methods for the analysis of survival data. Suppose we have a clinical trial with two arms, Arm 1 and Arm 2. Under the proportional hazards assumption, the hazard of death in one arm is proportional to the hazard of death in the other arm for all times, $t$. Let this proportion be denoted by $\gamma$, and the hazard functions for the two arms of the trial by $h_1(t)$ and $h_2(t)$. Then

$$h_2(t) = \gamma h_1(t). \tag{3.1}$$

So it does not matter what time $t$ is, if you are in Arm 1 of the trial, your instantaneous chance of dying is $\gamma$ times the instantaneous chance of dying for a patient in Arm 2 of the trial. Likewise, a patient in Arm 2 of the trial has an instantaneous chance of dying, $1/\gamma$ times your instantaneous chance of dying at all times, $t$. Proportional hazards is a reciprocal relationship.

From Equation (3.1),

$$\int \frac{f_2(t)}{S_2(t)}\, dt = \gamma \int \frac{f_1(t)}{S_1(t)}\, dt,$$

and so

$$\log\{S_2(t)\} = \gamma \log\{S_1(t)\}. \tag{3.2}$$

Exponentiating,

$$S_2(t) = \{S_1(t)\}^{\gamma}, \quad \text{or equivalently} \quad S_1(t) = \{S_2(t)\}^{1/\gamma}.$$

So proportional hazards gives a simple relationship between the two survival curves. If $\gamma > 1$ then the survival curve for Arm 2 lies below that of Arm 1, but if $\gamma < 1$, it lies above that for Arm 1. If $\gamma = 1$, then the survival curves are identical. The parameter $\gamma$ is called the *hazard ratio* and is widely quoted as a measure of difference in survival between two groups.

### 3.1.3 The cumulative hazard function

Life is hazardous! Each moment in time we are subjected to hazards, for instance catching a cold, being run over by a bus, dying from a heart attack. Let us concentrate on the ultimate hazard of dying. As we age, we are continually avoiding this hazard, accumulating the risk (probability) of dying at any one moment in time, until the hazard actually happens. The *cumulative hazard function*, $H(t)$, is essentially the accumulation of the risk of the hazard, as measured by $h(t)$, for the event of interest, from time $= 0$, until time $= t$. So

$$H(t) = \int_0^t h(s)\, ds.$$

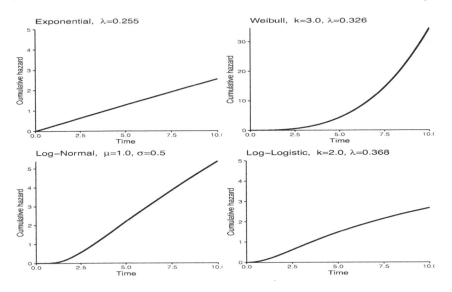

**FIGURE 3.6**
Cumulative hazard functions for the exponential, Weibull, log-normal and log-logistic distributions

Thus

$$H(t) = \int_0^t \frac{f(s)}{S(s)}\, ds = -\log\{S(T)\},$$

or equivalently,

$$S(t) = \exp\{-H(t)\}.$$

Figure 3.6 shows the cumulative hazard functions obtained from the hazard functions in Figure 3.3. Clearly, the cumulative hazard is non-decreasing and will usually go to infinity as $t$ goes to infinity, but not always.

That is the last of our survival equations. The survival equations for you to become familiar with are summarised in the box below.

---

**The Amazing Survival Equations**

$$S(t) = -\int f(s)\, ds, \qquad h(t) = \frac{f(t)}{S(t)}$$

$$H(t) = \int h(s)\, ds \qquad S(t) = \exp\{-H(t)\}.$$

---

Remember, we may have four equations and four functions, $f(t)$, $S(t)$, $h(t)$ and $H(t)$, but in reality, we only have one function, from which the other three functions can be derived. We could start with $f(t)$ and then derive $S(t)$, $h(t)$ and then $H(t)$, or more likely, we would start with $S(t)$ or $h(t)$ and derive the others from that. So why are these survival equations so amazing for me? My answer: they are basically one function which, in its various guises, conveys different insights into survival, with a neat simple structural relationship between the component parts.

**Alternative parameterisations of survival distributions**

There is more than one way of parameterising a survival distribution. For example, the exponential survival distribution may be parametrised as $S(t) = \exp(-t/\alpha)$, instead of $S(t) = \exp(-\lambda t)$ as in Table 3.1. For the parameterisation in the table, $\lambda$ is a "rate" parameter and for the alternative parameterisation, $\alpha$ is a "scale" parameter. For the Weibull distribution, an alternative parameterisation is, $S(t) = \exp(-\lambda t^k)$. In the parameterisation in Table 3.1, $S(t) = \exp\{-(\lambda t)^k\}$, $\lambda$ is a "rate" parameter and $k$ is a "shape" parameter. Yet another parameterisation is $S(t) = \exp\{-(t/\alpha)^k\}$, and $\alpha$ is a "scale" parameter. The terms "scale" and "rate" are used since the parameter $\alpha$ has dimension, time, and so "scales time", whilst, $\lambda$ has dimension time$^{-1}$ and represents an amount per unit time, i.e. a rate. There can be much confusion with the parameterisations of the Weibull and other distributions, with scale and rate terms used loosly. For instance, the $\lambda$ of the parameterisation, $S(t) = \exp\{-\lambda t^k\}$, is often referred to as a scale parameter, but technically, it is a mixture of scale and shape. We will use appropriate parameterisations depending upon what we are trying to achieve.

### 3.1.4   Design your own survival function

Before moving on, just for fun, let us see how we can design our own survival function, $S(t)$. It is not easy choosing a function for $S(t)$ – it has to be defined on zero to infinity, equal to one at time zero, non-negative, non-increasing and drop down to zero, either reaching zero at some time point or approaching zero as time tends to infinity. That is a lot conditions on the function. But we don't give up. We start by choosing a hazard function, $h(t)$. It has to be non-negative and we must be able to integrate it from zero to time $t$. That's so much easier now. First, we will choose

$$h(t) = t.$$

Then, using our survival equations,

$$H(t) = \int_0^t u \, du = t^2/2,$$

and so

$$S(t) = \exp(-H(t)) = e^{-t^2/2}.$$

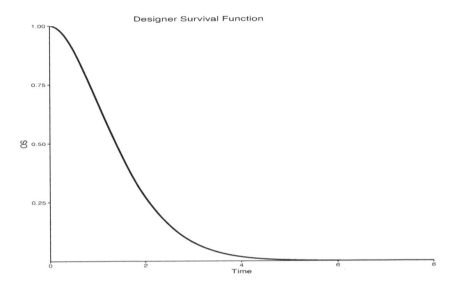

**FIGURE 3.7**
Our own design of a survival function

But that's not new; it is a special case of the Weibull distribution. Let us try another, let

$$h(t) = \log(t+1).$$

So

$$H(t) = \int_0^t \log(u+1)\, du = (t+1)\log(t+1) - t,$$

and so

$$S(t) = \exp(-H(t)) = \frac{e^t}{(t+1)^{(t+1)}}.$$

So we have our own survival function – or is it used somewhere out there already? Figure 3.7 shows a graph of it.

## 3.2   Non-parametric estimation of survival curves

Next we see how survival curves can be fitted to data. Usually, with a set of univariate data, you would plot a histogram before thinking about fitting a parametric distribution. Having chosen an appropriate distribution, maximum likelihood methods are used to fit the distribution, i.e. estimate the parameters of the distribution. For survival data where there is censoring, plotting a histogram, pretending the censored survival times are true survival times is not a good idea. Nor is the plotting of a histogram excluding the censored survival times. In both cases, the mean and median survival times would be underestimated. The usual approach is to plot an empirical survival curve instead. We do this and leave the parametric estimation of survival curves to a later section.

### 3.2.1 The Kaplan-Meier survival curve

Edward Kaplan and Paul Meier independently submitted manuscripts to the Journal of the American Statistical Association, proposing the same method of plotting an empirical survival curve. The editor suggested they write a joint-author paper and so now we have the ubiquitous Kaplan-Meier survival curve. At the time of writing, when "Kaplan-Meier" is entered into the topic search field of the web-of-science, 56,307 entries are displayed, and for Google Scholar, over 527,000 entries.

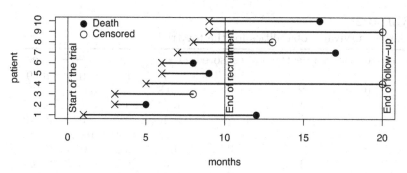

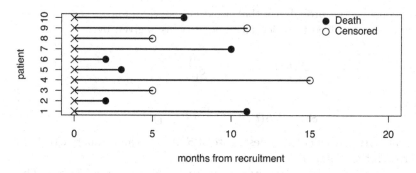

**FIGURE 3.8**
Clinical trial time-line for the toy trial

First, let us consider a toy clinical trial. Figure 3.8 shows the time-line of a trial from opening the trial for recruitment, to the end of the recruiting period and then on to the end of the follow up period. There are only ten patients in the trial, recruited between months 1 and 10. Patient 1 was recruited at month 1 and died at month 12, death being signified by a solid circle, patient 2 was recruited at month 3 and died at month 5, patient 3 was also recruited at month 3 but their survival time was censored at month 8, signified by an open circle, etc. The lower graph shows the actual survival times for the patients' month of death/censoring minus the month of recruitment (i.e. we are lining up the patient start times); note, for simplicity, we are assuming patients are recruited at the start of a month and that death/censoring date is recorded as the end of a month. so, for instance, patient 1 was recruited at the start of month 1 and died just before month 12 started.

The time axis is partitioned by the death times, $t_1, t_2, \ldots t_n$, giving the time intervals $(0, t_1], (t_1, t_2], \ldots (t_{n-1}, t_n]$, which are open on the left and closed on the right, i.e. the first interval contains all times $t$, such that $0 < t \le t_1$, the second interval contains all times $t$, such that $t_1 < t \le t_2$ etc.

## TABLE 3.2
Calculating survival probabilities for the toy trial

| time | no. at risk | no. deaths | no. censored | $(1 - \frac{d_i}{n_i})$ | $\hat{S}(t)$ | $\frac{d_i}{n_i(1-n_i)}$ | s.e. |
|---|---|---|---|---|---|---|---|
| 0 | 10 | 0 | 0 | 1 | 1.0 | 0 | 0.59 |
| 2 | 10 | 2 | 0 | 0.8 | 0.8 | 0.025 | 0.468 |
| 3 | 8 | 1 | 0 | 0.875 | 0.7 | 0.018 | 0.410 |
| 5 | 7 | 0 | 2 | 1 | 0.7 | 0 | 0.410 |
| 7 | 5 | 1 | 0 | 0.8 | 0.56 | 0.05 | 0.328 |
| 10 | 4 | 1 | 0 | 0.75 | 0.42 | 0.083 | 0.246 |
| 11 | 3 | 1 | 1 | 0.667 | 0.28 | 0.167 | 0.164 |
| 15 | 1 | 0 | 1 | 1 | 0.28 | 0 | 0.164 |
| 16 | 0 | 0 | 0 | | 0.28 | | |

Now, consider the probability that a patient survives longer than time $t_i$, given they have survived to at least time $t_{i-1}$. This is given by

$$S(t_i | T > t_{i-1}) = \frac{S(t_i)}{S(t_{i-1})},$$

and hence

$$S(t_i) = S(t_i | T > t_{i-1})S(t_{i-1}).$$

So we can recursively calculate an estimate of $S(t)$ if we can estimate $S(t_i | T > t_{i-1})$ for each time interval, and start with $S(0) = 1$.

Table 3.2 shows various quantities calculated from our toy trial data. The first column shows event times, $0, 2, 3, \ldots, 16$, and we have included the censored times too. The third column shows the number of deaths, $d_i$, in each interval, $(0, 2], (2, 3], \ldots$, and the fourth column the number of censored times.

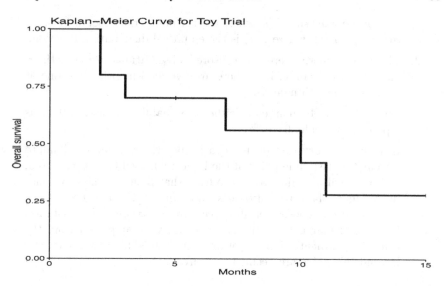

**FIGURE 3.9**
Kaplan-Meier curve for the toy trial

We make the assumption that censoring is independent of survival and so a patient with a censored time would have followed the same survival distribution, after their censoring, as a patient who does not have their survival time censored. Also it is assumed that the censoring distribution is the same throughout the trial. The second column of the table shows the number of patients at risk of dying, $n_i$, at the start of each interval, calculated as the number of patients at risk in the preceding interval minus the number of deaths and censored survival times in that interval.

An estimate of $S(t_i|T > t_{i-1})$ is,

$$\hat{S}(t_i|T > t_{i-1}) = 1 - \frac{d_i}{n_i},$$

i.e. one minus the proportion of patients who died during the interval. Now $\hat{S}(0) = 1$ and so $\hat{S}(t_1) = 1 \times (1 - d_1/n_1)$, $\hat{S}(t_2) = \hat{S}(t_1) \times (1 - d_2/n_2)$, etc., leading to the Kaplan-Meier (KM) estimate of the survival curve,

$$\hat{S}(t) = \prod_{t_i \leq t} \left(1 - \frac{d_i}{n_i}\right),$$

where "$t_i \leq t$" denotes the set of time intervals that have upper endpoints less than or equal to $t$. The KM curve for our toy trial is shown in Figure 3.9.

Here are some points to note about the KM survival curve.

1. The KM curve "steps" down at the death times and is horizontal between these. If there are no censored values, then the steps are

each of the same size, $1/n$, where $n$ is the number of observations, i.e. $\hat{S}(t) = 1 - \hat{F}(t)$, where $\hat{F}(t)$ is the empirical distribution function.

2. The times where there are censored observations are sometimes marked on the curve; here there are two at 5 months, and one at each of 11 and 15 months.

3. The y-axis can be marked as either a probability scale, 0–1.0, or as a percentage scale, 0–100%.

4. If the largest survival time is a death, then the KM curve will fall to zero at this last time point; if the largest survival time is censored, the curve will remain above zero from that point all the way to infinity. Hence the length of time to show on the time axis needs to be considered and chosen sensibly, according to context. For instance, for a particular cancer, survival over a five year period from the time of treatment might be of interest. It might be that survivors beyond this point are deemed as "cured".

**Example from a lung cancer trial**

We will use data from a lung cancer clinical trial[50] that tested the efficacy of maintenance therapy (treatment usually given after initial treatment) for small-cell lung cancer. Patients who had not progressed after initial chemotherapy were randomised to maintenance therapy by Sunitinib or Placebo. The primary endpoint was progression-free survival. The data are available from "Project Data Sphere", a platform for openly sharing cancer data (www.projectdatasphere.org; see doi.org/10.34949/zbwg-ef42 for this particular trial).

Figure 3.10 shows the KM curve of progression-free survival (PFS) for the Sunitinib arm of the trial. After one month, the gradient of the survival curve becomes steep and more or less constant, until six months when it levels off somewhat. Progression free survival beyond 12 months is very uncommon – nearly all patients will have progressed by 12 months from registration and more than 80% will have progressed by 6 months. Lung cancer is a nasty disease.

### 3.2.2  Confidence intervals for KM curves

Various estimates of the standard error of the Kaplan-Meier curve have been proposed, two of which are found by using *Greenwood's formula*:

$$\widehat{\text{se}(\hat{S}(t))} = \hat{\sigma}_G(t) = \hat{S}(t)\left\{ \sum_{t_i \leq t} \frac{d_i}{n_i(n_i - d_i)} \right\}^{1/2},$$

and by *Peto's formula*:

$$\widehat{\text{se}(\hat{S}(t))} = \hat{\sigma}_P(t) = \left\{ \hat{S}^2(t)\{1 - \hat{S}(t)\}/n_i \right\}^{1/2}, \quad t_i \leq t < t_{i+1}.$$

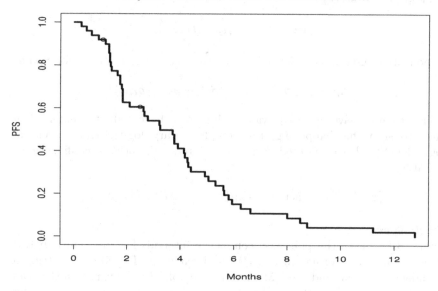

**FIGURE 3.10**
KM curve for Suninitib arm of the lung cancer trial

We will use Greenwood's formula for pointwise confidence intervals for the KM curve. Asymptotically, $\hat{S}(t)$ has a normal distribution with mean and standard error that we will estimate by $\hat{S}(t)$ and $\hat{\sigma}_G(t)$ respectively. Thus a confidence interval for $\hat{S}(t)$ is given by

$$\hat{S}(t) \pm z_{\alpha/2}\hat{\sigma}_G(t),$$

where $z_{\alpha/2}$ is the upper $100\alpha/2$ percentage point of the standard normal distribution.

One problem is that this confidence interval can have limits below zero, or above one, which is not sensible for survival curves which are bounded by zero and one. To stop negative limits, a log transformation can be applied. First, we need to know the approximate mean and variance of a transformation $Y = g(X)$ of a random variable, $X$, in terms of the mean, $\mu_X$ and variance, $\sigma_X^2$ of $X$. This is covered in the Appendix, where it is seen that,

$$E(Y) \approx g(\mu_X), \quad \sigma_Y^2 \approx \left\{\frac{dg(X)}{dX}\right\}^2 \sigma_X^2.$$

Our transformation will be $Y = \log(\hat{S}(t))$ and so $\frac{dY}{d\hat{S}(t)} = 1/\hat{S}(t)$. Hence $\log(\hat{S}(t))$ has, asymptotically, an approximate normal distribution with mean and standard error estimated by $\log(\hat{S}(t))$ and $\hat{\sigma}_G/\hat{S}(t)$, giving a confidence interval for $\log(\hat{S}(t))$ as

$$\log(\hat{S}(t)) \pm z_{\alpha/2}\hat{\sigma}_G(t)/\hat{S}(t).$$

Exponentiating these confidence limits, gives a confidence interval for $\hat{S}(t)$ as

$$[\hat{S}(t)\exp(-z_{\alpha/2}\hat{\sigma}_G(t)/\hat{S}(t)), \;\; \hat{S}(t)\exp(z_{\alpha/2}\hat{\sigma}_G(t)/\hat{S}(t)],$$

which cannot have negative values, but can have values greater than one. To stop this happening, the complementary log-log transformation, $\log\{-\log(\hat{S}(t))\}$, can be used. Similarly to above, the derived confidence interval for $\hat{S}(t)$ is

$$[\hat{S}^{1/\theta}, \; \hat{S}^{\theta}], \quad \text{where } \theta = \exp\{z_{\alpha/2}\hat{\sigma}_G(t)/[\hat{S}(t)\log(\hat{S}(t))]\}.$$

*Sunitinib arm*
Figure 3.11 shows the pointwise confidence intervals for the Sunitinib KM curve. A log transformation of $\hat{S}(t)$ has been used. The idea of putting a confidence band around the KM curve is to emphasise the variation that can occur when estimating the survival function. If another trial was run, under exactly the same conditions for this lung cancer trial, the KM curve obtained would be similar to the one we have now, and within these confidence bands (or most of the curve should be within these confidence bands). Of course, this is not guaranteed to be the case. As with any confidence interval, the greater the sample size, the narrower the confidence interval. Also, note that the confidence band is made up from pointwise 95% confidence intervals. This is not the same as an overall 95% confidence band for the whole survival curve.

### 3.2.3 Mean and median survival times

**Mean survival time**
There is a useful formula for calculating the population mean from a non-negative random variable, $X$. You just integrate $1 - F(x)$ over $[0, \infty]$, where $F(x)$ is the cumulative distribution function, or equivalently, the survival function $S(x)$. This is not widely known as a standard result. It is easily shown using integration by parts,

$$\int_0^\infty S(t)\,dt = [tS(t)]_0^\infty + \int_0^\infty tf(t)\,dt = \int_0^\infty tf(t)\,dt, \qquad (3.3)$$

as $[tS(t)]_0^\infty = 0$, assuming $\lim_{t\to\infty} tS(t) = 0$. The last term in the equation is the population mean, $\mu_T$ of the survival time $T$.

The problem of calculating the mean survival time from a survival dataset, is that there are censored observations. These cannot simply be ignored. Because of the censoring, the best we can do regarding the mean survival time, is to define the *restricted mean survival time*, up to time $t_k$, RMST$(t_k)$, where

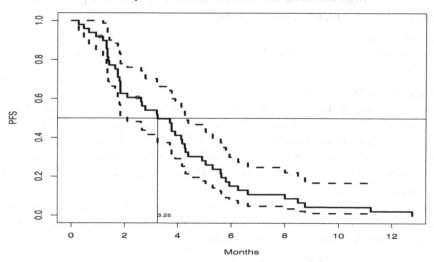

**FIGURE 3.11**
KM curve with confidence intervals for the Suninitib arm of the lung cancer trial

$t_k$ has to chosen, e.g. the time of the last event. To find $\mathrm{RMST}(t_k)$, define the restricted survival time, $T_{\text{res}}$, as $T_{\text{res}} = \min(T, t_k)$. Then

$$\mathrm{E}(T_{\text{res}}) = \mathrm{E}(T|T \leq t_k) + t_k \Pr(t > t_k) = \int_0^{t_k} t f(t)\, dt + t_k S(t_k)$$

$$= [-tS(t)]_0^{t_k} + \int_0^{t_k} S(t)\, dt + t_k S(t_k) = \int_0^{t_k} S(t)\, dt,$$

which is just the integral of the survival function of $T$ from 0 to $t_k$, and can be estimated by the area under the KM curve over the same range. (See Royston and Parmar[53] for further details.)

*Sunitinib arm*
The area under the Sunitinib KM curve is calculated up to 12 months which gives an estimate, $\widehat{\mathrm{RMST}}(12)$, for PFS as 3.80 months.

**Confidence interval for the restricted mean survival time**
The variance of $T_{\text{res}}$ can be shown to be,

$$\mathrm{var}(T_{\text{res}}) = 2 \int_0^{t_k} t S(t)\, dt - \left\{ \int_0^{t_k} S(t)\, dt \right\}^2,$$

where the first term can be estimated by integrating the KM curve multiplied by $t$, and the second term is the square of the estimated mean above. Other methods of estimation can be used such as perturbation-resampling, but we cannot go further into this here. Once obtained, a confidence interval for RMST($t_k$) can be found in the usual way.

*Sunitinib arm*
The 95% confidence interval for RMST(12) for the Sunitinib arm is (3.03, 4.57) months.

## Median survival time
The median survival time is very widely used, a lot more that the restricted mean survival time. It is estimated by the point on the x-axis of the KM curve, where a horizontal line through the point 0.5 on the y-axis meets the curve. It is the largest ordered death time for all ordered death times for which the KM curve is less than 0.5, i.e. median $= \max\{t_{(i)}|S(t_{(i)}) < 0.5\}$.

*Sunitinib arm*
The median PFS, $m$ is estimated as $\hat{m} = 3.25$ months as can be seen in Figure 3.11. It is less than the restricted mean PFS as would be expected. Other percentiles can be found in a similar way.

## Confidence interval for the median
It is not straightforward to find a confidence interval for the median survival time. We need an estimate of the standard error, se($\hat{m}$) of $\hat{m}$. One method is to use $\{\hat{f}(\hat{m})\}^{-2}$se$\{\hat{S}(\hat{m})\}$, where $\hat{f}(\hat{m})$ is an estimate of the survival time pdf at the estimated median value, based on $\hat{S}(t)$. See Collett[6] for the details. From this, the confidence interval is, $\hat{m} \pm z_{\alpha/2}$ se($\hat{m}$). The 95% confidence interval for the median survival time for the Sunitinib Arm is [2.10, 4.30]. Often, the points on the x-axis where the horizontal line giving the median on the KM curve cuts the lower and upper confidence intervals for the KM curve, will be approximately equal to this confidence interval.

## 3.2.4  The Nelson-Aalen cumulative hazard curve
Next, we look at estimating the cumulative hazard function. This can be done using the *Nelson-Aalen estimator* (NA), $\hat{A}(t)$, given by

$$\hat{A}(t) = \sum_{t_i \leq t} \frac{d_i}{n_i},$$

where $d_i$ and $n_i$ are the number of deaths and the number at risk in the $i$th interval, just as for calculating the KM curve. The NA curve is a non-decreasing step function. Figure 3.12 shows the NA cumulative hazard curve for the Sunitinib arm. The cumulative hazard increases at a lower rate in the first month than the rest of the time. After one month the cumulative risk increases at a more or less constant rate.

The variance of the NA estimator is given by

$$\hat{\sigma}^2_{NA}(t) = \sum_{t_i \leq t} \frac{(n_i - d_i)d_i}{n_i^2(n_i - 1)},$$

and has, asymptotically, a normal distribution. Hence an approximate confidence interval for $A(t)$ is

$$\hat{A}(t) \pm z_{1-\alpha/2}\hat{\sigma}_{NA}(t).$$

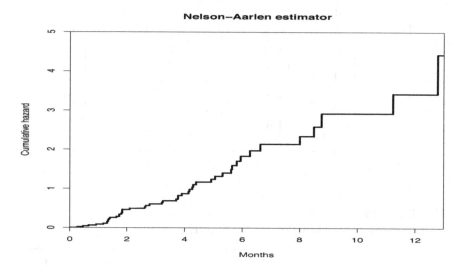

**FIGURE 3.12**
Nelson-Aalen estimate of the cumulative hazard function for the Sunitinib arm

## 3.2.5   Estimating the hazard function

The above step function estimates of the survival curve and the cumulative hazard curve give good representations of the true survival curve and cumulative hazard curve. Estimation of the hazard function in a similar way does not lead to such a nice curve unless some smoothing is carried out.

The Kaplan-Meier type estimate, $\hat{h}(t)$, of the hazard function is given by,

$$\hat{h}(t) = \frac{d_i}{n_i \tau_i},$$

where, as before, $d_i$ and $n_i$ are the number of deaths and the number at risk at the $i$th death time, $t_{(i)}$, and $\tau_i = t_{(i+1)} - t_{(i)}$. The $d_i/(n_i\tau_i)$ is estimating the risk of death per unit of time in the $i$ time interval and is constant over the interval.

Assuming a binomial distribution, Bin$(N, p)$, for $d_i$ with $N = n_i$ and $p$ estimated by $d_i/n_i$, an approximate confidence interval for $\hat{h}(t)$ is,

$$\hat{h}(t) \pm z_{\alpha/2}\hat{h}(t)\sqrt{\frac{n_i - d_i}{n_i d_i}},$$

which will be very wide when $d_i$ is small.

*Survival Analysis*

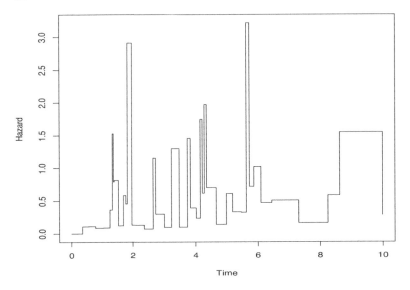

**FIGURE 3.13**
Estimate of the hazard function for the Sunitinib arm

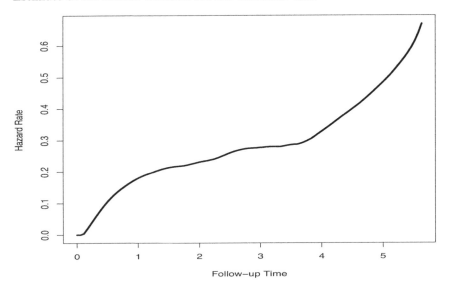

**FIGURE 3.14**
Smoothed estimate of the hazard function for the Sunitinib arm

Figure 3.13 shows the estimated hazard function for the Sunitinib arm. It is not very illuminating, looking like the outline of the tower blocks in a city skyline. The figure was produced using the R functions kphaz.fit() and kphaz.plot() from the muhaz package0. The package has a function muhaz() that smooths the hazard function estimate using kernel based methods and this has been used to produce Figure 3.14. This smoothed estimate is much more informative – we can now see that the hazard appears to increase rapidly near the start, then flatten off and then starts to increase rapidly again. Note though, the smoothed curve depends on various parameters that have to be set in the smoothing program, and caution in interpretation is needed near the end of the curve where data can be sparse.

## 3.3 Fitting parametric survival curves to data

A parametric survival curve can be fitted using maximum likelihood. First, we consider the exponential distribution with no censoring. The likelihood is

$$L = \prod_{i=1}^{n} \lambda \exp(-\lambda t_i),$$

and the log-likelihood is

$$l = \log(L) = \sum_{i=1}^{n} \{\log(\lambda) - \lambda t_i\}$$

$$= n \log(\lambda) - \lambda \sum_{i=1}^{n} t_i.$$

Differentiating $l$ with regard to $\lambda$ to obtain the likelihood equation and then equating to zero, gives the maximum likelihood estimate, $\hat{\lambda}$, as

$$\hat{\lambda} = \frac{n}{\sum_{i=1}^{n} t_i}.$$

For many distributions, the likelihood equations do not have explicit solutions and so numerical methods need to be used to find the parameter estimates.

Suppose now that we have some right censored observations. We cannot use the above estimator for $\lambda$, pretending the censored observations are observed observations – we have to take account of the censoring. The following is a way we can do this.

Each individual follows a time-to-event process and a censoring process. Let the $i$th individual have an event time, $Y_i$. and a censoring time, $C_i$. Only one of these will be observed – either the patient has the event (death say) or their event time is censored. Let $\Delta_i$ be an indicator variable for which is observed, event or censoring, and so

$$\Delta = 1 \quad \text{if the } i\text{th individual has died}$$
$$= 0 \quad \text{if the } i\text{th individual has a censored time.}$$

The observed value of $\Delta_i$ is denoted by $\delta_i$. The survival time, $T$, is the time actually recorded for a patient, either the time of the event or the time of censoring, and so,

$$T_i = \min(Y_i, C_i).$$

The observed data are denoted by $(t_1, \delta_1), \ldots, (t_n, \delta_n)$. Now we assume the censoring process is independent of the time-to-event process and hence the pdf of $(T_i, \Delta_i)$ can be written as,

$$f_{T_i, \Delta_i}(t_i, \delta_i) = f_{T_i|\Delta_i}(t_i, \delta_i = 1)\Pr(\Delta_i = 1) + f_{T_i|\Delta_i}(t_i, \delta_i = 0)\Pr(\Delta_i = 0).$$

Now $\Pr(\Delta_i = 1) = S_C(t_i)$, i.e. the probability that the censoring time is greater than the event time, and $\Pr(\Delta_i = 0) = S_T(t_i)$, i.e. the probability that the event time is greater than the censoring time. Hence

$$f_{T_i, \Delta_i}(t_i, \delta_i) = f_Y(t_i)S_C(t_i) + f_C(t_i)S_Y(t_i).$$

Conditioned on $\Delta$,

$$f_{T_i|\Delta_i}(t_i, \delta_i = 1) = f_Y(t_i)S_C(t_i)$$
$$f_{T_i|\Delta_i}(t_i, \delta_i = 0) = f_C(t_i)S_Y(t_i),$$

which can be combined as

$$f_{T_i|\Delta_i}(t_i, \delta_i) = \{f_Y(t_i)S_C(t_i)\}^{\delta_i}\{f_C(t_i)S_Y(t_i)\}^{1-\delta_i},$$

and hence the likelihood is

$$\text{L} = \prod_{i=1}^{n}\{f_Y(t_i)S_C(t_i)\}^{\delta_i}\{f_C(t_i)S_Y(t_i)\}^{1-\delta_i}$$

$$= \prod_{i=1}^{n}\{f_Y(t_i)\}^{\delta_i}\{S_Y(t_i)\}^{1-\delta_i} \prod_{i=1}^{n}\{f_C(t_i)\}^{1-\delta_i}\{S_C(t_i)\}^{\delta_i}.$$

But, the second product term only involves the censoring distribution and so has no parameters involved in the survival distribution since the two processes are independent. As far as the likelihood is concerned, the second product term is a constant. Hence

$$\text{L} \propto \prod_{i=1}^{n}\{f_Y(t_i)\}^{\delta_i}\{S_Y(t_i)\}^{1-\delta_i}.$$

So an actual survival time contributes the density $f_Y(t_i)$ to the likelihood and a censored survival time contributes the survival probability $S_Y(t_i)$.

Now we change the notation and replace $Y$ by our more favoured $T$. So now

$$L \propto \prod_{i=1}^{n} \{f_T(t_i)\}^{\delta_i} \{S_T(t_i)\}^{1-\delta_i}. \tag{3.4}$$

As an example, let us revisit the exponential distribution but with censoring. The likelihood is

$$L = \prod_{i=1}^{n} \{\lambda \exp(-\lambda t_i)\}^{\delta_i} \{\exp(-\lambda t_i)\}^{1-\delta_i}.$$

The log-likelihood is

$$l = \log(\lambda) \sum_{i=1}^{n} \delta_i - \lambda \sum_{i=1}^{n} \delta_i t_i - \lambda \sum_{i=1}^{n} (1 - \delta_i) t_i$$

$$= \log(\lambda) \sum_{i=1}^{n} \delta_i - \lambda \sum_{i=1}^{n} t_i.$$

Differentiating with respect to $\lambda$ and equating to zero,

$$\hat{\lambda} = \frac{\sum_{i=1}^{n} \delta_i}{\sum_{i=1}^{n} t_i} = \frac{\text{no. of deaths}}{\text{sum of all survival times (censored or not)}}.$$

In practice, for most survival distributions the likelihood equations cannot be solved explicitly and so numerical methods have to be used, and so we have to revert to a statistical package, allowing it to do the hard work.

### 3.3.1 Parametric survival curves fitted to the lung cancer trial data

The lung cancer data set, described in Section 3.2.1 is used to illustrate the fitting of parametric survival curves. Firstly, a Weibull survival curve is fitted to the progression free survival times for each of arm of the trial. The R function flexsurvreg() from the flexsurv package is used to fit the distributions, but note it uses the parameterisation, $S(t) = \exp(-(t/\lambda)^k)$. The fitted survival curves are shown in Figures 3.15(a) and (b). Corresponding Kaplan-Meier curves are also plotted on the graphs, which allows a judgement on how well the parametric survival curves fits the data.

The estimated parameters for the two fitted Weibull distributions are shown in Table 3.3. The fit of the Weibull distribution for the Sunitinib arm looks good. The estimate of the median, $\hat{\lambda} (\log(2))^{1/\hat{k}}$, is 3.28 months compared to 3.25 months from the Kaplan-Meier curve. However, the fit of the

**TABLE 3.3**

Parameter estimates for the fitted Weibull
distributions

|  |  | Estimate | 95% CI | se |
|---|---|---|---|---|
| Suninitib | $\hat{\lambda} = 4.224$ | [3.434, 5.195] | 0.446 |
|  | $\hat{k} = 1.449$ | [1.167, 1.800] | 0.160 |
| Placebo | $\hat{\lambda} = 3.490$ | [2.566, 4.748] | 0.548 |
|  | $\hat{k} = 1.010$ | [0.826, 1.135] | 0.104 |

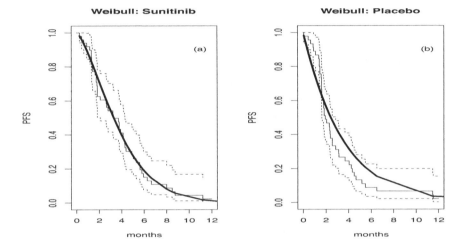

**FIGURE 3.15**
Weibull survival curves fitted to each arm of the lung cancer trial

Weibull distribution for the Placebo arm is not good. The estimate of the median is 2.43 months compared to 1.93 months from the Kaplan-Meier curve.

Log-logistic distributions were tried next. The survival function used is, $S(t) = 1/\{1 + (t/\lambda)^k\}$. The estimated parameters are shown in Table 3.4 and the fitted distributions in Figure 3.16.

The fits of the log-logistic curves look good for both arms. The estimates for the median, $\hat{\lambda}$, is 3.00 months for the Suninitib arm and 2.20 months for the Placebo arm, compared to 3.25 months and 1.93 months from the KM curves.

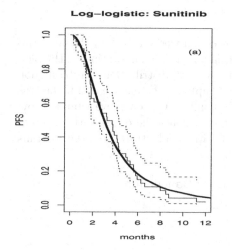

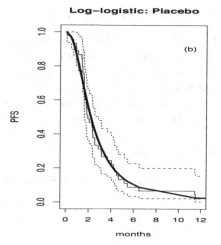

**FIGURE 3.16**

Log-logistic survival curves fitted to each arm of the lung cancer trial

**TABLE 3.4**

Parameter estimates for fitted log-logistic distributions

|  | Estimate | 95% CI | se |
|---|---|---|---|
| Suninitib | $\hat{\lambda} = 2.998$ | [2.379, 3.778] | 0.354 |
|  | $\hat{k} = 2.151$ | [1.703, 2.718] | 0.257 |
| Placebo | $\hat{\lambda} = 2.203$ | [1.755, 2.765] | 0.256 |
|  | $\hat{k} = 2.163$ | [1.679, 2.786] | 0.279 |

## 3.4 Comparing two survival distributions

When we have a two arm study, the first question that springs to mind is whether the two arms have, theoretically, the same survival distribution, and in terms of a null hypothesis, we would like to test $H_0 : S_1(t) \equiv S_2(t)$, i.e. equal for all $t$, against the two survival distributions not being equal for all $t$. A weaker hypothesis is to test just one summary quantity of the survival distributions, such as $H_0 : \mu_1 = \mu_2$, or $H_0 : \text{median}_1 = \text{median}_2$, testing whether the means or medians are equal. We can use a non-parametric approach or a parametric approach. Firstly, we look at a non-parametric approach and describe the *log-rank test*, which is a very widely used test in survival analysis and then we will consider the parametric approach.

### 3.4.1 The log-rank test

Figure 3.17 shows the KM curves of progression-free survival for the two arms of the lung cancer trial. For the moment, ignore all the annotation on the plot. Do you think the two arms have the same survival distribution? Probably not, but has that gap between the two curves happened by chance and if the trial was repeated would that become much smaller? Of couse, we cannot prove this either way, we can only weigh up the evidence, here, using the *log-rank test*. First, we use the lung cancer trial data to illustrate how the log-rank statistic is calculated.

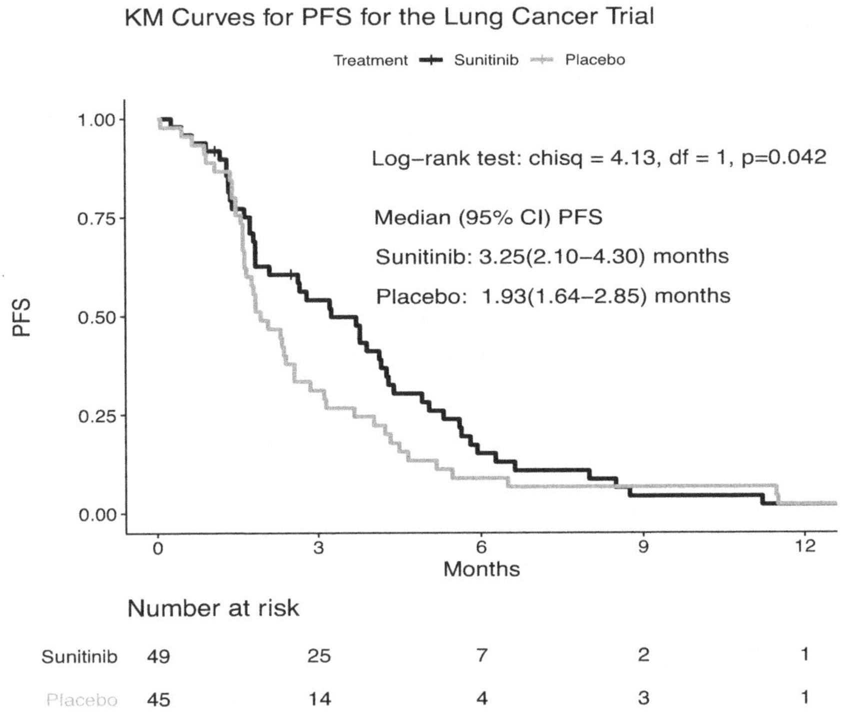
KM Curves for PFS for the Lung Cancer Trial
Treatment Sunitinib Placebo
Log–rank test: chisq = 4.13, df = 1, p=0.042
Median (95% CI) PFS
Sunitinib: 3.25(2.10–4.30) months
Placebo: 1.93(1.64–2.85) months
PFS
1.00
0.75
0.50
0.25
0.00
0
3
6
9
12
Months
Number at risk
Sunitinib 49 25 7 2 1
Placebo 45 14 4 3 1

**FIGURE 3.17**
KM curves for progression-free survival for the arms of the lung cancer trial

**TABLE 3.5**
Ordered survival times

| | | | | | | | | | | | | | |
|---|---|---|---|---|---|---|---|---|---|---|---|---|---|
| st | 1.08 | 1.34 | 2.49 | 2.49 | 2.56 | 2.66 | 2.75 | 2.82 | 2.85 | ... | 28.75 | 29.93 | 40.36 |
| cen | 0 | 1 | 1 | 0 | 1 | 1 | 1 | 1 | 1 | ... | 0 | 0 | 1 |
| arm | 1 | 1 | 2 | 1 | 1 | 2 | 1 | 1 | 2 | ... | 1 | 2 | 1 |

First, order all the survival times, regardless of arm (remember we are using PFS here, but still call it survival time). These are partly shown in Table 3.5, together with the censored times (cen=1 for an event, cen=0 for a censored time) and arm (arm=1 denotes Suninitib, arm=2 denotes Placebo). There are 49 Suninitib patients and 45 Placebo patients. Let the number of distinct event times be $r$, so here $r = 75$. and these are,

$$1.34 \quad 2.49 \quad 2.56 \quad 2.66 \quad 2.75 \quad 2.82 \quad 2.85 \ldots 40.36,$$

which we will denote by $t_{(1)}, t_{(2)}, \ldots, t_{(r)}$, the brackets on subscripts denoting that the event times have been ordered, so $t_{(1)}$ is the smallest event time, $t_{(2)}$ the second smallest, etc.

**TABLE 3.6**

Counts of number of deaths (events), number at risk and number surviving

| Death time | Suninitib | | | Placebo | | |
|---|---|---|---|---|---|---|
| | no. at risk | no. events | no. surv | no. at risk | no. events | no. surv |
| 1.34 | 48 | 1 | 47 | 45 | 0 | 45 |
| 2.49 | 47 | 0 | 47 | 45 | 1 | 44 |
| 2.56 | 46 | 1 | 45 | 44 | 0 | 44 |
| 2.66 | 45 | 0 | 45 | 44 | 1 | 43 |
| 2.75 | 45 | 1 | 44 | 43 | 0 | 43 |
| 2.82 | 44 | 1 | 43 | 43 | 0 | 43 |
| 2.85 | 43 | 0 | 43 | 43 | 1 | 42 |
| ⋮ | ⋮ | ⋮ | ⋮ | ⋮ | ⋮ | ⋮ |
| 4.10 | 42 | 0 | 42 | 36 | 1 | 35 |
| 4.23 | 42 | 0 | 42 | 35 | 2 | 33 |
| 4.26 | 42 | 2 | 40 | 33 | 0 | 33 |
| 4.62 | 40 | 0 | 40 | 33 | 1 | 32 |
| ⋮ | ⋮ | ⋮ | ⋮ | ⋮ | ⋮ | ⋮ |
| 16.56 | 10 | 0 | 10 | 9 | 1 | 8 |
| 17.74 | 8 | 1 | 7 | 7 | 0 | 7 |
| 18.62 | 7 | 0 | 7 | 6 | 1 | 5 |
| ⋮ | ⋮ | ⋮ | ⋮ | ⋮ | ⋮ | ⋮ |
| 21.84 | 6 | 1 | 5 | 1 | 0 | 1 |
| 24.82 | 3 | 1 | 2 | 1 | 0 | 1 |
| 40.36 | 1 | 1 | 0 | 0 | 0 | 0 |

We now count the number of patients that have tumour progression, within each arm, in each of the time intervals $(0 - t_{(1)}], (t_{(1)} - t_{(2)}], \ldots, (t_{(r-1)} - t_{(r)}]$. We also count the number of patients at risk of tumour progression at the start of each event time, $t_i$, and the number surviving beyond that time interval. Note, event times that are censored in the interval, before the event time are not in the risk group at the event time. All these counts are placed in a table, which for space reasons, we show only a limited part as Table 3.6.

Note that in most intervals there are only one or zero PFS events. This is often the case unless there are a large number of patients or the event times are measured coarsely, e.g. to the nearest month. As time progresses, patients with censored event times drop out from the "at risk" groups.

Our lung cancer example is about progression free survival. But now, in describing the theory for the log-rank test, we will revert back to "deaths" at the event of interest, and use "treatment arm" and "control arm" for our two arms. Let the number at risk at the start of the $i$th interval be $n_{1i}$ for the treatment arm and $n_{2i}$ for the control arm. Let $n_i = n_{1i} + n_{2i}$, the total number at risk at the start of the $i$th interval. Let the number of deaths in the $i$th interval for the treatment arm be $d_{1i}$ and the number for the control arm be $d_{2i}$. Let $d_i = d_{1i} + d_{2i}$, the total number of deaths in the $i$th interval. The number of patients in the treatment arm surviving past $t_i$ is $n_{1i} - d_{1i}$, and the number in the control arm is $n_{2i} - d_{2i}$. Now, consider the number of deaths and the numbers surviving past the end of the $i$th time interval for each arm, as a $2 \times 2$ contingency table, as shown in Table 3.7.

**TABLE 3.7**

Counts of deaths and survivors within the $i$th time interval

| Arm | No. deaths in $(t_{(i-1)} - t_{(i)}]$ | No. surviving past $t_{(i)}$ | No. at risk at $t_{(i-1)}$ |
|---|---|---|---|
| treatment | $d_{1i}$ | $n_{1i} - d_{1i}$ | $n_{1i}$ |
| control | $d_{2i}$ | $n_{2i} - d_{2i}$ | $n_{2i}$ |
| Total | $d_i$ | $n_i - d_i$ | $n_i$ |

If $H_0$ is true, all of the patients in the overall risk group, i.e. both risk groups combined, have the same chance of dying. The hypergeometric distribution gives us the probability that $d_{1i}$ die within the treatment arm and $d_{2i}$ die in the control arm, conditioned on the marginal totals of the table, as,

$$\binom{d_i}{d_{1i}}\binom{n_i - d_i}{n_{1i} - d_{1i}} \Big/ \binom{n_i}{n_{1i}}.$$

The expected number of deaths in the treatment and control arms are, respectively,

$$e_{1j} = n_{1i}d_i/n_i, \quad e_{2j} = n_{2i}d_i/n_i.$$

The variances of the number of deaths in the two arms are the same, given by

$$v_{1i} = v_{2i} = v_i = \frac{n_{1i}n_{2i}d_i(n_i - d_i)}{n_i^2(n_i - 1)}.$$

It may seem odd that the variances are the same, until you realise that $d_{2i} = d_i - d_{1i}$, and as $d_i$ is considered as fixed, $d_{1i}$ and $d_{2i}$ are the same variable, apart from a negative sign and the addition of a constant. Hence, we only need to consider one of these.

Our test statistic, $U$, is the sum of the difference between $d_{1i}$ and its expected value under $H_0$,

$$U = \sum_{i=1}^{r} (d_{1i} - e_{1i}),$$

which is the difference in the total number of deaths in the treatment arm and the expected number of deaths in the treatment arm, under the null hypothesis. As we have conditioned on the fixed death times, the $d_{1i}$'s are independent and hence the variance, $V$, of $\sum_{i=1}^{r}(d_{1i} - e_{1i})$ is $V = \sum_{i=1}^{r} v_i$, and as we are summing $r$ random variables, it is no surprise that the central limit theorem applies, giving

$$U = \sum_{i=1}^{r} (d_{1i} - e_{1i}) \;\dot\sim\; N\left(0, \sum_{i=1}^{r} v_i\right), \text{ i.e. } N(0, V), \qquad (3.5)$$

where $\dot\sim$ means "distributed approximately as". Normalising by the standard deviation to give a $N(0, 1)$ random variable, and then squaring gives the log-rank statistic,

$$\text{LR} = \left\{ \sum_{i=1}^{r} (d_{1i} - e_{1i}) \right\}^2 \Big/ \sum_{i=1}^{r} v_i \sim \chi_1^2.$$

(Recall that the square of a $N(0, 1)$ random variable has a chi-squared distribution with one degree of freedom, and so LR has a chi-squared distribution, but note, p-values can be found using the standard normal distribution instead.)

*Lung cancer trial*
The log-rank test for PFS gives, $\text{LR} = 1.70$, with $p = 0.193$, giving no evidence that the PFS curves for the two arms are different. But, read the following section – things will change!

### 3.4.1.1 Stratified log-rank test

Stratification, as mentioned in Chapter 1 and to be discussed in Section 6.4.1.3, is when patient allocation to arms is balanced over one or more factors, such as sex and resection margin. When this is the case, the *stratified log-rank test* should be used.

Let there be $k$ strata, then, essentially, the log-rank statistic is calculated for each stratum and combined to give the test statistic. For the $j$th stratum, let $U_j = \sum_{i=1}^{r(j)} (d_{1i} - e_{1i})$, where the $d_{1i}$'s and $e_{1i}$'s are based on only the data for those patients within the $j$th stratum and $r(j)$ is the number of distinct death times in the $j$th stratum. Let $V_j = \sum_{i=1}^{r(j)} v_i$. Then the stratified log-rank test statistic is

$$\text{LR}_S = \frac{\left(\sum_{j=1}^{k} U_j\right)^2}{\sum_{j=1}^{k} V_j},$$

which has a chi-squared distribution with one degree of freedom under $H_0$.

### 3.4.1.2   Log-rank tests for the lung cancer trial

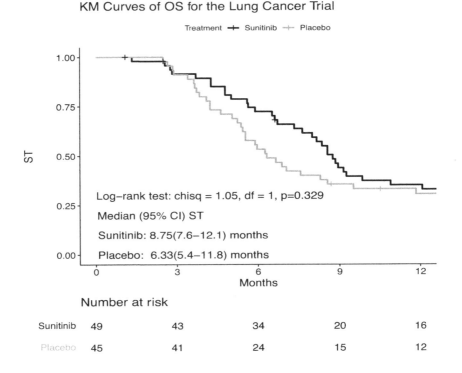

**FIGURE 3.18**
KM curves for overall survival for the arms of the lung cancer trial

We now look at the analysis of the lung cancer trial data in more detail. The patients were stratified on two factors: (i) whether treatment began with cisplatin or carboplatin; (ii) less than six cycles of chemotherapy were received, against the full treatment of six cycles of chemotherapy being received. Thus there are four strata in total. Carrying out a stratified log-rank test gave $LR_S = 4.13$, with $p = 0.042$. So, having taken stratification into account we now have a statistically significant result, indicating a difference in the progression-free survival curves.

We now return to Figure 3.17 and look in more detail. We have added more information to the KM curves than previously. This detail is what is expected when trial results are reported or published. The number at risk table is shown below the time axis. The numbers of patients in the two arms decreases as time progresses, due to events or from censoring. By three months, 24 patients have been lost in the Sunitinib arm and 31 in the Placebo arm. By twelve months there is only one patient left in each arm.

The plot is also annotated with the median progression-free survival times for each arm, together with confidence intervals. The result of the log-rank test is also shown. A coloured plot is particularly useful for distinguishing arms for a multi-arm trial.

This is a good place to show the results for overall survival (OS), which was a secondary endpoint of the trial. Figure 3.18 shows the KM curves for OS together with the median survival times and the results of the stratified log-rank test. There are 16 Sunitinib patients and 12 Placebo patients surviving (or have not been censored) at 12 months. The median survival times are 8.77 months and 6.33 months respectively. The stratified log-rank statistic has a value of 1.05, with $p = 0.329$, showing no evidence of a difference in the two survival curves. Do you believe this to be true? Do we need a larger sample size?

## 3.4.2    Other tests

The log-rank test is ubiquitous, but there are other tests of the equality of survival distributions including tests based on the log-rank test, but using various weightings in its calculation. Before describing these, first let us look at the survival curves for two arms in Figure 3.19. They have been produced from simulated data. For the upper plot, eighty observations were simulated for each arm from exponential distributions, where for Arm 1, $\lambda = 4.0$ and for Arm 2, $\lambda = 2.22$. Random censoring of 10% has been applied. The median survival times, read from the Kaplan-Meier curves are 3.11 months and 1.66 months respectively. The log-rank test has LR $= 6.43$ with a p-value of $p = 0.011$. Visually, the survival curves are different, and as the statistician of the trial team, you would be happy with the result, if this were real data.

In the lower plot, the survival curves cross at about three months. The data for Arm 1 have been simulated from an exponential distribution with $\lambda = 4.0$ and for Arm 2, from a Weibull distribution with $\lambda = 3.33$ and $k = 2.0$. Patients in Arm 2 do better than patients in Arm 1 early on from the time of randomisation, but then patients in Arm 1 do better than patients in Arm 2 later. The median survival times are fairly close at 2.39 and 2.85 months. Clearly the survival curves are different, but we cannot reject the null hypothesis of no difference in the survival curves, based on the log-rank test, as LR $= 1.10$ with $p = 0.294$. So, we must be careful when using the log-rank test. What is happening when these survival curves cross, is that the $(d_{1i} - e_{1i})$'s will all tend to be positive for the first part of the survival curves and then all tend to be negative for the second part of the survival curves, thus cancelling each other out and producing a low value of LR. It can be shown that the log-rank test is optimal when the proportional hazards assumption holds, i.e. $S_2(t) \equiv S_1(t)^\gamma$, is when it is most powerful.

Kaplan–Meier Curves for Exp/Exp Data

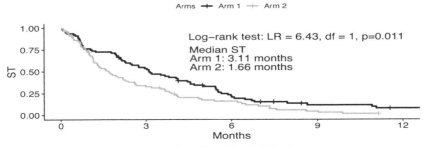

Kaplan–Meier Curves for Exp/Weibull Data

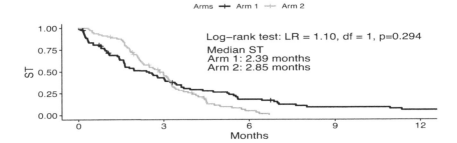

**FIGURE 3.19**
Log-rank tests for (i) a proportional hazards situation and (ii) a non-proportional hazards situation

### 3.4.2.1   Checking for proportional hazards

We have said that the log-rank test is optimal for the case of proportional hazards, although you do not need proportional hazards in order to carry out the test. We can use Equation (3.2) as a basis for checking for proportional hazards. Now $\log\{S(t)\}$ is always negative, and so, negating the equation and taking logs again,

$$\log\big(-\log\big(S_2(t)\big)\big) = \log(\gamma) + \log\big(-\log\big(S_1(t)\big)\big).$$

The function $\log(-\log(.))$ is the *complementary log-log function*. If $\log\big(-\log\big(S_1(t)\big)\big)$ and $\log\big(-\log\big(S_2(t)\big)\big)$ are both plotted against $t$, then, for proportional hazards, the two curves will be parallel, separated by a vertical distance of $\log(\gamma)$ along all of the time axis.

Figure 3.20 shows the log-log survival plots for the simulated data of Figure 3.19. The curves are mostly parallel for the exponential/exponential data but not at all so for the exponential/Weibull data.

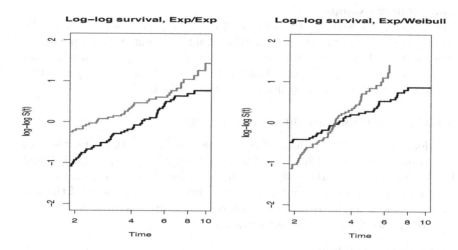

**FIGURE 3.20**
Testing for proportional hazards for the simulated data of Figure 3.19

### 3.4.2.2 Weighted log-rank tests

Various weights, $\{w_i\}$, have been suggested for the $(d_{1i} - e_{1i})$ in the log-rank tests, so the test statistic becomes

$$\left\{ \sum_{i=1}^{r} w_i(d_{1i} - e_{1i}) \right\}^2 \Bigg/ \sum_{i=1}^{r} w_i^2 v_i.$$

Table 3.8 shows the appropriate weights that lead to various alternative tests to the log-rank test.

**TABLE 3.8**
Alternative tests to the log-rank test

| Test | weights $w_i$ |
|---|---|
| Log-rank | $w_i = 1$ |
| Wilcoxon | $w_i = n_i$ |
| Tarone-Ware | $w_i = \sqrt{n_i}$ |
| Peto | $w_i = \hat{S}(t_i)$ |
| Fleming-Harrington | $w_i = \{\hat{S}(t_i)\}^\rho$ |

For the Peto and Fleming-Harrington tests, $\hat{S}(t_i)$ is the KM estimate of the combined sample, i.e. ignoring arm. For the Fleming-Harrington test, $\rho$ has to be chosen. Clearly, $\rho = 0$ gives the log-rank test and $\rho = 1$ gives the Peto test. The tests can be carried out using the R function survdiff() in the survival package and ggsurvplot() in the survminer package, or comp() in the survMisc package, the latter being the most comprehensive.

**Lung cancer example**

The log-rank test for overall survival for the lung cancer trial had LR $= 0.95$, $p = 0.955$ when stratification was not taken into account. Using some of the other tests in Table 3.8: Wilcoxon: test statistic $= 1.92$, $p = 0.166$; Tarone-Ware: test statistic $= 1.43$, $p = 0.231$; Peto: test statistic $= 1.99$, $p = 0.160$. These tests all give more weight to the earlier death times.

### 3.4.2.3   Log-rank test: three or more arms

The log-rank test can be generalised for the case where there are three or more arms. Let there be $K$ arms. Then essentially, $\sum_{i=1}^{r}(d_{ki} - e_{ki})$ is found for the $k$th arm. The variance for the $k$th arm is found and also the covariance for each pair of arms ($k$th and $k'$th). From these a generalisation of the log-rank statistic is found. It has a chi-squared distribution with $K - 1$ degrees of freedom. See Collett[6] for further details.

**Lung cancer example**

As an example, Arm and Gender were made into one variable, Group, with:

$$Group = SM : Treatment = Sunitinib, Gender = Male$$
$$Group = SF : Treatment = Sunitinib, Gender = Female$$
$$Group = PM : Treatment = Placebo, \ \ Gender = Male$$
$$Group = PF : Treatment = Placebo, \ \ Gender = Female$$

Figure 3.21 shows the KM curves for all four groups. The log-rank test gives LR=1.20 on 3 degrees of freedom with p=0.753. Again a non statistically significant result at which we are probably not surprised.

   If we want to make pairwise comparisons of the four groups (probably not with these data!), we could use one of a variety of methods. For the readers who have not met *multiple comparisons* before, here is a very brief outline. If you have several treatments, $K$ in number say, then if you wanted to compare every treatment with every other treatment, you would have $N = K(K-1)/2$ comparisons to make. For $K$ large, this is a lot of comparisons. Here we have six. The problem is that if you make a lot of comparisons, calculating a test statistic and p-value each time, the overall Type I error rate (or experimentwise error rate) is seriously affected. If the significance level is $\alpha$ for each comparison, the overall significance level is not $\alpha$. For example, if $K = 3$, there would be three tests to carry out. Suppose there were no differences between the three treatments, and suppose you could make your three tests independent, e.g. new random samples for each test. Then the probability of a Type I error for each test is $\alpha$, i.e. the probability of rejecting $H_0$ when it is true. But the probability of rejecting at least one of the three tests is $1 - (1 - \alpha)^3$. This is the overall Type I error rate. So for $\alpha = 0.05$, the overall Type I error rate is 0.143. For $K = 4$, there are $N = 6$ comparisons and the overall Type I error rate is 0.265. Multiple comparison procedures adjust the tests so that the overall Type I error rate is closer to $\alpha$. Note, the

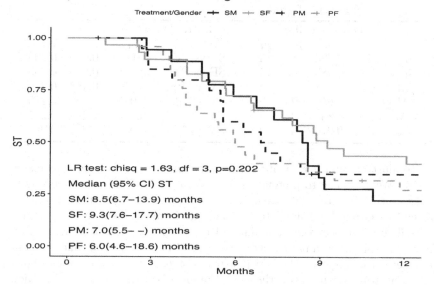

Kaplan–Meier Curves for Lung Cancer Trial

Treatment/Gender — SM — SF — PM — PF

LR test: chisq = 1.63, df = 3, p=0.202
Median (95% CI) ST
SM: 8.5(6.7–13.9) months
SF: 9.3(7.6–17.7) months
PM: 7.0(5.5– –) months
PF: 6.0(4.6–18.6) months

**FIGURE 3.21**
KM curves for the four treatment/gender groups of the lung cancer trial

multitude of tests that are carried out are not pairwise independent usually, and so there are approximations going on with the calculated significances.

A common multiple comparison method is the Bonferroni method, which uses a significance level of $\alpha/N$ for each test. To make the *Bonferroni correction* for each test made, simply multiply the p-value, $p$, obtained for each test by $N$, to give the adjusted p-value, $p_{adj}$ for the test. Then compare this with the significance level, $\alpha$. The Bonferroni correction is rather conservative and only cautiously rejects a null-hypothesis.

The first three columns of Table 3.9 show the comparisons (tests) made, the p-values, $p_i$, for the individual tests before adjustment, and the Bonferroni adjusted p-value, $p_{Bi}$. The latter are all recorded as "1" since $6p_i > 1$ for all six tests. The fourth column gives the adjusted p-values using the Holm method, which is similar to the Bonferroni method, but is less conservative. First the tests are placed in order of the increasing $p_i$ values, as in the table. The smallest $p_i$ is multiplied by $N$, the second smallest by $N-1$, the third by $N-2$, etc. Any adjusted p-values greater than 1 are replaced by 1. These Holm adjusted p-values are denoted by $p_{Hi}$. However, going down the list of $p_H$ values, if a $p_H$ value is less than the preceding $p_H$ value, it is changed to the preceding value. So, in row six of the table, the Holm adjusted p-value is $1 \times 0.976 = 0.976$, but the preceeding one is 1, and so the 0.976 is replaced by 1. Not surprisingly, no unadjusted p-value or adjusted p-value is significant.

**TABLE 3.9**

Multiple comparison of groups

| True data (n=94) | | | | Extra data (n=1316) | | | |
|---|---|---|---|---|---|---|---|
| Test | $p_i$ | $p_{Bi}$ | $p_{Hi}$ | Test | $p_i$ | $p_{Bi}$ | $p_{Hi}$ |
| SF vs PF | 0.293 | 1 | 1 | SF vs PF | < 0.001 | 0.000 | 0.000 |
| SF vs PM | 0.485 | 1 | 1 | SF vs PM | 0.008 | 0.046 | 0.039 |
| SM vs PF | 0.637 | 1 | 1 | SM vs PF | 0.070 | 0.418 | 0.279 |
| SM vs SF | 0.656 | 1 | 1 | SM vs SF | 0.087 | 0.523 | 0.279 |
| PM vs PF | 0.858 | 1 | 1 | PM vs PF | 0.494 | 1 | 0.987 |
| SM vs PM | 0.976 | 1 | 1 | SM vs PM | 0.907 | 1 | 0.987 |

To illustrate multiple testing further, the lung cancer data set is increased in size. We are going to pretend we have data for 1316 patients. One way to make up this fictitious data would be to fit a parametric distribution to each of the four Groups and simulate survival data from the distribution. But, no, we will simple repeat the same data 14 times, giving us 1316 patients. This may seem silly, as each patient will give rise to 14 equal survival times, but the data serve the purpose. The KM plot will look exactly like the one in Figure 3.21. Can you see why this is the case? Columns 5-8 of Table 3.9 give the new multiple comparison results. It is easier to see how the Holm adjusted p-values are calculated now. There are two significant unadjusted p-values, which are also still significant after using the two adjustment methods.

The log-rank test for the enhanced data set has LR = 17.3, with 3 degrees of freedom and $p = 0.001$. This is a highly significant result, showing that by increasing the sample size, a non-significant result can be turned into a significant one. That is why the *clinically significant difference* should always be stated before data are collected. If the data analysis finds a statistically significant difference, it can only be usefully used to say there is evidence of a difference between groups if the clinical significant difference has also been achieved. Chapter 6 covers sample size calculations for clinical trials and other studies.

## 3.5   The ESPAC4-Trial

One of the main data sources we use in this text is the series of ESPAC trials. To further illustrate the methods of this chapter, the ESPAC4 trial is used which compared two chemotherapy treatments for pancreatic cancer. The title of the trial is,

"European study group for pancreatic cancer —Trial 4. Combination versus single agent chemotherapy in resectable pancreatic ductal and peri-ampullary cancers".

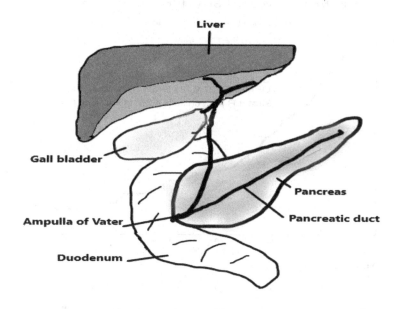

**FIGURE 3.22**
The pancreas

This was the fourth trial in a series of five trials to date, for pancreatic cancer by the European study Group for Pancreatic Cancer (ESPAC), although Trial number 2 never came to fruition. Figure 3.22 shows a diagram of the pancreas. The pancreas has two functions, (i) to produce enzymes and secrete them into the small intestine for breaking down food and (ii) to produce insulin which regulates glucose in the body. The most common type of pancreatic cancer is "ductal" where the tumour develops around the pancreatic duct. Ampullary cancers occur in the "Ampulla of Vater" where the pancreatic and bile ducts meet the small intestine. Peri-ampullary cancers occur around the area. Pancreatic cancer has a notoriously poor survival rate with only 5% of patients surviving more than 5 years, but longer for those diagnosed at an early stage. Patients with peri-ampullary cancers can fare better.

The trial was a phase III, two arm, open-label, multi-centre randomised clinical trial. The treatments were: combination gemcitabine and capecitabine (GEMCAP) therapy, and therapy by gemcitabine (GEM) alone, both treatments used as an adjuvant therapy following resection (i.e. chemotherapy used after surgery). Approximately 120 centres in approximately 8 countries were used for recruiting 722 patients with pancreatic ductal adenocarcinoma (361 in each arm) and 740 patients with peri-ampullary carcinoma (370 in each arm). Patients were randomised to arm, but stratified by country and resection margin. Patients received up to 24 weeks of chemotherapy and were

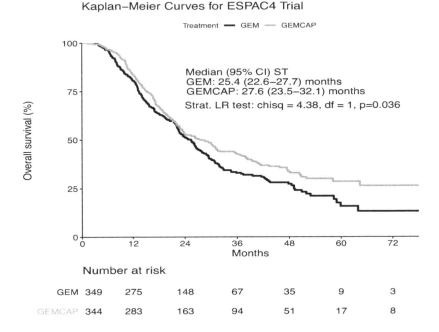

FIGURE 3.23

KM curves for overall survival for the arms of the ESPAC4 trial

followed up every three months for a minimum of 5 years, or until death or withdrawal from the trial.

There were 15 criteria for inclusion in the study, e.g. able to attend for long-term follow-up, life expectancy greater than 3 months, fully informed written consent given. There were 13 criteria for exclusion, e.g. patients with TNM Stage IV disease, patients younger than 18 years and pregnancy.

The trial was essentially two trials in one – the first for ductal carcinoma which would recruit patients faster and report results earlier, the second for peri-ampullary carcinoma where it would take much longer to recruit patients. The primary outcome for both sub-trials was *overall survival*, OS. For the ductal carcinoma, the secondary outcomes were: toxicity, quality of life, two year survival, five year survival and relapse free survival. For the peri-ampullary carcinoma, the secondary outcomes were: toxicity, quality of life, four year survival and relapse free survival. See Neoptolemos[44] for further details.

For the ESPAC4 ductal trial, overall survival, measured from the time of randomisation of the patient, was compared between the two treatment arms, GEM and GEMCAP. The number of deaths and patients in the final analysis was less than that in the trial design, because the Independent Data and Safety Monitoring Committee (IDSMC) had recommended early publication

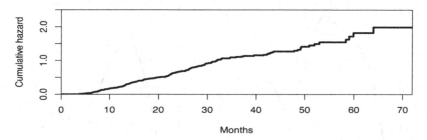

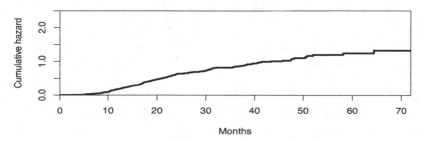

**FIGURE 3.24**
Nelson-Aalen estimates of the cumulative hazards for the ESPAC4 trial

based on efficacy which was accepted by the Trial Steering Committee (TSC)
– see Chapter 4 for details about these two committees.

We use the ESPAC4 data here to illustrate the survival analysis methods.
We use data for 693 patients; the paper uses data for 730 patients. Figure 3.23
shows the Kaplan-Meier curves for the two arms.

The two survival curves are close for the first 24 months with the GEMCAP
curve being slightly above that for GEM, but then the curves diverge, with
GEMCAP having a survival advantage. The median survival time for the GEM
patients was 25.4 months, with 95% confidence interval, [22.6, 27.7]. For the
GEMCAP arm the median survival time was 27.6 months with 95% confidence
interval [23.5, 32.1]. The stratified log-rank test statistic had a value of 4.38
on 1 df, giving $p = 0.036$, a significant result.

Figure 3.24 shows the Nelson-Aalen plots of the cumulative hazard functions. This would not usually be used in a statistical report or an academic
paper reporting the trial, but we include it here for interest. For both arms,
the cumulative hazard increases very slowly to start with, which is not surprising as one of the entry criteria was life expectancy greater than three months.
The cumulative hazard functions then increase steadily till about 36 months
and then the rate of increase drops slightly. The cumulative hazard for GEM
attains a higher level than that for GEMCAP.

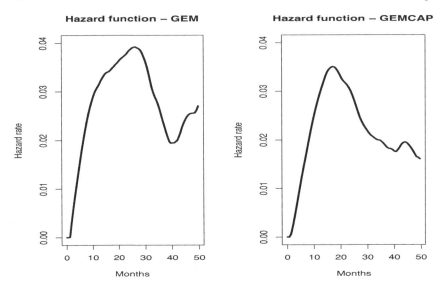

**FIGURE 3.25**
Smoothed estimates of the hazard functions for the ESPAC4 trial

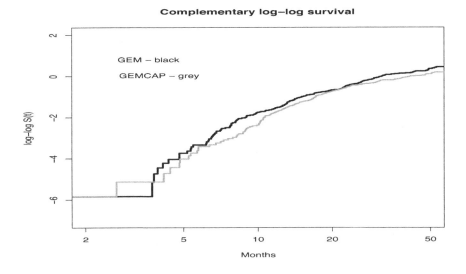

**FIGURE 3.26**
Plot for assessing the proportional hazards assumption for the ESPAC4 trial

Next, Figure 3.25 shows smoothed estimates of the hazard functions. The hazard functions for both arms increase rapidly until about 25 months for GEM and 18 months for GEMCAP. Then the hazard rates fall off, concurring with the changes of slope seen in the cumulative hazard functions. The estimates of the hazard functions are not reliable past 50 months.

Finally, Figure 3.26 shows a plot of the complementary log-log transformations of the survival curves where proportional hazards can be assessed. Note the R function survfit() puts the x-axis on a log-scale. Are the two plots parallel? Its too hard to say – probably not. The vertical distance between the two curves is small for most of the range.

## 3.6   Comparing two parametric survival curves

How can we test for a difference in two fitted parametric curves, such as those in Figure 3.16 for the lung cancer trial? Well, just looking at them tells us they are not exactly the same, but what we really mean is can we fit the same parametric curve to the two arms because there is no difference in efficacy for the two treatment regimes, and the difference we see is just due to random variation. There are two ways we can do this. Firstly, we can fit a particular distribution to each of the arms and then test the hypothesis that the parameters of the two distributions are identical, and secondly, we can use the generalised log-likelihood ratio test which compares the likelihoods of the case where two separate distributions (both from the same family) are fitted to the data, one for each arm, to the case where one distribution is fitted to the combined data. We will use the lung cancer trial data with the two fitted log-logistic distributions for illustration.

**Testing equality of parameters**

For the log-logistic distribution, there are two parameters, $\lambda$ and $k$. For the Sunitinib arm let these be denoted by $\lambda_S$ and $k_S$ and let $\hat{\lambda}_S$ and $\hat{k}_S$ be the maximum likelihood estimates of these. Similarly, for the Placebo arm, the parameters are $\lambda_P$ and $k_P$ with maximum likelihood estimates, $\hat{\lambda}_P$ and $\hat{k}_P$. We test the null hypothesis,

$$H_0 : \lambda_S = \lambda_P \text{ and } k_S = k_P,$$

against the alternative hypothesis

$$H_1 : \lambda_S \neq \lambda_P \text{ and/or } k_S \neq k_P.$$

We cannot test $H_0 : \lambda_S = \lambda_P$ and $H_0 : k_S = k_P$ separately using the maximum likelihood estimators, $\hat{\lambda}_S$, $\hat{\lambda}_P$, $\hat{k}_S$ and $\hat{k}_P$, because $\hat{\lambda}_S$ and $\hat{k}_S$ are not independent and have a covariance, and similarly for $\hat{\lambda}_P$ and $\hat{k}_P$. We have to carry out a multivariate test. The null and alternative hypotheses are best written in vector form as,

$$H_0 : (\lambda_S, k_S) = (\lambda_P, k_P),$$

and
$$H_1 : (\lambda_S, k_S) \neq (\lambda_P, k_P).$$

Asymptotically (i.e. large sample size), $(\hat{\lambda}_S, \hat{k}_S)$ has a bivariate normal distribution with covariance matrix, $\mathbf{V}_S$, the inverse of the information matrix which in turn is the negative of the expected value of the Hessian matrix. Similarly for $\mathbf{V}_P$. Now we look at the difference $\mathbf{d} = ((\hat{\lambda}_S, \hat{k}_S) - (\hat{\lambda}_P, \hat{k}_P))$, which has covariance matrix $\mathbf{V}_S + \mathbf{V}_P$ and hence

$$D = \mathbf{d}^{\mathsf{T}}(\mathbf{V}_S + \mathbf{V}_P)^{-1}\mathbf{d}$$

has a chi-squared distribution with 2 df under $H_0$.

For the fitted log-logistic distributions for the lung cancer trial, we have,

$$\text{Sunitinib}: \quad \hat{\lambda}_S = 2.998, \hat{k}_S = 2.151, \quad \mathbf{V}_S = \begin{bmatrix} 0.1252 & 0.0018 \\ 0.0018 & 0.0658 \end{bmatrix}$$

$$\text{Placebo}: \quad \hat{\lambda}_P = 2.203, \hat{k}_P = 2.163, \quad \mathbf{V}_P = \begin{bmatrix} 0.0653 & -0.0025 \\ -0.0025 & 0.0780 \end{bmatrix}.$$

Then $D = 3.32$ with $p = 0.190$ and so there is no evidence to reject the null hypothesis of identical parameters. (Note flexsurvreg() works on the log scale for parameters, whereas we have worked with the parameters directly. The results are very similar.)

### Generalised likelihood-ratio test
The likelihood ratio test compares the likelihood under the null hypothesis to the likelihood under the alternative hypothesis, the comparison being their ratio.

Under $H_1$, the likelihood is the product of the two separate likelihoods for the two arms (because the arms are independent). Under $H_0$, the likelihood is obtained by combining the data from the two arms and fitting one distribution. As the likelihoods are extremely small, the log-likelihoods, $LL$, are given:

$$\text{Sunitinib}: - 107.091, \text{ Placebo}: - 89.223, \text{ Combined}: - 198.002.$$

The likelihood ratio, $LR$, is thus $LR = \exp(-198.002 + 107.091 + 89.223)$, which is equal to 0.18.

Asymptotically, $-2\log(LR)$ has a chi-squared distribution with degrees of freedom the difference in the number of parameters in $H_1$ and $H_0$, i.e $4 - 2 = 2$ here. Now $-2\log(LR) = 3.377$, with $p = 0.185$, again a non-significant result.

We have seen how we can parametrically test the equality of two survival distributions, but this is not the route usually taken. The problem is, if you reject the hypothesis of equality of distributions, it is then awkward to further explore the differences in the two distributions. What does it mean, in practical terms, if $k_S$ is greater than $k_P$, while $\lambda_S$ is smaller than $\lambda_P$? Instead, inference is based on useful quantities of interest, such as the hazard ratio $\gamma$.

# 4

## Designing and Running a Clinical Trial

If you want to run a clinical trial, you need a lot of money because, generally, clinical trials are costly. Many people can be involved: physicians, nurses, clinical trial centre staff and of course, patients. Drugs to be used in the trial might be expensive. In the UK, trials are usually funded by the National Health Service (NHS), the Medical Research Council (MRC), charities such as Cancer Research UK (CRUK) and drug companies.

Clinical Trials Units (CTUs) and Contract Research Organisations (CROs) run clinical trials. Drug companies run their own clinical trials or contract with CTUs and CROs. Many universities have their own CTU and are heavily involved in medical research, as would be expected. A trial might only be run at a single centre or site, e.g. a particular hospital, or several centres. The terms "centre" and "site" are used interchangeably.

## 4.1 Types of trials and studies

Here, we take time to briefly discuss various types of trial and studies. It is a large area with many books dedicated to trial design, e.g. Hackshaw[20], Green[18], Jung[29] and Machin[38].

Figure 4.1 shows a categorisation of medical studies, clinical trials being just a subset of medical studies. Studies are either *observational* or *experimental*. Patients recruited for observational studies do not have any intervention – they are not given drugs or other treatments, they are just "observed", which might involve the patient completing questionnaires, being interviewed, or information about the patient might be gleaned from their medical records. Patients recruited to clinical trials do have an intervention of some kind, for example the administration of a chemotherapy drug, or surgery, or both. Simply put, a clinical trial is an experiment on humans.

Studies can be *retrospective* or *prospective*. Experimental studies are always prospective. Observational studies can be either. A *cross-sectional* study looks at data at one moment in time, or within a short time interval for practical reasons. A *longitudinal* study collects data at various time points.

A *case-control* study selects patients, some with the disease in question and some without, and looks back in time at factors that might be related

DOI: 10.1201/9781003041931-4

to the disease. For example, patients are selected, some with lung cancer and some without, and then past smoking habits are established with the view of finding a link between smoking and lung cancer.

A *cohort* study is a prospective study, where subjects are selected according to some criteria, and then followed for a period of time to establish who succumbs to the disease in question. For example, a group of people are selected, some of whom smoke and others do not, and during the next twenty years records are kept of who develop lung cancer and who do not. One problem with this is that you have to wait twenty years! A retrospective study can use data that was collected in the past and this will give a much speedier result. For example, patients needing frequent repeat interventions for chronic lymphocytic leukaemia (CLL) are selected from hospital records. Then their medical notes are scrutinised from the start of their treatment until ten years later, in order to map the progress of the disease. Other examples are given in Figure 4.1.

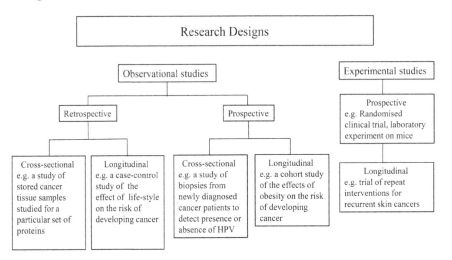

**FIGURE 4.1**
Categorisation of medical study designs

There is a hierarchy of evidence for clinical trials and studies. It is

1. *Systematic Review*
2. *Randomised Controlled Trial (RCT)*
3. *Non-Randomised Controlled Trial*
4. *Cohort Study*
5. *Case-control Study*
6. *Case Reports, Case Series*
7. *Expert Opinion*

The higher up the list, the more evidence there is for the conclusions being made. A systematic review gives the best evidence for the intervention in question and expert opinion gives the least evidence.

A systematic review looks at relevant publications relating to the intervention in question and summarises their findings, see Chapter 8 – Meta-Analyses. Evidence is based on several studies, whereas for a single RCT, evidence is based on just one study – and, as always with statistical methods, errors do occur such as Type I and Type II errors in hypothesis testing which can lead to wrong conclusions. The US Food and Drug Administration (FDA) usually requires two well-designed clinical trials supporting the use of a new treatment in order to gain drug approval. Case-control studies and cohort studies are described in detail in Chapter 7 – Cancer Epidemiology. A case report gives the details of a particular patient's demographics, diagnosis, treatment and outcome data. A case series is the case reports of several patients who have received similar treatments. Expert opinion can often be wrong!

## 4.2 Clinical trials

We now concentrate on clinical trials giving a brief overview of trial set up and the running of a trial. Clinical trials are generally categorised into four phases. These are,

- *Phase I Trials* are "first in man" trials. Drugs will have been developed in a laboratory and then tested on animals, before trying them out on humans. They involve a small number of subjects, often healthy volunteers.

- *Phase II Trials* are relatively small trials, generally ranging from 30 to 120 patients, that glean more information about the safety and efficacy of the drug or treatment being tested, and help set the dose level for a Phase III trial.

- *Phase III Trials* are confirmatory trials, in that they are designed to give a definitive answer to the hypothesis being tested, for example "Does this new cancer drug increase the median survival time of patients compared to the existing treatments?" A sample size of 500 to 1000 patients is not uncommon, with many centres involved in collecting the patient data.

- *Phase IV Trials* are run after a drug has been licenced to collect data on the long-term effects of the drug.

There are other types of clinical trials that are relevant to cancer, including,

1. *Pilot Studies* are run when basic information is lacking. A pilot study might be run in order to obtain an estimation of the variance

for a primary outcome, so that a sample size calculation can be made for further studies.

2. *Feasibility Studies* A feasibility study is run to see if a particular trial could be run in the future. It would ascertain whether enough patients could be recruited and whether centres were willing to participate in a future trial.

3. *Prevention Trials* As the name suggests, these trials investigate whether a treatment or intervention can prevent cancer or other disease from developing. For example, can an advertising campaign manage to reduce alcohol and tobacco consumption, which is then sustained into the future, hence reducing lifetime cancer risk.

4. *Biomarker Trials* Biomarker trials are similar to drug trials, but biomarkers are tested to see whether they can help diagnose cancer or predict the course of the disease. See Chapter 9 – Biomarkers.

### 4.2.1   Regulatory and other bodies

This is a good place to mention that clinical trials need to be run according to regulations and guidelines. The following are some of the important things to consider. See Hackshaw[20] for further details.

The *International Council for Harmonisation of Technical Requirements for Pharmaceuticals for Human Use* (ICH) (https://www.ich.org) has guidelines for *Good Clinical Practice* (GCP). There are 13 core principles for GCP, which we will not repeat here, they include ethics, risks, safety and well-being for subjects, information to be made available, etc.

In the UK, the *Medicines and Healthcare products Regulatory Agency* (MHRA) regulates clinical trials that involve an *Investigational Medicinal Product* (IMP) and hence called a CTIMP. A CTIMP has to be authorised by the MHRA. If a trial does not have an IMP, then the MHRA does not have to be involved. The MHRA inspects CTUs and other sites where CTIMPs are run, to check they are being conducted according to GCP.

The *European Medicines Agency* (EMA) has a role to guarantee the scientific evaluation, supervision and safety monitoring of human and veterinary medicines in the EU. The *European Union Drug Regulating Authorities Clinical Trials Database* (EudraCT) is the European database for all interventional clinical trials on medicinal products authorised in the EU.

All trials have to be registered in an appropriate registry, for example the *International Standard Randomised Controlled Trial Number* (ISRCTN) Registry, or the *US National Library of Medicine* (ClinicalTrials.gov).

All trials have to be approved by a *Research Ethics Committee* (REC), which is an independent committee that assesses the ethics of a trial and is there to protect the safety and other interests of the participants.

## 4.2.2 Trial setup

Before a clinical trial is set up, the initial idea for a trial has to be developed. A design for the trial has to be established, sample sizes chosen and resources needed to run the trial have to be assessed. For example, a meeting consisting of the person destined to lead the trial (Chief Investigator), a statistician and a senior member of the CTU or CRO might take place to discuss taking the trial forward.

There is a lot of work involved in setting up a clinical trial, from the initial idea for the trial, until the recruitment of the first patient, participant or subject, three terms we will use interchangeably. The NHS has a useful website (http://www.ct-toolkit.ac.uk/routemap) that covers all aspects of setting up a clinical trial, cancer or otherwise.

A typical clinical trial needs the following people:

- *Sponsor*: The sponsor is responsible for the finance, quality assurance and insurance for the trial. The sponsor can be a company, e.g. a pharmaceutical company, an organisation, e.g. a CRO, an institution, e.g. a university, or, rarely, an individual.

- *Chief Investigator* (CI): The person who has overall responsibility for the overall scientific integrity, the ethical aspects and the administration of the trial. The trial might have been their idea or they are acting as the most senior person of a research group.

- *Trial Coordinator*: The trial coordinator facilitates and coordinates the running of the trial, reporting to the CI. They liaise with many others, including the sponsor, PPI, trial monitors, centres, etc.

- *Data Manager*: The data manager is responsible for the recording of the trial data, setting up a database and ensuring data quality.

- *Trial Statistician*: The trial statistician is responsible for all statistical aspects on the trial and is involved from start to finish of the trial. A more detailed description of the work for the trial statistician is give in Chapter 6.

- *Trial Monitor*: The trial monitor checks that the trial is being conducted according to the protocol and Standard Operating Procedures (SOPs) (see later). They check centre (hospital) records, data, master files, etc. Much can be done centrally from the CTU, but site visits will occur when appropriate.

- *Principal Investigators* (PI): Each centre participating in the trial has a PI who is responsible for the conduct of the clinical trial at that particular centre. The PI must be a health care professional.

- *Research nurses, pharmacists, plus others*: These are the people who carry out the day-to-day running of the trial at the centres, recruitment of patients, administering treatments, etc.

- *Patient and Public Involvement* (PPI): PPI is an important aspect of the clinical trial. Patients and the public are involved at every stage of a trial, from helping with initial design to attending trial committee meetings. The members of the PPI group can give the patient's view of the trial, for instance a PPI member may point out that the treatment regime proposed within a trial might not be acceptable to most trial subjects, e.g. they may see that the many proposed visits to a hospital for tests may be too onerous for most patients. The PPI group can help write patient information leaflets and keeping over ambitious clinicians in line with patients' general knowledge.

- *Trial Management Group* (TMG): The TMG has responsibility for running the trial. It is chaired by the CI and its members consist of relevant researchers, the trial manager, the trial statistician and other staff and a PPI representative as appropriate. The Group usually meets regularly.

- *Trial Steering Committee* (TSC): The TSC provides executive oversight to the trial and advises the Sponsor on the trial's conduct. Members include the CI, some members of the TMG and independent members. The Committee reviews safety and efficacy data, resolves higher level problems within the trial and can terminate the trial prematurely before the planned end of the trial, if for example, it is already clear that one treatment is superior to the other. Meetings are scheduled as appropriate, for instance, a meeting might only be necessary every 12 months.

- *Independent Safety and Data Monitoring Committee (ISDMC or DMC)*: The main remit of the ISDMC is to guard the safety of patients participating in the trial. It assesses safety and efficacy data as the trial progresses, by holding *interim analyses* at pre-specified intervals. It looks at the frequency of adverse events and checks whether one arm accumulates these at a faster rate than the other. It looks at the efficacy data and can recommend that a trial be stopped for safety reasons, if one treatment is clearly superior to the other, or for futility, if there is clearly no difference between the treatments and there is no point in continuing with the trial. The committee consists mainly of independent members. It will have an independent statistician as well as a statistician from the trials centre, possibly the study statistician.

- *Patients*: The patients who kindly agree to participate in trials, without whom the trials could not take place.

Once the outline design of the trial has been established and a risk assessment made, a sponsor and funding needs to be found. A pharmaceutical company may be sponsoring and funding the trial, or an academic research group will approach a charity or a government funded body. Examples of charities funding cancer clinical trials are Cancer Research UK (https://www.cancerresearchuk.org), Blood Cancer UK (https://bloodcancer.org.uk), Kidney Cancer UK (https://www.kcuk.org.uk) and the Roy Castle Lung Cancer Foundation (https://roycastle.org/). The sponsor and/or funding body will probably require peer review. Proposed cancer trials in the UK will often be peer reviewed by one of the NCRI clinical studies groups, each one concerned with a particular cancer type. Government organisations funding clinical trials are the Medical Research Council

(MRC) (https://mrc.ukri.org/) and the National Institute for Health Research (NIHR) (https://www.nihr.ac.uk/). Universities that have clinical trial centres sponsor their own clinical trials. NHS Trusts also sponsor clinical trials.

The next step is to write the protocol, a document that covers all aspects of the clinical trial and how it will be run. Protocols often run to well over 100 pages. There is a Health Research Authority (HRA) document , entitled "Protocol guidance and template for use in a Clinical Trial of an Investigational Medicinal Product (CTIMP)" (https://www.hra.nhs.uk/planning-and-improving-research/research-planning/protocol/) that would be a good starting point for writing a protocol from scratch for a CTIMP. There is also a similar one for qualitative studies.

The *ICH Guideline for good clinical practice*, E6(R1), states that a protocol should include the following topics:

1. General Information
2. Background Information
3. Trial Objectives and Purpose
4. Trial Design
5. Selection and Withdrawal of Subjects
6. Treatment of Subjects
7. Assessment of Efficacy
8. Assessment of Safety
9. Statistics
10. Direct Access to Source Data/Documents
11. Quality Control and Quality Assurance
12. Ethics
13. Data Handling and Record Keeping
14. Financing and Insurance
15. Publication Policy
16. Supplements

The protocol might also detail how samples of blood and tumour tissue are to be taken from patients and stored in tissue banks. These will be used in future *translational research.*

The protocol is a very important document as it tells everybody involved with the clinical trial what the trial is about, how it will be conducted, what data are to be collected and stored, and how the data will be analysed and reported. Above all, it is there to ensure the safety of the patient. People working on the trial will also use *standard operating procedures* (SOPs), some of which are centre specific and some are trial specific.

All CTIMP trials must be registered on the *European Clinical Trials Database* (EudraCT). A EudraCT number will be issued for the trial. The *Integrated Research Application System* (IRAS) is a system for applying for health and social care research in the UK and so this is another number to be secured. Then the trial will need *clinical trial authorisation* (CTA) from the MHRA. Clinical trials run within the NHS need permission from the local NHS Research and Development Office.

Once the protocol is finalised, submission to a research ethics committee can be made. The committee will scrutinise the proposed research looking at all aspects of patient safety and well-being. A trial cannot start without ethics approval.

Documentation has to be being sorted out. There has to be a *Trial Master File* (TMF), which is kept at the trial coordinating site, e.g at the Clinical Trials Unit, and a copy is kept at each participating site in the *Investigator Site File* (ISF). The ICH E6(R1) guideline lists the essential documentation that should be stored for the trial.

*Pharmacovigilance* is the detection, assessment and reporting of adverse effects of medicines. Plans have to be drawn up as to how this will happen within the trial. During the course of a trial, patients can experience one or more of the following,

- *Adverse Event* (AE), e.g. mild nausea when the patient does not usually suffer from nausea
- *Adverse Reaction* (AR), e.g. a rash which is related to the drug being taken
- *Serious Adverse Event* (SAE), e.g. severe nausea leading to hospitalisation
- *Serious Adverse Reaction* (SAR), e.g. kidney damage caused by the drug
- *Suspected Unexpected Serious Adverse Reaction* (SUSAR), e.g. death of a patient when this was unexpected; SUSARs have to be reported promptly to the MHRA and the relevant ethics committee.

Patients invited to join a clinical trial must be given full information about the trial, the treatment regimes, possible side effects and if the trial is a randomised one, then it has to be explained that they may or may not be given the treatment on test, but receive the placebo or the treatment that is current standard practice. Because of equipoise, it is explained that until the results of the trial are known, then a trial patient is no better off whether they are allocated to the test or control treatment.

### 4.2.3   Running the trial

When all the documentation and authorisations are in place, the clinical trial can start. The start date can be defined as the date the first patient is recruited, or some other date as described in the protocol. Patients are enrolled continuously over a recruitment period until the sample size requirements are met, whereupon recruitment stops. For large trials with many centres, the centres are not usually opened all at once, but one after another over the course of the first part of the recruitment period. It makes sense to open the larger centres first, where it is expected that the rate of recruitment will be high. When recruitment ends, a follow-up period starts.

Throughout the recruitment and follow-up period, trial management continually happens and meetings of the TMG take place regularly, as does trial monitoring where monitors use a mixture of central monitoring and onsite monitoring. Central monitoring takes place at the trial centre and is to check and quality control patient data. For onsite monitoring, a monitor will visit a centre (site) and assure that the centre is following the protocol, safety reporting is active, AEs, SAEs ARs and SUSARs are recorded and reported as appropriate.

The DMC meets regularly, studies the safety and efficacy data and issues reports. The TSC meets and reviews progress and takes an overview of the whole trial.

The MHRA might inspect the trial. Hopefully, this is not because of a serious breach of GCP or the trial protocol, but because the trial has been chosen by the MHRA as part of their routine inspection of the clinical trials centre.

At the end of the trial, once the data are cleaned and checked, the database is locked. The data are analysed and a statistical report written. This is amalgamated into the end of study report. Research papers are written and submitted to medical journals for publication.

At the end of the trial, an EudraCT, *Declaration of End of Trial form* has to be submitted to the research ethics committee to the MHRA. All data has to be archived by law. A pharmaceutical company may prepare documents for application to the EMA for licensing of the test drug in the trial.

# 5

## Regression Analysis for Survival Data

Regression analysis is a very widely used statistical technique to analyse data and fit models. There are three types of regression models: parametric regression models, non parametric regression models and semiparametric regression models. These are described in the Appendix. In this chapter we look at some parametric and semi-parametric regression models for survival data. The dependent variable will be survival time, the independent variables will be various patient demographic and clinical variables. Note, we will sometimes refer to the independent variables as factors or covariates or explanatory variables or prognostic variables. We will use the terms interchangeably. Collett[6] gives a more comprehensive coverage of the area.

### 5.1   A Weibull parametric regression model

First, we look at a Weibull regression model. The dependent variable is survival time, $T$ and $p$ dependent variables (covariates) are $X_j$, for example, age, treatment, lymph node involvement, etc.

Recall that the Weibull distribution can be parameterised in various ways and the one we will use is $S(t) = \exp(-\lambda t^k)$, $f(t) = \lambda k t^{k-1} \exp(-\lambda t^k)$ giving the hazard function as, $h(t) = \lambda k t^{k-1}$.

We could model $\lambda$ and $k$ on our covariates, $x_j$, for example,

$$k = \exp(\alpha_0 + \alpha_1 x_1 + \alpha_2 x_2 + \ldots + \alpha_p x_p),$$
$$\lambda = \exp(\beta_0 + \beta_1 x_1 + \beta_2 x_2 + \ldots + \beta_p x_p),$$

and then use maximum likelihood to estimate the alphas and betas, giving us $\hat{\alpha}_j$, $\hat{\beta}_j$ and hence $\hat{\lambda}$ and $\hat{k}$. But how would we jointly interpret $\hat{\lambda}$ and $\hat{k}$? We could substitute these into the equation for the hazard function, but better still, why not start by modelling the hazard function from the start.

To do this, we model $\lambda$ for the $i$th patient as,

$$\lambda_i = \exp(\beta_1 x_{i1} + \beta_2 x_{i2} + \ldots + \beta_p x_{ip})\lambda,$$

and so their hazard function is,

$$h_i(t) = \exp(\beta_1 x_{i1} + \beta_2 x_{i2} + \ldots + \beta_p x_{ip})\lambda k t^{k-1} = \exp(\boldsymbol{\beta}^\mathsf{T}\mathbf{x_i})\lambda k t^{k-1},$$

DOI: 10.1201/9781003041931-5

where we use the vector form, $\boldsymbol{\beta}^{\mathsf{T}}\mathbf{x_i}$, for $\beta_1 x_{i1} + \beta_2 x_{i2} + \ldots + \beta_p x_{ip}$ for convenience.

Then the corresponding survival function and probability density function, evaluated at time $t_i$, are,

$$S(t_i|\mathbf{x_i}) = \exp[-\exp(\boldsymbol{\beta}^{\mathsf{T}}\mathbf{x_i})\lambda t_i^k],$$
$$f(t_i|\mathbf{x_i}) = \exp(\boldsymbol{\beta}^{\mathsf{T}}\mathbf{x_i})\lambda k t_i^{k-1}\exp[-\exp(\boldsymbol{\beta}^{\mathsf{T}}\mathbf{x_i})\lambda t_i^k].$$

The likelihood can be found using Equation 3.4,

$$\mathrm{L} \propto \prod_{i=1}^{n}\{f(t_i|\mathbf{x_i})\}^{\delta_i}\{S(t_i|\mathbf{x_i})\}^{1-\delta_i} = \prod_{i=1}^{n}\left\{\frac{f(t_i|\mathbf{x_i})}{S(t_i|\mathbf{x_i})}\right\}^{\delta_i} S(t_i|\mathbf{x_i})$$
$$= \prod_{i=1}^{n}\{h_i(t_i|\mathbf{x_i})\}^{\delta_i} S(t_i|\mathbf{x_i}),$$

maximisation of which with respect to $\lambda$, $k$ and $\beta_1, \beta_2, \ldots, \beta_p$, will fit the model.

Basically, the model is,

$$h_i(t) = \exp(\beta_1 x_{i1} + \beta_2 x_{i2} + \ldots + \beta_p x_{ip}) h_0(t),$$

where $h_0(t) = \lambda k t^{k-1}$ is the *baseline hazard function*. Each patient "adjusts" the baseline hazard function according to the values of their covariates.

The hazard ratio of the hazard functions, $\gamma_{ij}(t)$, for two patients, $i$ and $j$, is

$$\gamma_{ij}(t) = \frac{\exp(\boldsymbol{\beta}^{\mathsf{T}}\mathbf{x_i})h_0(t)}{\exp(\boldsymbol{\beta}^{\mathsf{T}}\mathbf{x_j})h_0(t)} = \frac{\exp(\boldsymbol{\beta}^{\mathsf{T}}\mathbf{x_i})}{\exp(\boldsymbol{\beta}^{\mathsf{T}}\mathbf{x_j})}.$$

We see that $\gamma_{ij}$ does not depend on time $t$, i.e. it is a constant that only depends on $\boldsymbol{\beta}$, $\mathbf{x_i}$ and $\mathbf{x_j}$; the model is called a *proportional hazards* regression model.

### Example: ESPAC4 trial

The R function weibreg() from the package eha was used to fit a Weibull proportional hazards regression model using one covariate, $x_1$, treatment. Let $x_1 = 0$ for treatment by GEM and $x_1 = 1$ for treatment by GEMCAP. The results are:

$$\hat{\beta}_1 = -0.255, \ \hat{k} = 1.352, \ \hat{\lambda} = 0.0080, \ \text{log-likelihood} = -2049.6,$$

and so $h(t) = \exp(-0.255x_1) \times 0.0080t^{1.352}$. The hazard ratio for the two treatments, GEMCAP relative to GEM, is $\exp(-0.255)/\exp(0) = 0.77$.

Now, two covariates are fitted – treatment and lymph node status (negative/positive). The results are:

$$\hat{\beta}_1 = -0.234, \ \hat{\beta}_2 = 0.850, \ \hat{k} = 1.390, \ \hat{\lambda} = 0.0035, \ \text{log-likelihood} = -2027.9.$$

The model fits better than the first one since the log-likelihood is larger. The likelihood ratio test statistic, minus twice the difference in log-likelihoods, is LR = 43.3, $p < 0.001$. The hazard ratio for the treatments is now, 0.79, for fixed lymph node status, and that for lymph node status, 2.34, for fixed treatment. Lymph node status is a significant covariate.

We will see some more parametric regression models later.

## 5.2 Cox proportional hazards model

We continue on the theme of proportional hazards models and give substantial attention to the Cox proportional hazards model. It is a semi-parametric model, which is widely used to analyze survival and other time-to-event data. The model for the hazard function is,

$$h_i(t) = h_0(t) \exp(\boldsymbol{\beta}^\mathsf{T} \mathbf{x}_i), \tag{5.1}$$

where again, $\mathbf{x}_i$ is the vector of explanatory variable values for the $i$th individual, $\boldsymbol{\beta}$ is the vector of regression coefficients and $h_0(t)$ is a *non-specified* baseline hazard function.

As we do not have the probability distribution for the data, we cannot write down the likelihood. We use a partial likelihood approach instead. Repeating the set up of survival times and risk sets that we saw for the log-rank test in Chapter 3, let there be $n$ individuals, $r$ of whom die in the time interval considered, the rest having censored survival times. Let the death times, $t_1, \ldots, t_r$ be ordered to give the ordered death times $t_{(1)}, \ldots, t_{(r)}$ and to start with, suppose there are no ties in the death times. So, if the 10th individual dies first, they are labelled (1), and we will label the 10th individual's explanatory variable vector, as $\mathbf{x}_{(1)}$. The probability that the individual (1) dies first, is proportional to their hazard rate $h_0(t_{(1)}) \exp(\boldsymbol{\beta}^\mathsf{T} \mathbf{x}_{(1)})$ at time $t_{(1)}$. The constant of proportionality is the reciprocal of the sum of all the similar hazard rates for all the individuals at risk at time $t_{(1)}$, i.e. the risk set containing all individuals except those with censored times before time $t_{(1)}$. Hence

$$\Pr(\text{individual (1) dies at } t_{(1)} | \text{a death at } t_{(1)}) = \frac{h_0(t_{(1)}) \exp(\boldsymbol{\beta}^\mathsf{T} \mathbf{x}_{(1)})}{\sum_{j \in R_{t(1)}} h_0(t_{(1)}) \exp(\boldsymbol{\beta}^\mathsf{T} \mathbf{x}_{(j)})},$$

where $R_{t(1)}$ is the risk set at time $t_{(1)}$. But, we see $h_0(t_{(1)})$ cancels from this expression and so the baseline hazard function does not enter the probability.

Similar probabilities are calculated for all the $r$ individuals who die, the risk set decreasing at each $t_{(i)}$, by the previous death and individuals with censored times. The *partial likelihood* for the data is the product of all the probabilities,

$$L(\beta) = \prod_{i=1}^{r} \frac{\exp(\beta^{\mathsf{T}}\mathbf{x}_{(i)})}{\sum_{j \in R_{t(i)}} \exp(\beta^{\mathsf{T}}\mathbf{x}_j)}.$$

This can be written in terms of all the survival times, censored or not, as

$$L(\beta) = \prod_{i=1}^{n} \left\{ \frac{\exp(\beta^{\mathsf{T}}\mathbf{x}_i)}{\sum_{j \in R_{t(i)}} \exp(\beta^{\mathsf{T}}\mathbf{x}_j)} \right\}^{\delta_i}. \tag{5.2}$$

Note the partial likelihood does not involve the actual survival or censored times, only their rank order. The partial likelihood, $L(\beta)$, or the log-partial likelihood has to be maximised numerically. If there are ties in the data, the partial likelihood is more complicated and the reader is referred to Collett[6] for further details.

A nice property of a proportional hazards model is that the hazard function in Equation 5.1 can be written as,

$$h(t) = h_0(t)\exp(\beta^{\mathsf{T}}\mathbf{x}_i) = h_0(t) \times \exp(\beta_1 x_1) \times \exp(\beta_2 x_2) \times \ldots \times \exp(\beta_p x_p),$$

and the hazard ratio for two observed sets of $x$ values, say, $x_1^1, x_2^1, \ldots, x_p^1$ and $x_1^2, x_2^2, \ldots, x_p^2$ is,

$$\text{HR} = \exp[\beta_1(x_1^1 - x_1^2)] \times \exp[\beta_2(x_2^1 - x_2^2)] \times \ldots \times \exp[\beta_p(x_p^1 - x_p^2)],$$

showing the multiplicative nature of the hazard function and the hazard ratio. For the special case of an indicator variable, $x_k$ say, taking the values 0 and 1, then when all the other variables are equal in the two sets of values, i.e. $x_j^1 = x_j^2$ $(j \neq k)$, then $\text{HR} = \exp(\beta_k)$, which is estimated by $\exp(\hat{\beta}_k)$.

## Cox PH regression for ESPAC4

We use the ESPAC4 pancreatic cancer trial data again, this time to illustrate Cox proportional hazards regression. First we just use arm (treatment) as a univariariate explanatory variable. The R function coxph() is used to fit the model. The fitted coefficient for arm is $\hat{\beta} = -0.210$ with standard error, 0.096 giving the z-value as $-2.197$ and $p = 0.028$. (Or equivalently, the Wald test statistic is $W = \hat{\beta}^2/\text{var}(\hat{\beta})) = 4.827$. Under $\text{H}_0 : \beta = 0$, $W \sim \chi_1^2$, and gives the same p-value.)

The hazard function is,

$$h_i(t) = h_0(t)\exp(-0.210\,x_i).$$

The hazard ratio of treatment by GEMCAP, relative to treatment by GEM, is $\text{HR} = \exp(-0.210) = 0.81$. A 95% confidence interval for HR is given by,

$$[\exp\{\hat{\beta} - 1.96 \times \text{se}(\hat{\beta})\}, \exp\{\hat{\beta} + 1.96 \times \text{se}(\hat{\beta})\}] = [0.67, 0.98].$$

In conclusion, there is evidence that treatment by GEMCAP gives a survival advantage, compared to treatment by GEM.

There is a direct connection between this Cox proportional hazards model and the log-rank test. First, going back to maximum likelihood estimation in general, the *efficient score statistic*, $u(\beta)$, is $u(\beta) = \frac{\partial l}{\partial \beta}$, where $l$ is the log-likelihood. Then $u(\beta)$ has variance equal to the Fisher information, $I(\beta) = -\mathrm{E}\left(\frac{\partial^2 l}{\partial \beta^2}\right)$. The *score test* is a test of the null hypothesis, $H_0 : \beta = 0$, using the statistic, $u(0)/I(0)$, which has an approximate chi-squared distribution. Now, the connection between the Cox proportional hazards model and the log-rank test, is that the score test for the proportional hazards model is equivalent to the log-rank test. See Collett[6] for further details.

The ESPAC4 paper, reporting the results of the clinical trial, has a table with eighteen univariate Cox PH analyses using demographic, clinical and biomarker variables. Here, we just analyse eleven of the variables, Table 5.1 showing the results. We use *Wald* tests to test the hypotheses that parameter values are zero.

From the Wald tests, Sex, Age, WHO status and Country are not significant, i.e. the HRs are not significantly different from unity. We could have looked at the significance of each factor level within those factors with more than two levels – none were significant. The significant factors are resection margin, tumour grade, lymph node status, tumour stage, maximum tumour diameter, postoperative carbohydrate antigen (CA19-9) and treatment. Note, the number of individuals with a recorded value, N, for each variable is shown. For CA19-9, N = 628, that is 65 missing values.

Next, we look at multivariate Cox PH models. We could throw all eighteen of the original variables used into a model, but that would invite a massive overfitting problem. Instead we will use forward selection to select a "good-fitting" model. The R function stepAIC() is used to carry this out.

Now, because country and resection margin were stratification variables when patients were recruited to the trial, this stratification ought to be allowed for in the multivariate model. To see this, suppose we had just one stratification variable, country, with just two countries. We allow the two countries to have different baseline hazard functions, and so the two models are

$$h_1(t) = h_{01}(t)\exp(\boldsymbol{\beta}^\mathsf{T}\mathbf{x}), \quad h_2(t) = h_{02}(t)\exp(\boldsymbol{\beta}^\mathsf{T}\mathbf{x}), \tag{5.3}$$

where $\mathbf{x}$ does not include the variable country. The partial likelihood for the Cox stratified PH model is simply the product of the partial likelihoods found for each stratum separately. In the coxph() function, our country stratification variable is allowed for, using the term $+$ strata(country) in the model formula. The results given do not include anything for the stratification variable – the stratification variable has only combined the individual partial likelihoods, it has done nothing else. Resection margin is also a stratification variable and that combined with country gives ten combined strata, from which we

**TABLE 5.1**
Univariate Cox PH models for ESPAC4

| Variable | | HR [95% CI] | Wald test | p-value |
|---|---|---|---|---|
| Sex (N=693) | Male | 1 | | |
| | Female | 0.85 [0.71, 1.03] | 2.65 | 0.103 |
| Age (years) (N=693) | | 1.003 [0.993, 1.014] | 0.35 | 0.553 |
| WHO status (N=693) | 0 | 1 | | |
| | 1 | 1.09 [0.90, 1.32] | 1.17 | 0.556 |
| | 2 | 1.26 [0.70, 2.26] | | |
| Resection margin (N=693) | Negative | 1 | | |
| | Positive | 1.54 [1.27 1.88] | 18.40 | <0.001 |
| Country (N=693) | England | 1 | | |
| | Wales | 0.80 [0.59, 1.08] | | |
| | Scotland | 1.06 [0.71, 1.59] | 6.56 | 0.256 |
| | France | 0.95 [0.62, 1.45] | | |
| | Germany | 0.99 [0.62, 1.58] | | |
| | Sweden | 1.46 [0.84, 2.53] | | |
| Tumour grade (N=681) | Well | 1 | | |
| | Moderate | 1.29 [0.87, 1.91] | 33.82 | <0.001 |
| | Poorly | 2.16 [1.45, 3.21] | | |
| Lymph node status (N=693) | Negative | 1 | | |
| | Positive | 2.20 [1.66, 2.91] | 30.56 | <0.001 |
| Tumour stage (N=693) | I and II | 1 | | |
| | III and IV | 1.56 [1.09, 2.22] | 5.99 | 0.014 |
| Max. tum. size* (N=679) | | 1.12 [1.06, 1.18] | 1.58 | <0.001 |
| Post CA19-9** (N=628) | | 1.37 [1.28, 1.46] | 9.09 | <0.001 |
| Treatment (N=693) | Gem | 1 | | |
| | GemCap | 0.81 [0.67, 0.98] | 4.83 | 0.028 |

* mm and log(x+1) transformation, ** square root transformation

would calculate the product of the individual strata partial likelihoods, losing both country and resection margin for the results. But, we have a problem. Resection margin is a vital variable needed in calculating HRs as it has a significant influence on survival.

To overcome this, we will assume proportionality within the strata baseline hazard functions. So, for our simple two country example, let

$$h_{0i}(t) = \exp(\beta_0 z_i) h_0(t),$$

where $z_i = 0$ if an individual is in the first stratum and $z_i = 1$ if in the second stratum. Hence the overall model is now,

$$h(t) = h_0(t) \exp(\beta_0 z + \boldsymbol{\beta}^\mathsf{T} \mathbf{x}),$$

and so the stratification variable is now just another factor in the Cox PH model. So, if we assume proportionality between strata, we just have to make sure we include the stratification variables within the Cox model.

We can now do the forward selection of variables. The R function stepAIC() will carry out the procedure. The function needs to be told which

**TABLE 5.2**

Multivariate Cox PH model using forward selection

| Variable | | HR [95% CI] | Wald test | p-value |
|---|---|---|---|---|
| Resection margin (N=604) | Negative | 1 | | |
| | Positive | 1.34 [1.07 1.66] | 6.82 | 0.009 |
| Country (N=604) | England | 1 | | |
| | Wales | 0.90 [0.63, 1.27] | | |
| | Scotland | 0.98 [0.53, 1.80] | 5.63 | 0.344 |
| | France | 0.78 [0.45, 1.35] | | |
| | Germany | 0.99 [0.52, 1.88] | | |
| | Sweden | 2.02 [0.85, 4.78] | | |
| Tumour grade (N=604) | Well | 1 | | |
| | Moderate | 1.73 [1.08, 2.78] | 29.40 | <0.001 |
| | Poorly | 2.77 [1.72, 4.47] | | |
| Lymph node status (N=604) | Negative | 1 | | |
| | Positive | 1.59 [1.16, 2.17] | 8.35 | 0.004 |
| Max. tum. size* (N=604) | | 1.12 [1.03, 1.23] | 6.86 | 0.009 |
| Post CA19-9** (N=604) | | 1.33 [1.24, 1.42] | 66.00 | <0.001 |
| Treatment (N=604) | Gem | 1 | | |
| | GemCap | 0.82 [0.67, 1.01] | 3.63 | 0.057 |

* mm and log(x+1) transformation, ** square root transformation

variables must be kept in the model and what is the full model, i.e. the one with all possible variables included. If you have many variables for inclusion, then some pre-screening of them should be carried out. For example, maybe only keep those for which $p < 0.1$ in the series of univariate analyses. Now, one problem is "missing data". Some of the patients have data missing on some of the variables. The easiest way of dealing with this is to delete all the data for a patient that has any missing data whatsoever on any of the regressor variables, even if they have just one piece of missing data. This is *casewise deletion* and this is what stepAIC expects. We return to the problem of missing data later.

Table 5.2 shows the results for the forward selection. The variables selected for the final model, apart from the stratifying variables are: tumour grade, lymph node status, maximum tumour diameter, post CA19-9 and treatment. If stepwise selection is selected in the method option for stepAIC(), then the same model is arrived at.

The model contains seven variables. In many medical papers these are referred to as "independent prognostic factors" or "independent risk factors" for survival. I am not sure why. They are certainly the "independent variables", i.e. the explanatory variables in the regression model, but are they referred to as independent because their effect on the hazard rate is multiplicative? To back up the notion of independence, interaction terms between the variables should be included in the model and the interaction coefficients tested to be equal to zero. But, there is usually a problem with this, because of the large number of interaction terms; for our model there would be 55.

There is criticism of using stepwise methods in regression generally. Basically, the model you end up with using stepwise methods, is conditioned on all the previous models that were fitted in the preceding steps. Several authors have written about the problems that occur with stepwise methods; parameter estimates are biased away from zero, the standard errors are too small and so their confidence intervals are too narrow, p-values are too low, collinearity problems are worse and other problems[60]. But, stepwise regression methods are still commonly used. See the article by Flom and Cassell[13]. Some alternative methods to stepwise regression are: best subsets regression, model averaging, lasso regression and cross-validation. We will use best subsets regression on the ESPAC4 data and look at cross-validation in Chapter 9 on biomarkers.

The R function bess() from the BeSS package carries out best subsets regression, maximising the log-partial likelihood, using search algorithms, and so it is not guaranteed to find the optimal solution. Complete enumeration would be better, but for a large number of variables, this becomes impossible. Using this on our data, the same factors as for our forward selection analysis were chosen, but with the extra variables, age, WHO and tumour. For this model the log-partial likelihood is $-2204.9$ and the AIC is 4443.9. For the stepwise model, the log-partial likelihood is $-2191.3$ and the AIC 4406.567 – so the best subsets model found is not actually the true best subsets model!

The R function glmulti() from the glmulti package can carry out a complete enumeration of all possible subsets. It uses AIC and BIC as criteria for finding the best subset. For generalised linear models it is possible to force variables into all the subsets to be considered, but for Cox PH models, this appears not to be straightforward. However, you can look at the full list of fitted models, ordered by the criterion used to find the best model and find the one nearest the top that includes the particular variables you need in the model. When this was done for our data, the best subsets model that included country and resection margin was the same as that for the forward selection model. That's good!

## 5.2.1   Multiple imputation for missing data

One problem with the analyses we have just carried out is that we have had to throw away valuable information. We only used 604 rows of our 693 row dataset. If there is only variable with its observation missing for a particular individual, all the remaining data are disregarded in the analysis – that seems rather harsh. It should be noted that excluding missing data can introduce bias into the analysis and reduce precision and power.

Missing data are classified according to the nature of the "missingness", see Sterne[61]. The classifications are:

- *Missing completely at random* (MCAR): whether data points are missing or not, does not depend on any of the observed and unobserved data. MCAR data are rather rare. For example, a blood test to measure CA19-9 goes wrong or the blood sample is lost.

- *Missing at random* (MAR): whether data points are missing or not does depend on the observed data, but not the unobserved data, and the effect of the missing data can be explained by the observed data. For example, those who drop out of a trial due to drug-related adverse events.

- *Missing not at random* (MNAR): whether a data point is missing or not is systematically related to an unobserved variable. For example, an individual who has a low quality of life, might be less likely to fill in a quality of life questionnaire than a patient who has a high quality of life.

To overcome the problem of missing data, we can impute the missing data. There are various ways of doing this, for example,

- For a particular variable, replace a missing value with the mean, median or mode of all the observed values.

- Use regression, regressing a variable with missing data on other variables in the dataset. Then predicted values replace the missing values.

- Use an indicator variable to show which of the values of a particular variable are missing and replace the missing values by zeros. Then carry out a regression incorporating the indicator variable as an extra explanatory variable.

- For an individual with missing data, find its nearest neighbour using a similarity/dissimilarity measure (see Appendix) and replace the missing values by those of its nearest neighbour.

Generally, when data are imputed, standard errors of estimated parameters are too small. For casewise deletion, the standard errors are too large.

Multiple imputation imputes data multiple times – as the name suggests, producing several data sets, each containing the original observed data, and with the missing data replaced by randomly chosen observations from a probability model for the missing data. Each imputed dataset is analysed separately, as if it was the true full dataset and then the results are pooled, giving one set of parameter estimates, together with estimates of standard errors. Under appropriate conditions, the pooled estimates are unbiased and have the correct statistical properties.

So for example, we might use simple linear regression to predict a missing value for variable $X_1$ using variable $X_2$ as a predictor variable, i.e. the model is $X_1 = \beta_0 + \beta_1 X_2$. Then if an individual with a missing $X_1$ value has $X_2$ value of $x_2$, then their estimated value of $X_1$ is $\hat{\beta}_0 + \hat{\beta}_1 x_2$. But we can do better, adding in noise and parameter variation into the model. Then we can use Bayesian methods or the bootstrap to impute the missing value several times, producing multiple datasets. Then we would carry out our analysis, e.g.

multiple regression, Cox PH regression, on each of these datasets, and then the final step is to pool the results using Rubin's rules. These rules say how parameter estimates and test results are to be combined. For those who wish to take this further, see van Buuren[69, 70].

The above example considered a single variable with missing data. You could consider each variable in turn and impute its missing values, but it would be better to treat all the variables together and do multivariate imputation. A popular method for this is imputation by chained equations (MICE), also known as sequential regressions and fully conditioned specification. It is essentially Gibbs sampling (see Appendix) of conditional probabilities of the missing data in each variable, $X_j^{\text{miss}}$, in turn, on the observed data on the variable, $X_j^{\text{obs}}$, all the other variables and the indicator matrix of missing data. See van Buuren[70, 69] for further details.

The R mice package carries out MICE. It has built-in models for various types of data, continuous, categorical, etc. and provides various plots to assess the imputations. Figure 5.1 shows the missing data patterns for the ESPAC4 data. The first column shows number of individuals and the final column shows the number of variables in the particular patterns. So, 604 individuals had all data present, 64 individuals had only post CA19-9 missing, 12 had only maximum tumour dimension missing, one individual had both these variables missing, etc. The data here have only missing data on three variables. The listing and diagram can become unwieldy when there is a lot of missing data.

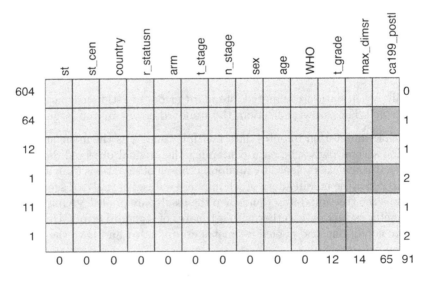

**FIGURE 5.1**
Patterns of missing data in the ESPAC4 data

The md.pairs() function gives four tables relating the number of individuals with observed/missing data for each pair $(X_j, X_k)$ of variables. The first table (labelled rr) gives the number of individuals with data observed for both $X_j$ and $X_k$, the second table (rm), the number where $X_j$ is observed and $X_k$ is missing, the third table (mr), the number where $X_j$ is missing and $X_k$ is observed, and the fourth table (mm) where $X_j$ and $X_k$ are both missing. We show just the table (rm), Table 5.3. With our simple patterns of missing data, it is easy to relate the numbers in Table 5.3 back to Figure 5.1.

**TABLE 5.3**
Number of individuals with data for $X_j$ present and $X_k$ missing (rm)

|  | st | st_cen | country | r_status | arm | ca199_postl | t_grade | t_stage | n_stage | max_dimsr | sex | age | WHO |
|---|---|---|---|---|---|---|---|---|---|---|---|---|---|
| st | 0 | 0 | 0 | 0 | 0 | 65 | 12 | 0 | 0 | 14 | 0 | 0 | 0 |
| st_cen | 0 | 0 | 0 | 0 | 0 | 65 | 12 | 0 | 0 | 14 | 0 | 0 | 0 |
| country | 0 | 0 | 0 | 0 | 0 | 65 | 12 | 0 | 0 | 14 | 0 | 0 | 0 |
| r_status | 0 | 0 | 0 | 0 | 0 | 65 | 12 | 0 | 0 | 14 | 0 | 0 | 0 |
| arm | 0 | 0 | 0 | 0 | 0 | 65 | 12 | 0 | 0 | 14 | 0 | 0 | 0 |
| ca199_postl | 0 | 0 | 0 | 0 | 0 | 0 | 12 | 0 | 0 | 13 | 0 | 0 | 0 |
| t_grade | 0 | 0 | 0 | 0 | 0 | 65 | 0 | 0 | 0 | 13 | 0 | 0 | 0 |
| t_stage | 0 | 0 | 0 | 0 | 0 | 65 | 12 | 0 | 0 | 14 | 0 | 0 | 0 |
| n_stage | 0 | 0 | 0 | 0 | 0 | 65 | 12 | 0 | 0 | 14 | 0 | 0 | 0 |
| max_dimsr | 0 | 0 | 0 | 0 | 0 | 64 | 11 | 0 | 0 | 0 | 0 | 0 | 0 |
| sex | 0 | 0 | 0 | 0 | 0 | 65 | 12 | 0 | 0 | 14 | 0 | 0 | 0 |
| age | 0 | 0 | 0 | 0 | 0 | 65 | 12 | 0 | 0 | 14 | 0 | 0 | 0 |
| **WHO** | 0 | 0 | 0 | 0 | 0 | 65 | 12 | 0 | 0 | 14 | 0 | 0 | 0 |

Let $\mathbf{R}$ be the indicator matrix of missing values, so $r_{ij} = 1$ if the $i$th individual has observed data on the $j$th variable and $r_{ij} = 0$ otherwise. There are two indices that measure the ability to impute the missing data from the observed data. The *proportion of usable cases* or *inbound* statistic, $I_{jk}$, for imputing $X_j$ from $X_k$, and the *outbound* statistic, $O_{jk}$, are defined as,

$$ I_{jk} = \frac{\sum_{i=1}^{n}(1 - r_{ij})r_{ik}}{\sum_{i=1}^{n}(1 - r_{ij})}, \quad O_{jk} = \frac{\sum_{i=1}^{n} r_{ij}(1 - r_{ik})}{\sum_{i=1}^{n} r_{ij}}. $$

The denominator in $I_{jk}$ counts the number of missing values of $X_j$ and the numerator counts the number of these where $X_k$ is observed. The higher the value of $I_{jk}$, the greater the potential for $X_k$ to help impute $X_j$'s missing values. The denominator in $O_{jk}$ counts the number of observed values in $X_j$ and the numerator counts the number of these where $X_k$ is missing. The higher the value of $O_{jk}$, the greater the potential for $X_j$ to help impute $X_k$'s missing values. For our data, where missing values are not a substantial problem, these statistics do not give us much information. For better illustration, we look at the missing data patterns within six biomarker variables, ca199, crp (c-reactive protein), bil (bilirubin), pre and post surgery, within the ESPAC4 database and assess whether missing data for any of them can be imputed from the others.

Figure 5.2 shows the missing data patterns for the six variables, where we have had to transpose the table outputted from the R function in order to save space. The variable names are on the left, the number of missing values for each variable are on the right, the number of variables in each pattern are on the top and the number of individuals with the particular patterns are on the bottom. The patterns are more interesting than those in Figure 5.1. We leave the reader to peruse the patterns.

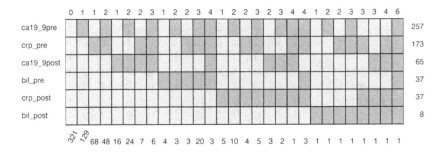

**FIGURE 5.2**
Patterns of missing data in ESPAC4 biomarker variables

Table 5.4 shows the number of missing values for each variable and the $O_{jk}$ values. looking for the largest values of $O_{jk}$, shows ca199_pre missing values can be imputed using all of the other five variables and similarly crp_pre missing values can be imputed using all of the other five variables. The variables, ca199_post, crp_post and bil_post are best suited to imputing missing values for bil_pre. The variable ca199_post can be imputed from all other variables.

**TABLE 5.4**
Values of the outbound statistic, $O_{jk}$

| Msg | Vbl | ca199_pre | ca199_post | crp_pre | crp_post | bil_pre | bil_post |
|-----|-----|-----------|------------|---------|----------|---------|----------|
| 257 | **ca199_pre** | 0.000 | 0.064 | 0.195 | 0.034 | 0.016 | 0.011 |
| 65 | **ca199_post** | 0.350 | 0.000 | 0.245 | 0.045 | 0.053 | 0.008 |
| 173 | **crp_pre** | 0.325 | 0.088 | 0.000 | 0.040 | 0.013 | 0.006 |
| 37 | **crp_post** | 0.358 | 0.085 | 0.239 | 0.000 | 0.050 | 0.006 |
| 37 | **bil_pre** | 0.346 | 0.093 | 0.218 | 0.050 | 0.000 | 0.011 |
| 8 | **bil_post** | 0.371 | 0.091 | 0.245 | 0.048 | 0.053 | 0.000 |

The mice() function has default methods for the imputation of the missing data for the various types of variable in the dataset, but you can override these and tell the function which methods you wish to use, for example, predictive mean matching, random forest imputation, linear regression, logistic regression for binary variables and polytomous logistic regression for categorical variables.

**TABLE 5.5**
Multivariate Cox PH model using forward selection after multiple imputation

| Variable | | HR [95% CI] | Wald test | p-value |
|---|---|---|---|---|
| Resection margin (N=604) | Negative | 1 | | |
| | Positive | 1.37 [1.11 1.68] | 8.74 | 0.016 |
| Country (N=693) | England | 1 | | |
| | Wales | 0.91 [0.66, 1.24] | | |
| | Scotland | 0.85 [0.56, 1.30] | 1.81 | 0.111 |
| | France | 0.79 [0.51, 1.22] | | |
| | Germany | 1.15 [0.71, 1.87] | | |
| | Sweden | 2.05 [1.14, 3.67] | | |
| Tumour grade (N=693) | Well | 1 | | |
| | Moderate | 1.55 [1.00, 2.38] | 14.74 | < 0.001 |
| | Poorly | 2.48 [1.59, 3.85] | | |
| Lymph node status (N=693) | Negative | 1 | | |
| | Positive | 1.57 [1.17, 2.10] | 9.17 | 0.039 |
| Max. tum. size* (N=693) | | 1.11 [1.04, 1.20] | 8.87 | 0.040 |
| Post CA19-9** (N=693) | | 1.32 [1.23, 1.41] | 59.20 | < 0.001 |
| Treatment (N=693) | Gem | 1 | | |
| | GemCap | 0.78 [0.64, 0.95] | 5.96 | 0.071 |

* mm and log(x+1) transformation, ** square root transformation

Using mice() on our ESPAC4 data, predictive mean matching was used for the variables, ca199_postl and max_dimsr, and polytomous logistic regression for t_grade. Five sets of imputed data were generated and each subjected to Cox PH regression using forward selection. The variables chosen to be in the final model were those that were in at least three of the five regressions. Then five more sets of imputed data were generated and Cox PH regression carried out on these and the results pooled using Rubin's rules.

Table 5.5 shows the results. The variables in the model are the same as those in the forward selection model earlier. The estimates and confidence intervals are similar.

Murray[43] gives an overview of multiple imputation. Sterne[61] discusses its potential and pitfalls' .

## 5.2.2 Assessing the fit of the Cox model

Returning to our Cox PH models, we need to assess the fit of a model, looking at outliers, influential observations, non-linearity of continuous predictor variables and checking the proportional hazards assumption. A lot of this can be accomplished by analysing residuals of the model.

There are four types of residual for survival regression: *Cox-Snell residuals*, *martingale residuals*, *deviance residuals* and *Schoenfeld residuals*. We will briefly describe what they are and how they are derived. More detail can be found in Collett's text.

**Cox-Snell residuals**

To define these, we first find the distribution of the variable $Y$ defined as $Y = S(T)$, where $T$ is the usual time variable with survival function, $S(t)$. As $Y$ is a probability, the range of $Y$ is $[0, 1]$. Letting the distribution function of $Y$ be $G(y)$, and the inverse function of $S(t)$ be $S^{-1}(t)$, we have,

$$G(y) = \Pr(Y \leq y) = \Pr(S(T) \leq y) = \Pr(T \geq S^{-1}(y)) = S(S^{-1}(y)) = y,$$

and hence the pdf of $Y$ is $g(y) = 1$, a uniform distribution on $[0, 1]$. Note, this result is usually shown in terms of the transformation, $Y = F(X)$, where $F(x)$ is a distribution function.

Next, the distribution of $Z = -\log(Y)$ is found in the same manner, and that is left for the reader to do. You will see $-\log(Y)$ has an exponential distribution with rate, $\lambda = 1$. Hence, $-\log(S(T))$ has an exponential distribution with mean unity and variance unity – and it is true for any continuous survival function. When a model fits well, the set of values, $\{-\log(\hat{S}(t_i)\}$, will be a sample from a unit exponential distribution. These are calculated as,

$$\hat{e}_i = \exp(\hat{\boldsymbol{\beta}}^{\mathsf{T}}\boldsymbol{x_i})\hat{H}_0(t_i),$$

where $\hat{H}_0(t_i)$ is an estimate of the baseline cumulative hazard function. These are the Cox-Snell residuals.

We also have *modified Cox-Snell residuals* which have 1 or 0.693 added to each residual corresponding to a censored observation, the value depending on which of the mean or median is chosen to account for the extra non-observed survival time for a censored observation.

Cox-Snell residuals can be used to check the *proportional hazards assumption*.

**Martingale residuals**

Martingale residuals are defined as,

$$\hat{m}_i = \delta_i - \exp(\hat{\boldsymbol{\beta}}^{\mathsf{T}}\boldsymbol{x_i})\hat{H}_0(t_i).$$

The definition comes from the theory of the counting process approach to survival analysis, which we will not enter into here. The $\delta_i$ is the observed number of deaths in the time interval $(0, t_i)$ for the $i$th individual (either 0 or 1 depending on whether the individual's survival time is censored or not). The $\exp(\hat{\boldsymbol{\beta}}^{\mathsf{T}}\boldsymbol{x_i})\hat{H}_0(t_i)$ is the expected number of deaths for the $i$th individual. That sounds rather odd – an individual can have 0.58 expected deaths in time interval $(0, t_i)$! Clearly, $\hat{m}_i = \delta_i - \hat{e}_i$. Martingale residuals lie between $-\infty$ and 1, have mean zero and are approximately uncorrelated.

Martingale residuals can be used to check *linearity of continuous predictors* and for *outliers*.

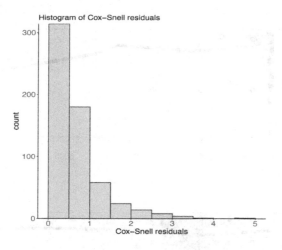

**FIGURE 5.3**
Histogram of the Cox-Snell residuals for the ESPAC4 multivariate Cox PH model

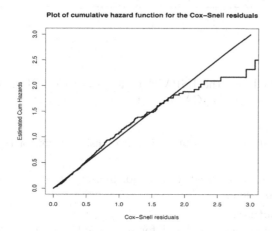

**FIGURE 5.4**
Estimated cumulative hazard function of the Cox-Snell residuals

**Deviance residuals**

As the name suggests, the deviance residuals come from the deviance, $D = -2\{l(\hat{\beta}) - l(\hat{\beta}_{sat})\}$, where $l(\hat{\beta})$ is the log-likelihood for the fitted model and $l(\hat{\beta}_{sat})$ is the log-likelihood for the saturated model (see Appendix). Without describing the derivation, the deviance can be written as, $D = \sum \hat{d}_i^2$, where,

$$\hat{d}_i = \text{sgn}(\hat{m}_i)[-2\{\hat{m}_i + \delta_i \log(\delta_i - \hat{m}_i)\}]^{\frac{1}{2}},$$

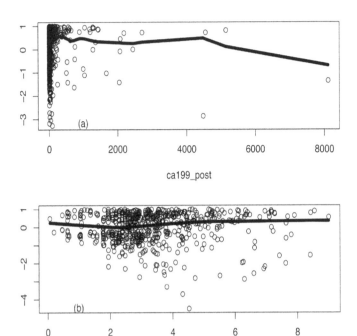

**FIGURE 5.5**
Martingale residuals for the ESPAC4 multivariate Cox PH models plotted against CA19-9post and log(CA19-9post)

$\hat{d}_i$ being the $i$th deviance residual. It is a transformation of the martingale residual.

Deviance residuals can be used to check for *outliers*, with large residuals suggesting possible outliers.

**Schoenfeld residuals**
Schoenfeld residuals are calculated for each predictor. The $i$th Schoenfeld residual is essentially the difference in the value of the predictor for the $i$th individual and the expected value of the predictor at the time of the $i$th individual's death. Censored individuals do not have a Schoenfeld residual. The $i$th individual's contribution to the score equation for the $j$th variable is,

$$\hat{s}_{ij} = \delta_i \left\{ x_{ij} - \frac{\sum_{k \in R(t_i)} x_{kj} \exp(\hat{\boldsymbol{\beta}}^\mathsf{T} \boldsymbol{x_k})}{\sum_{k \in R(t_i)} \exp(\hat{\boldsymbol{\beta}}^\mathsf{T} \boldsymbol{x_k})} \right\},$$

$\hat{s}_{ij}$ being the Schoenfeld residual for the $i$th individual for the $j$th predictor. The Schoelfeld residuals sum to zero and are approximately uncorrelated.

The Schoenfeld residuals can be used to detect *time dependence of predictors*.

Now let us find and plot the four types of residual for our multivariate Cox PH model for the ESPAC4 data, where we have removed all individuals with missing data. Figure 5.3 shows a histogram of the Cox-Snell residuals. They look as if they follow a unit exponential distribution. A plot of the estimated cumulative hazard function of the Cox-Snell residuals should be a straight line with slope unity, since, for the exponential distribution, $H(t) = t$. Figure 5.4 shows this – the tail of the distribution wanders off the straight line, but we will accept that overall the residuals more or less follow an Exp(1) distribution and so the model fit is satisfactory so far.

Note, some studies have shown that Cox-Snell residuals are not very good at showing violations in the model assumptions, and some suggest they should be avoided.

**Deviance residuals**

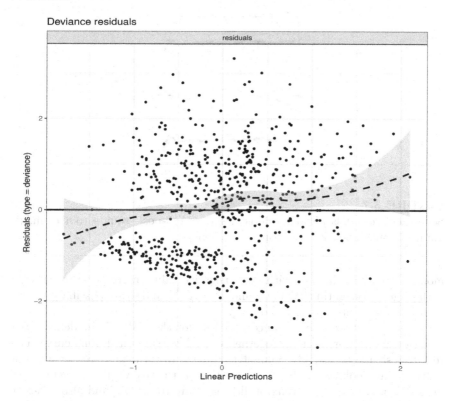

**FIGURE 5.6**
Deviance residuals plotted against linear predictions for the ESPAC4 multivariate Cox PH model

The upper plot in Figure 5.5 shows the martingale residuals for post CA19-9. There seems to be a problem. The lower figure shows the martingale residuals for the log transformed CA19-9 values. This plot shows that transforming the CA19-9 data in this way should be done before using it in the Cox PH

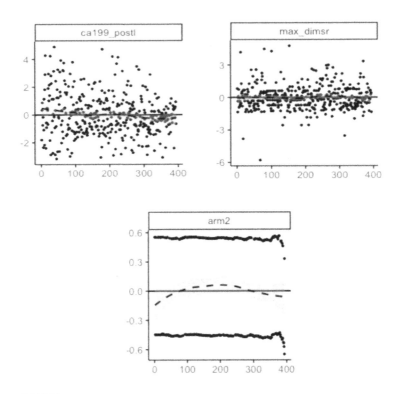

**FIGURE 5.7**
Schoenfeld residuals plotted against log of CA19-post, square root of maximum tumour dimension and arm (treatment)

models. Points in the plot with large negative values correspond to patients with longer survival times than the model suggests and so are possible outliers. Similarly for points very close to unity

Figure 5.6 shows the deviance residuals for the ESPAC4 model, plotted against the linear predictions, together with a loess smoothed trend curve. The residuals should be roughly symmetric about zero with standard deviation of unity. We are looking for very large positive or negative values, or a cluster of points around zero, suggesting outliers that are too small, and also whether there is a pattern relating the residuals to the linear predictions. We see that the maximum absolute value of the residuals is 3.32 – anything above 4 might give us cause for concern. We also notice there does look to be a pattern in the residuals with two groups of them appearing, the first with points mostly above the zero line and the second below towards the left with a region where there is a dearth of points. The trend curve is probably not a problem, it has a slight slope and we can ignore both ends where there are few points. Comparing the predictor values and overall survival time for the two groups of residuals shows the second group has much longer survival times and much more favourable predictor variable values. Overall, I would accept the model, despite the slight misfitting; statistical models are never perfect! The figure was drawn using the R function, ggcoxdiagnostics(), where the x-axis can be the linear predictions or observation identity.

Figure 5.7 shows the Schoenfeld residuals plotted against three of the predictors, post CA19-9 (log-scale), maximum tumour diameter and treatment (arm). The plots for the two continuous variables show no disturbing patterns and that the parameters can be constant rather than time-dependent. The plot for arm shows the type of plot obtained for categorical variables.

## 5.2.3   Estimating the baseline hazard function

With the Cox PH model, we can compare survival between individuals in terms of hazard ratios, but this does not give us any idea as to the overall survival function for individuals. Recall that if we have the hazard function, $h(t)$, and hence the cumulative hazard function, $H(t)$, we can find the survival function as $S(t) = \exp(-H(t))$. So, in order to obtain an estimate of the survival function, we need an estimate of the baseline hazard function, $h_0(t)$. There are two commonly used estimates, the *Breslow estimator* and the *Kalbfleisch-Prentice estimator*.

We give an indication of how the Breslow estimate is derived. As before, let $t_{(i)}$ be the ordered death times. The $j$th individual in the risk set, $R_t(i)$, has probability, $h_0(t_{(i)}) \exp(\boldsymbol{\beta}^\mathsf{T} \mathbf{x}_j)\delta t$ of dying in the time interval, $(t_{(i)}, t_{(i)} + \delta t)$. Hence the approximate expected number of deaths, $d_i$, in the time interval, $(t_{(i)}, t_{(i)} + \delta t)$, is

$$d_i \approx \sum_{j \in R_{t(i)}} h_0(t_{(i)}) \exp(\boldsymbol{\beta}^\mathsf{T} \mathbf{x}_j) \times \delta t,$$

as so

$$h_0(t_{(i)}) \times \delta t \approx \frac{d_i}{\sum_{j \in R_{t(i)}} \exp(\boldsymbol{\beta}^\mathsf{T} \mathbf{x}_j)}.$$

Hence, summing over the death times, the estimate of the cumulative hazard function is

$$\hat{H}_0(t) = \sum_{i:t_i \leq t} h_0(t_{(i)}) \times \delta t = \sum_{i:t_i \leq t} \left( \frac{d_i}{\sum_{j \in R_{t(i)}} \exp(\boldsymbol{\beta}^\mathsf{T} \mathbf{x}_j)} \right). \qquad (5.4)$$

Figure 5.8(a) shows a plot of the estimated baseline cumulative hazard function using Equation 5.4. Figure 5.8(b) shows a plot of the estimated baseline hazard function and, as expected, the plot is very spiky and for appearances, the y-axis has been truncated at 0.2. Also plotted is a loess trend curve, which is a more sensible estimate of $h_0(t)$. This shows the hazard increasing from zero to about 20 months and then decreasing and levelling off up to 60 months. After 60 months the estimate is not reliable. Figure 5.8(c) shows a plot of the estimated baseline survival curve, using, $\hat{S}_0(t) = \exp(-\hat{H}_0(t))$, and

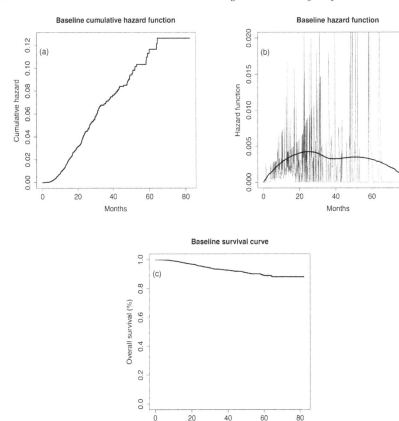

**FIGURE 5.8**
Estimate of the baseline cumulative hazard function, the baseline hazard function and the baseline survival function

because the baseline covariate values are all zero, survival in this case will be very, very good and impossible in practice.

Now, for the $i$th patient, $\hat{h}_i(t) = \hat{h}_0(t)\exp(\hat{\boldsymbol{\beta}}^{\mathsf{T}}\mathbf{x}_i)$, from which we can find the estimated survival curve for the $i$th patient as,

$$\hat{S}_i(t) = \exp\{-\hat{H}_i(t)\} = \exp\{-\hat{H}_0(t)\exp(\hat{\boldsymbol{\beta}}^{\mathsf{T}}\mathbf{x}_i)\} = \hat{S}_0(t)^{\exp(\hat{\boldsymbol{\beta}}^{\mathsf{T}}\mathbf{x}_i)}.$$

Figure 5.9 shows the estimated survival curves for three patients, one with very good survival predictor values, close to the values for the baseline survival curve, one with very poor survival predictor values and one with moderate values.

The baseline hazard has $\mathbf{x} = \mathbf{0}$ in the model, $h(t) = h_0(t)\exp(\boldsymbol{\beta}^{\mathsf{T}}\mathbf{x})$, and is found using the R function survfit(). But, as a default, survfit() uses the

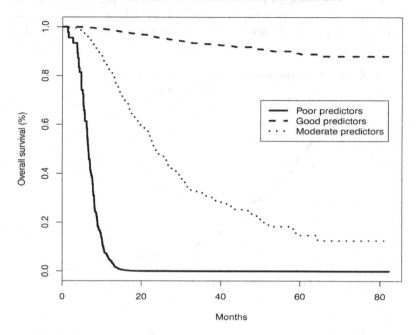

**FIGURE 5.9**
Estimates of the survival curves for three patients: one with good predictors, one with moderate predictors and one with poor predictors

mean of the predictor values as the baseline. This is usually not very sensible. For example, the mean of the treatment variable for our data is 0.47, with treatment coded as "0" for Gem and "1" for GemCap. Which patients get treatment 0.47? So, survfit() needs to be told what is the baseline patient, i.e. what values of the predictors constitute the baseline, and then survfit() will centre its calculations around these. For instance, we could have chosen the patient with the moderate predictor values in Figure 5.9 as the baseline.

## 5.2.4 Time dependent predictors

We will have a brief look at predictors in the Cox PH model that depend on time. For instance, CA19-9 might be measured every month for each patient in a pancreatic cancer trial, with reducing CA19-9 levels indicating that the tumour is responding to treatment. In chronic lymphocytic leukaemia (CLL), mostly found in people over the age of sixty, progression is usually slow, and remission can be achieved with treatments such as chemotherapy, radiotherapy and targeted drugs. When the cancer progresses, more treatment is given. The age-standardised 5-year survival rate is about 70% in the UK. Thus patients can be monitored to check on progress of their cancer using biomarkers and other variables, which can be used in prognostic modelling. These biomarkers can change with time. A simple prognostic variable related to time is a patient's age!

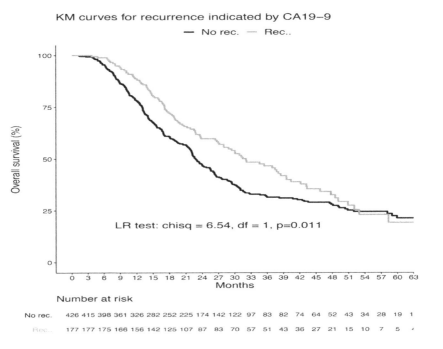

**FIGURE 5.10**
KM curves for pancreatic cancer patients; patients with recurrence vs patients with no recurrence. (This is an erroneous analysis!)

How do we incorporate time-dependent predictors into a Cox PH model? Firstly, suppose some or all of the predictors are time-dependent and can be written as a function of $t$, for example, $age(t) = age_0 + t/12$, where $age_0$ is the age at which a patient joins a clinical trial, and time is measured in months. Then the model for the hazard function in Equation 5.1 becomes

$$h_i(t) = h_0(t) \exp(\boldsymbol{\beta}^\mathsf{T} \mathbf{x}_i(t)).$$

The model is no longer a PH model as $h_i(t)/h_0(t)$ now depends on $t$. The model is fitted maximising the partial likelihood as before. Alternatively, if there are not "nice" functions for the time-dependent predictors, such as we have for age, the actual values $\mathbf{x}_i(t_k)$, of the predictors for the patients at various time points can be used. But, at each death time, every patient in the risk set must have the value of their time-dependent predictors known. This is not always feasible and so approximations can be made, such as using the last known value.

Another way of incorporating a time-dependent predictor is to use intervals of time. We start with an example, using the ESPAC4 data, but inventing a scenario where CA19-9 is measured every three months after surgery for each patient, up to a maximum of 18 months. It has been shown that CA19-9 levels can detect recurrence or metastasis of pancreatic cancer. We will simulate the CA19-9 data for each patient and determine if a recurrence or metastasis has occurred. Not all patients will be tested six times, as some will die or their survival time will be censored before the course of tests can be completed. If a patient has a positive test, the recurrence is deemed to have occurred one and a half months after the previous test, or at one and a half months if positive at the first test.

Figure 5.10 shows KM curves for the two groups of patients: those patients with recurrence versus those patients with no recurrence. The log-rank test value is 6.54, with p-value, $p = 0.011$, indicating a significant difference in survival. But, look closely, the patients with recurrence have better survival prospects than those who did not have recurrence! This cannot be right, unless the CA19-9 values were simulated in a perverse manner. They were not. Each patient had the same probability, 0.07, of recurrence at each CA19-9 test. So what has gone wrong? The patients who died or were censored early, could not have had as many tests as those patients who survived longer. So, the longer surviving patients were more likely to have a recurrence than those who died early. Hence the difference. There should be no difference as all patients were treated equally in the simulation.

**TABLE 5.6**

Counting process layout of ESPAC4 data

| pat.id | $x_1$ | $x_2$ | ... | st | st_cen | REC | tstart | tstop | death |
|--------|-------|-------|-----|-------|--------|-----|--------|-------|-------|
| 1 | 1 | 3 | ... | 7.91 | 1 | 0 | 0 | 7.91 | 1 |
| 2 | 0 | 2 | ... | 10.21 | 1 | 0 | 0 | 1.5 | 0 |
| 2 | 0 | 2 | ... | 10.21 | 1 | 1 | 1.5 | 10.21 | 1 |
| 3 | 0 | 3 | ... | 13.60 | 1 | 0 | 0 | 13.60 | 1 |
| 4 | 1 | 1 | ... | 12.62 | 0 | 0 | 0 | 12.62 | 0 |
| 5 | 1 | 2 | ... | 16.30 | 0 | 0 | 0 | 10.50 | 0 |
| 5 | 1 | 2 | ... | 16.30 | 0 | 1 | 10.5 | 16.30 | 0 |
| 6 | 0 | 3 | ... | 14.29 | 1 | 0 | 0 | 14.29 | 1 |
| 7 | 0 | 3 | ... | 7.75 | 1 | 0 | 0 | 7.75 | 0 |
| 8 | 0 | 2 | ... | 48.11 | 0 | 1 | 0 | 4.5 | 0 |
| 8 | 0 | 2 | ... | 48.11 | 1 | 1 | 4.5 | 48.11 | 1 |
| ⋮ | ⋮ | ⋮ | ⋮ | ⋮ | ⋮ | ⋮ | ⋮ | ⋮ | ⋮ |

A correct analysis can be had by splitting time into intervals, based on the time recurrence happens or not. The data are put into *counting process form*. Table 5.6 shows the form of this. The table shows patient id (pat.id), some

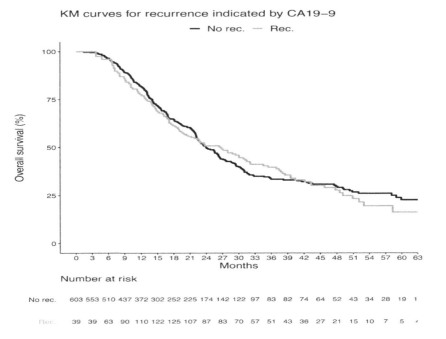

**FIGURE 5.11**
KM curves for pancreatic cancer patients; patients with recurrence vs patients
with no recurrence (correct analysis)

prognostic variables ($x_i$), survival time (st), the censoring indicator (st_cen),
recurrence (REC=1 yes; REC=0 no) and three new variables tstart, tstop and
death. For each patient who has a recurrence, an extra line of their data is
added to the dataset. This splits their time into two intervals, the first before
their recurrence and the second from their recurrence to their survival time.

Looking at the table, patients 1, 3 and 6 did not have a recurrence and
died. Patient 4 did not have a recurrence and their survival time was censored.
These four patients had tstart=0 and tstop=st and death=st_cen. Patient 2's
CA19-9 test at 3-months showed recurrence, and so their first line of data has
tstart=0, tstop=1.5 and as they were still alive, death=0. Their second line
of data has tstart=1.5 and tstop=10.21, their survival time and as they died,
death=1. Patient 5 has the same structure of their data as patient 2, but as
their survival time was censored, death=0 for both lines of data. Therneau [66]
shows how to turn a dataset into counting process form using the R function,
tmerge(). Figure 5.11 shows the KM curves derived from the counting process
form data. The curves are very close to each other, suggesting no survival
disadvantage for patients with recurrence – but as the recurrence or not was
assigned at random to patients, this is what we would expect to happen.

The number-at-risk table is interesting. At time, $t = 0$, all 603 patients are in the no-recurrence group, (The R program places 39 in the recurrence group, but really it should be 0.) Figure 5.12 shows how the at-risk numbers are calculated up to 21 months. At 3 months, 39 patients had their CA19-9 test show a recurrence which was recorded to be at time 1.5 months. They all survived to at least three months, otherwise they could not have had the CA19-9 test. So, 39 patients left the no-recurrence group to join the recurrence group. During the first 3 months, 11 patients in the no-recurrence group died or were censored. So at 3 months the number at risk in the no-recurrence group was $603 - 39 - 11 = 553$. Moving to the 3-month to 6-month time interval, 26 moved from the no-recurrence group to the recurrence group, 17 in the no-recurrence group died or were censored and 2 died or were censored in the recurrence group. Thus at 6 months, in the no-recurrence group, there were $553 - 26 - 17 = 510$ at risk, and in the recurrence group, $39 + 26 - 2 = 63$ were at risk, etc.

If you were to carry out a log-rank test using the counting process dataset, that would be a mistake. It is easy to see why, for a start, 177 of the patients appear in both groups! However, Cox PH regression can be done. You might think that Cox PH regression using the counting process dataset would not be appropriate since there are 177 extra lines of data than in the original dataset. But, it works because for each subject, only one row of their data is selected at each of the ordered death times for calculation of the partial likelihood in Equation 5.2. For example, patient 5 in Table 5.6 has two lines of data with intervals $(0, 10.5]$ and $(10.5, 16.3]$. If an ordered death time $t_{(i)}$ is before $t = 10.5$, the first line of data is used in the risk set; if an ordered death time is after $t = 10.5$, the second line of data is used in the risk set.

Using Cox PH univariate analysis on the counting process dataset, the results are:

|  | coef | exp(coef) | se(coef) | z | p-value |
|---|---|---|---|---|---|
| REC | −0.018 | 0.982 | 0.116 | −0.153 | 0.879, |

showing recurrence is not statistically significant.

Now, we turn our attention to the number of cycles of chemotherapy received by patients. For convenience, we assume each patient has up to six consecutive cycles of monthly treatment, starting after they join the ESPAC4 trial. For our 603 patients, 5 patients did not receive any chemotherapy, 35 received 1 cycle, 44 received 2 cycles, 42 received 3 cycles, 36 received 4 cycles, 36 received 4 cycles and 405 6 cycles. Clearly, as for the CA19-9 testing, you can only receive cycles of chemotherapy whilst you are still alive or you have not left the trial. If we were to do the erroneous analysis, Cox PH regression, treating the number of cycles as a continuous variable, gives the following result

|  | coef | exp(coef) | se(coef) | z | p-value |
|---|---|---|---|---|---|
| No. cycles | −0.119 | 0.888 | 0.028 | −4.184 | p<0.0001, |

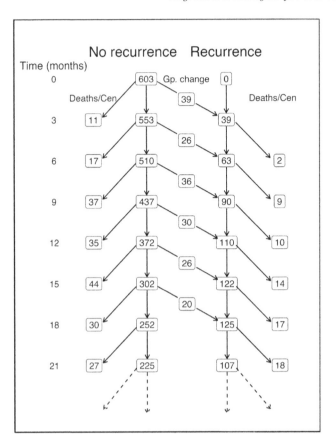

**FIGURE 5.12**
Calculation of the numbers at risk for the no-recurrence and recurrence groups

with HR (95%CI) equal to 0.89 [0.84, 0.94], a highly significant result.

Including number of cycles received in our previous multivariate model, the corresponding regression coefficient is,

|  | coef | exp(coef) | se(coef) | z | p-value |
|---|---|---|---|---|---|
| No. cycles | −0.091 | 0.913 | 0.030 | −3.064 | p=0.002, |

with HR (95%CI) equal to 0.91 (0.86, 0.97), still a significant result.

But we know this is the wrong analysis! Conducting it properly with the counting process approach, for the univariate regression,

|  | coef | exp(coef) | se(coef) | z | p-value |
|---|---|---|---|---|---|
| No. cycles | −0.106 | 0.899 | 0.029 | −3.653 | p<0.001, |

a highly significant result. We cannot quote the hazard ratio as it is not constant over time.

For the multivariate model,

|  | coef | exp(coef) | se(coef) | z | p-value |
|---|---|---|---|---|---|
| No. cycles | −0.078 | 0.925 | 0.030 | −2.580 | p=0.010, |

still a significant result.

When comparing deviances of the multivariate model, with and without the number of cycles received, $-2(l_1 - l_0) = 6.30$ with $p = 0.012$ and hence the number of cycles received is an important prognostic variable.

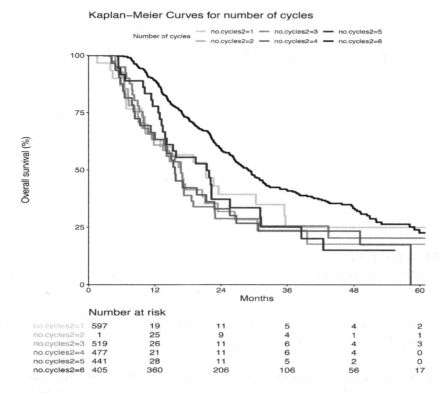

**FIGURE 5.13**

Calculation of the numbers at risk for the no-recurrence and recurrence groups

Figure 5.13 shows the KM type plot of the survival curves, grouped by number of chemotherapy cycles received, but not including the one for the five patients who did not receive any cycles. Coloured curves would have been better here, but the darkness of the curves increases with the number of cycles received. Clearly, those patients receiving the complete treatment of six cycles have better survival, with median survival 29.0 months, compared to the medians for the other groups which range from 15.6 months to 21.4 months. This is probably not surprising as those who manage all six cycles will generally be fitter (in some sense) than those whose chemotherapy treatment was cut short.

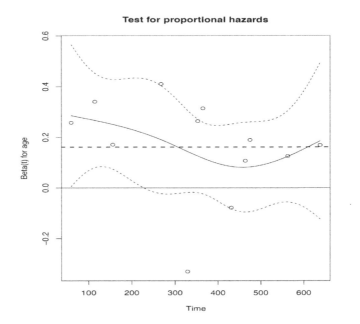

**FIGURE 5.14**
Test for proportional hazards for age in a univariate Cox PH model for ovarian cancer

Another approach is to have continuous time-dependent coefficients in the Cox model, so

$$h_i(t) = h_0(t) \exp(\boldsymbol{\beta}^{\mathsf{T}}(t)\mathbf{x}_i),$$

where some of the $\beta$s are a function of time, $t$. For an illustrative example, we will use the ovarian cancer dataset in the R survival package. The data are from a clinical trial comparing treatment of advanced ovarian cancer using cyclophosphamide compared to treatment with cyclophosphamide plus adriamycin. There are only 26 subjects. Survival time (*futime*) is in days, the censor variable is *fustat* and the treatment variable is *rx*. Prognostic factors are *age*, residual disease (*residual.ds*) and ECOG performance status (*ecog.ps*). Cox PH regression gives the results

|  | coef | exp(coef) | se(coef) | z | p-value |
|---|---|---|---|---|---|
| age | 0.125 | 1.133 | 0.047 | 2.662 | p=0.008 |
| residual.ds | 0.826 | 2.285 | 0.790 | 1.046 | p=0.295 |
| ecog.ps | 0.336 | 1.400 | 0.644 | 0.522 | p=0.602 |
| rx | −0.915 | 0.401 | 0.653 | −1.400 | p=0.162 |

Only age is significant and we will concentrate on that and use a univariate Cox model, the results being,

|  | coef | exp(coef) | se(coef) | z | p-value |
|---|---|---|---|---|---|
| age | 0.162 | 1.175 | 0.050 | 3.249 | p=0.001. |

The HR for age is 1.18 with 95% CI, [1.07, 1.30]; so for a one year increase in age, the hazard increases by a factor of 1.18. The log-likelihood is $l = -27.84$.

Figure 5.14 shows a test of proportional hazard using the function, cox.zph(). The coefficient for age could be replaced by one that is a function of time. The coefficient needs to be larger for small $t$ than for large $t$. The evidence is not strong for doing this and $n$ is only 26, but it will illustrate the theory.

We will replace $\beta$ by the function, $\beta(t) = \text{age}/\sqrt{t}$. The results are

|  | coef | exp(coef) | se(coef) | z | p-value |
|---|---|---|---|---|---|
| tt(age) | 2.515 | 12.369 | 0.807 | 3.115 | p=0.002, |

where tt(age) is the function we are using in place of the constant $\beta$. The log-likelihood is $l = -27.160$, slightly higher than that previously.

So, the model we have fitted is

$$h_i(t) = h_0(t)\exp(2.515 \times \text{age}_i/\sqrt{t}).$$

We do not have proportional hazards, as the hazard ratio for comparing age $a_1$ with age $a_2$ at time $t$ is $\text{HR}(a_1, a_2) = \exp[2.515(a_1 - a_2)/\sqrt{t}]$. For example, for ages 60 and 61 at time 100 days, HR = 1.29, but for the same ages at time 1000 days, HR = 1.08.

We can let all or some of the coefficients, $\boldsymbol{\beta}$, in the model for the hazard function in Equation 5.1 be dependent on time, and so

$$h_i(t) = h_0(t)\exp(\boldsymbol{\beta}^{\mathsf{T}}(t)\mathbf{x}_i).$$

An alternative to having a regression coefficient as a continuous function $\beta(t)$, is to use a step function, where time is split into intervals and $\beta$ is constant within each interval. This is easily accomplished using the R survSplit() function to create a new dataset, similar to the one in Table 5.6. We try it on the ovarian cancer data splitting time into the intervals $(0, 300]$ and $(300, \infty)$ and then use the coxph() function with just age as the predictor. The results are,

|  | coef | exp(coef) | se(coef) | z | p-value |
|---|---|---|---|---|---|
| age:strata(gp)Int=1 | 0.271 | 1.311 | 0.099 | 2.735 | p=0.006 |
| age:strata(gp)Int=2 | 0.084 | 1.088 | 0.057 | 1.461 | p=0.144, |

where Int is the group variable of observations in the two intervals. The HRs and 95% CIs within the first and second time intervals are, 1.31 [1.08, 1.59] and 1.09 [0.97, 1.22], only the first being statistically significant, so age only has an effect on survival for shorter survival times. The log-likelihood is $l = -26.23$, the largest of the three models we have fitted.

### 5.2.5 Frailty models

Subjects with the same values of the prognostic variables in a Cox PH regression model do not all have exactly the same survival time – there is *heterogeneity*. The prognostic variables account for the *observed heterogeneity*, but there is *unobserved heterogeneity*. A *frailty* model (or a model with frailty) models this unobserved heterogeneity as a random effect. The Cox PH model with frailty incorporates a multiplicative frailty variable, $Z$, as

$$h_i(t|Z = z_i) = z_i\, h_0(t) \exp(\boldsymbol{\beta}^\mathsf{T} \mathbf{x}_i), \qquad (5.5)$$

where $Z$ is a latent variable (i.e. unobservable) following a particular distribution.

Shortening $h_i(t|Z = z_i)$ to $h_i(t)$, we can rewrite Equation 5.5 as

$$h_i(t) = h_0(t) \exp(\boldsymbol{\beta}^\mathsf{T} \mathbf{x}_i + \log(z_i)) = h_0(t) \exp(\boldsymbol{\beta}^\mathsf{T} \mathbf{x}_i + u_i), \qquad (5.6)$$

where $u_i = \log(z_i)$. So, $u_i$ is acting as a variable accounting for all the relevant prognostic variables not contained in the Cox PH model. Commonly used distributions for $Z$ are the lognormal distribution and the gamma distribution. When $Z$ has a lognormal distribution, $U$ has a normal distribution, the mean of which is made to be equal to zero.

The partial likelihood of Equation 5.2 becomes

$$\mathrm{L}(\boldsymbol{\beta}, \boldsymbol{u}) = \prod_{i=1}^{n} \left\{ \frac{\exp(\boldsymbol{\beta}^\mathsf{T} \mathbf{x}_i + u_i)}{\sum_{j \in R_{t(i)}} \exp(\boldsymbol{\beta}^\mathsf{T} \mathbf{x}_j + u_j)} \right\}^{\delta_i} \times \prod_{i=1}^{n} f(u_i),$$

where $f(u)$ is the probability density function of the frailty variable, $U$, for example $f(u) = \exp(-u^2/2\sigma_u^2)/\sqrt{2\pi\sigma_u^2}$ for lognormal frailty. An iterative algorithm is used to obtain maximum likelihood estimates of $\boldsymbol{\beta}$ and $\sigma_u^2$. The way this is done is by using the log-likelihood to find estimates of $\boldsymbol{\beta}$, for fixed $\sigma_u^2$, and then the marginal log-likelihood, obtained by integrating the log-likelihood over the $u_i$'s, is maximised to find an estimate for $\sigma_u^2$. This process is repeated until convergence. See Collett for further details.

The R function coxph() can fit frailty models for $U$ having a gamma, normal or t-distribution. The function coxme() in the coxme package fits $U$ as normal and is more reliable with further model options available. Other R packages for frailty models include frailtyEM, frailtyHL, frailtypack and frailtySurv. They fit a variety of models using a variety of methods.

We first use coxme() to fit the model of Equation 5.6 to the ESPAC4 data, using the key prognostic variables selected in the multivariate Cox PH model of Table 5.2 and using patient identifier as the frailty variable. To save space, we will not show the table of coefficients, but only report the estimate of $\sigma_u^2$ and the log-likelihood. These are,

|  | log-likelihood | | | |
|  | null | integrated | penalised | $\hat{\sigma}_u^2$ |
| Model with pat.id frailty: | $-2265.6$ | $-2181.1$ | $-2097.3$ | $0.527$. |

The log-likelihood for the model without the frailty term had log-likelihood of $-2185.556$ and so the chi-square test for the significance of the frailty term is $X^2 = -2(-2185.556 - -2181.084) = 8.944$, on 1 d.f., with $p = 0.003$. Thus frailty is significant.

### 5.2.5.1 Shared frailty

A commonly used frailty model is the *shared frailty* model, where a group of patients share the same frailty. For instance, all patients within a country share the same frailty value. When country (site) is used as the frailty variable for the ESPAC4 data, the log-likelihood is,

|  | log-likelihood | | | |
|  | null | integrated | penalised | $\hat{\sigma}_u^2$ |
| Model with site shared frailty: | $-2265.6$ | $-2188.4$ | $-2187.9$ | $0.006$ |

The log-likelihood is smaller than that for the previous frailty model. The frailty estimates for countries are,

England: 0.046, Wales: 0.006, Scotland: $-0.013$, France: 0.021, Germany: $-0.024$, Sweden: $-0.036$

---

## 5.3 Accelerated failure time (AFT) models

Now for a different type of model – the accelerated failure time model. We start by considering a simple accelerated failure time model, with just two treatment groups. Suppose the first group follows a survival distribution, $S_0(t)$. We let the second group have a survival distribution, $S_1(t)$, that has the same form as $S_0(t)$, but events occur at a different pace. The model is,

$$S_1(t) = S_0(\phi t), \tag{5.7}$$

where $\phi$ is a constant. Basically, $S_1(t)$ is the same as $S_0(t)$, but with the time axis stretched ($\phi > 1$, survival times are shorter) or shrunk ($\phi < 1$, survival times are longer). The parameter $\phi$ is called the *acceleration parameter* or *acceleration factor*. From Equation 5.7,

$$f_1(t) = \phi f_0(\phi t), \text{ and hence, } h_1(t) = \phi h_0(\phi t). \tag{5.8}$$

We have to choose a survival distribution, for example an exponential, a Weibull, or a lognormal. The R function survreg() can fit AFT models. We will use the log-logistic distribution with $S(t) = 1/(1 + \lambda t^k)$ and

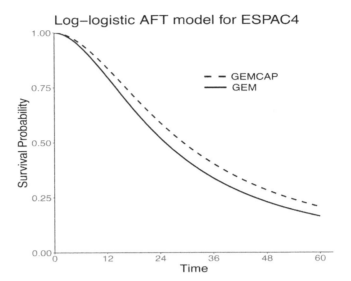

**FIGURE 5.15**
Log-logistic AFT model for treatment in the ESPAC4 data

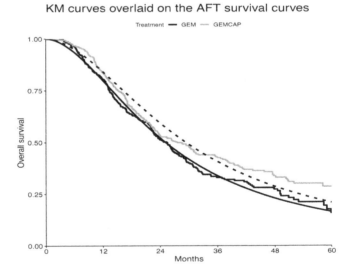

**FIGURE 5.16**
Kaplan-Meier curves overlaid on the log-logistic AFT model for the ESPAC4 data

$h(t) = \lambda k t^{k-1}/(1 + \lambda t^k)$ for the two arms in the ESPAC4 data. The like-lihood is maximised giving results which lead to the two fitted survival functions, $S_0(t) = 1/(1 + 0.00244t^{1.87})$ and $S_1(t) = 1/(1 + 0.00184t^{1.87}) = 1/(1 + 0.00244(0.86t)^{1.87})$. Hence the acceleration parameter is 0.86. Figure 5.15 shows the fitted survival functions. Figure 5.16 shows the Kaplan-Meier curves superimposed on the fitted log-logistic survival functions. The KM and log-logistic survival curves agree with each other quite well for the GEM arm, but not so well for the GEMCAP arm.

We would like to add some prognostic factors into the AFT model. To do this we make $\phi$ in Equations 5.7 and 5.8 into a function of them, choosing $\phi = \exp(\boldsymbol{\beta}^{\mathsf{T}}\mathbf{x})$. Hence, for the $i$th patient, we have,

$$S_i(t) = S_0\big(\exp(\boldsymbol{\beta}^{\mathsf{T}}\mathbf{x})t\big), \quad h_i(t) = \exp(\boldsymbol{\beta}^{\mathsf{T}}\mathbf{x})h_0\big(\exp(\boldsymbol{\beta}^{\mathsf{T}}\mathbf{x})t\big),$$

where $S_0()$ and $h_0()$ are baseline survival and hazard functions.

Now we have to choose a particular survival distribution, $S(t)$, but the way we will do it is to put the model in *log-linear form*. Consider making the transformation from survival random variable $T$ to random variable $\epsilon$, where $\log T = \mu + \boldsymbol{\beta}^{\mathsf{T}}\mathbf{x} + \sigma\epsilon$, or $T = \exp(\mu + \boldsymbol{\beta}^{\mathsf{T}}\mathbf{x} + \sigma\epsilon)$. Here $\mu$ is an *intercept* parameter and $\sigma$ a *scale* parameter. The distribution of $\epsilon$ is not the same as the distribution of $T$, for example, if $T$ is to have a Weibull distribution, $\epsilon$ has to have a Gumbel distribution, and for $T$ to have a log-logistic distribution, $\epsilon$ has to have a logistic distribution. The R function survreg() fits AFT models in the log-linear form.

As an example, the logistic cumulative distribution function with location parameter, $\mu$, and scale parameter, $\sigma$, both equal to 1, is $1/\{1+\exp(-\epsilon)\}$. The survival distribution, $S_i(t)$, is related to this cumulative distribution function by,

$$\begin{aligned} S_i(t) &= 1 - 1/[1 + \exp\{-(\log t - \mu - \boldsymbol{\beta}^{\mathsf{T}}\mathbf{x}_i)/\sigma\}] \\ &= 1/[1 + \exp\{(\log t - \mu - \boldsymbol{\beta}^{\mathsf{T}}\mathbf{x}_i)/\sigma\}] \\ &= 1/[1 + \exp\{(-(\mu + \boldsymbol{\beta}^{\mathsf{T}}\mathbf{x}_i)/\sigma\}\exp\{(\log t)/\sigma\}] \\ &= 1/\big[1 + \big(\exp\{-(\mu + \boldsymbol{\beta}^{\mathsf{T}}\mathbf{x}_i)\}t\big)^k\big], \end{aligned}$$

where $k = 1/\sigma$. Hence $S_i(t)$ has a log-logistic survival distribution with shape parameter, $1/\sigma$.

A log-logistic AFT model was fitted to the ESPAC4 data, using the same prognostic factors as before. Table 5.7 shows the estimated intercept, scale parameter and the $\beta$s. A Weibull AFT model was also fitted to the ESPAC4 data. The results for the two models are very similar. We will look further at the log-logistic model. How do we interpret the results?

The fitted model is,

$$S_i(t) = 1/\big[1 + \big(\exp\{-(0.138x_{1i} + 0.022x_{2i} + \ldots)\}\exp(-5.207)t\big)^{2.29}\big].$$

The log-likelihood for the model is $-1790.4$, the log-likelihood for the null model (intercept only) is $-1698.4$, leading to a chi-square test: $X^2 = 184.0$

**TABLE 5.7**
Two AFT models for the ESPAC4 data

|  | Log-logistic | | | | Weibull | | | |
|---|---|---|---|---|---|---|---|---|
|  | Coefft. | s.e. | z | p-value | Coefft. | s.e. | z | p-value |
| (Intercept) | 5.207 | 0.232 | 22.48 | <0.001 | 5.47 | 0.249 | 21.94 | <0.001 |
| Wales | 0.138 | 0.112 | 1.23 | 0.217 | 0.118 | 0.118 | 1.00 | 0.317 |
| Scotland | 0.022 | 0.177 | 0.12 | 0.901 | 0.071 | 0.204 | 0.35 | 0.726 |
| France | 0.170 | 0.167 | 1.02 | 0.309 | 0.127 | 0.184 | 0.69 | 0.487 |
| Germany | 0.042 | 0.193 | 0.22 | 0.828 | 0.036 | 0.215 | 0.17 | 0.868 |
| Sweden | −0.462 | 0.249 | −1.85 | 0.064 | −0.409 | 0.289 | −1.42 | 0.157 |
| Resec. margin | −0.208 | 0.070 | −2.97 | 0.003 | −0.225 | 0.072 | −3.11 | 0.002 |
| Post CA19-9 | −0.184 | 0.022 | −8.48 | <0.001 | −0.194 | 0.023 | −8.61 | <0.001 |
| Tumour grade 2 | −0.346 | 0.141 | −2.45 | 0.014 | −0.348 | 0.159 | −2.19 | 0.029 |
| Tumour grade 3 | −0.686 | 0.145 | −4.72 | <0.001 | −0.639 | 0.161 | −3.96 | <0.001 |
| Lymph node | −0.381 | 0.095 | −4.01 | <0.001 | −0.329 | 0.105 | −3.12 | 0.001 |
| Max. tum. size | −0.079 | 0.030 | −2.62 | 0.009 | −0.076 | 0.029 | −2.67 | 0.008 |
| Treatment | 0.124 | 0.066 | 1.87 | 0.061 | 0.142 | 0.067 | 2.12 | 0.034 |
| Log(scale) | −0.827 | 0.042 | −19.58 | <0.001 | −0.424 | 0.04 | −10.7 | <0.001 |
|  | Scale=0.437 | | | | Scale=0.655 | | | |

with 12 df and $p < 0.001$, showing the model is worth considering. The log-likelihood for the Weibull model is $-1730.9$, but we cannot compare this with the log-likelihood for the log-logistic model because the functional form of the survival distributions are different.

The baseline survival function is $S_0(t) = 1 / \left[ 1 + \left( \exp(-5.207)t \right)^{2.29} \right]$. The acceleration factor is $\hat{\phi} = \exp\{-(0.138x_{1i} + 0.022x_{2i} + \ldots)\}$. Each prognostic variable has a multiplicative contribution, $\exp(\hat{\beta}_i)$ to the overall value of the acceleration factor, which is interpreted as the amount of stretching or shrinking of the time axis, when all other prognostic variables are held constant. Table 5.7 shows the results for this log-logistic AFT model, as well as those for a Weibull AFT model. The acceleration factors for the key prognostic variables are: resection margin, 0.81; CA19-9, 0.83; tumour grade, stage two, 0.71; tumour grade, stage three, 0.50; lymph node status, 0.68, maximum tumour size, 0.92; GEMCAP, 1.13.

The survival function for a particular patient is found by placing their prognostic variable values into the estimated acceleration parameter. The hazard function for the $i$th patient is,

$$h_i(t) = \frac{2.29 \exp\{-2.29(5.207 + 0.138x_{1i} + 0.022x_{2i} + \ldots)\}t^{1.29}}{1 + \exp\{-2.29(5.207 + 0.138x_{1i} + 0.022x_{2i} + \ldots)\}t^{2.29}}.$$

The baseline hazard function is,

$$h_0(t) = \frac{\exp(-6.717)\, t^{1.29}}{1 + \exp(-11.924)\, t^{2.29}}.$$

Figure 5.17(a) shows the hazard function for two patients, both with the same values of prognostic variables, apart from arm, with one in the GEM arm and one in the GEMCAP arm. The GEMCAP patient's hazard function lies below that for the GEM patient. Figure 5.17(b) shows the ratio of the hazard functions for the two patients, GEMCAP compared to GEM. We can see that the hazard ratio starts at 0.75 at time zero and increases with time to nearly 1.0 at 60 months. Clearly, we do not have proportional hazards!

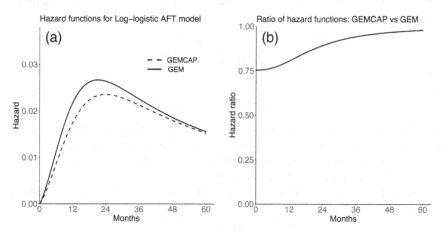

**FIGURE 5.17**
Hazard functions for two typical patients, one treated with GEM, the other with GEMCAP and the ratio of the hazard functions plotted against time

## 5.4    Proportional odds models

We will briefly look at proportional odds models. Recall, the odds of the event that your horse wins the next race, is $p_1/(1 - p_1)$, where $p_1$ is the probability that your horse does win the next race. The odds ratio of your horse winning the next race, compared to your friend's horse winning the race is $\{p_1/(1 - p_1)\} \div \{p_2/(1 - p_2)\} = \{p_1(1 - p_2)\}/\{(1 - p_1)p_2\}$, where $p_2$ is the probability that your friend's horse wins the next race. The odds for a survival function, $S(t)$, are $S(t)/(1 - S(t))$.

For two groups, the proportional odds model is,

$$\frac{S_2(t)}{1 - S_2(t)} = \theta \frac{S_1(t)}{1 - S_1(t)}, \tag{5.9}$$

where $\theta$ is a constant. Hence the name proportional odds model. Let us express $S_2(t)$ in terms of $S_1(t)$. After some easy algebra,

$$S_2(t) = \frac{\theta S_1(t)}{1 - (1 - \theta)S_1(t)}.$$

Now differentiate with respect to $t$ and multiply by $-1$, to obtain

$$f_2(t) = \frac{\theta f_1(t)}{\{1 - (1 - \theta)S_1(t)\}^2}.$$

Divide by $S_2(t)$, but using $S_2(t) = \theta S_1(t)/\{1 - (1-\theta)S_1(t)\}$ on the right hand side of the equation,

$$\frac{f_2(t)}{S_2(t)} = \frac{f_1(t)}{S_1(t)\{1 - (1 - \theta)S_1(t)\}}.$$

Hence,

$$h_2(t) = h_1(t)/\{1 - (1 - \theta)S_1(t)\}.$$

As $t$ tends to infinity, $S_1(t)$ tends to zero, and asymptotically, $h_2(t) = h_1(t)$, so the hazard ratio tends to 1. This can make the proportional odds model useful, for example, we have seen evidence above that for the ESPAC4 data, the hazard ratio approaches 1 at around 60 months.

Similarly, as for the proportional hazards models, we generalise Equation 5.9 to,

$$\frac{S_i(t)}{1 - S_i(t)} = \exp(\boldsymbol{\beta}^\mathsf{T} \mathbf{x}) \frac{S_0(t)}{1 - S_0(t)},$$

where $S_0(t)$ is the baseline survival function. Taking logs,

$$logit(S_i(t)) = \boldsymbol{\beta}^\mathsf{T} \mathbf{x} + logit(S_0(t)),$$

where *logit* is the log-odds function. We can use a parametric proportional odds approach by choosing an appropriate survival distribution, such as the log-logistic distribution, or we can use a semi-parametric approach and fit the model using a modified partial likelihood approach. We will do the latter using the R function prop.odds() from the timereg package.

**TABLE 5.8**
A semi-parametric proportional odds model for the
ESPAC4 data

| | coefft. | exp(coefft.) | s.e. | z | p-value |
|---|---|---|---|---|---|
| Wales | 0.258 | 1.29 | 0.236 | 1.09 | 0.275 |
| Scotland | 0.029 | 1.03 | 0.388 | 0.08 | 0.940 |
| France | 0.408 | 1.50 | 0.348 | 1.17 | 0.240 |
| Germany | 0.056 | 1.06 | 0.407 | 0.14 | 0.891 |
| Sweden | $-1.080$ | 0.34 | 0.546 | $-1.99$ | 0.047 |
| Resec. margin | $-0.461$ | 0.63 | 0.154 | $-3.00$ | 0.003 |
| Post CA19$-$9 | $-0.433$ | 0.65 | 0.055 | $-7.91$ | <0.001 |
| Tumour grade 2 | $-0.769$ | 0.46 | 0.310 | $-2.48$ | 0.013 |
| Tumour grade 3 | $-1.540$ | 0.21 | 0.325 | $-4.75$ | <0.001 |
| Lymph node | $-0.765$ | 0.47 | 0.217 | $-3.52$ | <0.001 |
| Max. tum. size | $-0.178$ | 0.84 | 0.072 | $-2.47$ | 0.014 |
| Treatment | 0.267 | 1.31 | 0.150 | 1.78 | 0.074 |

Table 5.8 shows the results. The standard errors of the estimates of the co-efficients are estimated using a re-sampling method. Prognostic variables with negative estimated coefficients decrease the odds function for a patient with respect to the baseline odds function, and variables with positive estimated coefficients, increase the odds function. This in turn implies that $S_i(t) < S_0(t)$ for negative coefficients and $S_i(t) > S_0(t)$ for positive coefficients. The results are similar to the other models we have fitted. (Note: the function prop.odds() models the odds as $F(t)/(1 - F(t))$ and so be be consistent with the other models we have looked at, the signs of the estimated coefficients have been reversed.)

## 5.5 Parametric survival distributions for PH and AFT models

Not all parametric survival distributions can be PH or AFT models. We see which can be used, by considering their hazard functions. To be a PH or an AFT model, the hazard function, $h(t)$, or cumulative hazard function, $H(t)$, for the survival distribution must be in one of the following forms:

$$\text{PH:} \quad h(t) = h_0(t)\exp(\boldsymbol{\beta}^\mathsf{T}\mathbf{x}), \qquad H(t) = H_0(t)\exp(\boldsymbol{\beta}^\mathsf{T}\mathbf{x}),$$
$$\text{AFT:} \quad h(t) = h_0\big(\exp(\boldsymbol{\beta}^\mathsf{T}\mathbf{x})t\big)\exp(\boldsymbol{\beta}^\mathsf{T}\mathbf{x}), \quad H(t) = H_0\big(\exp(\boldsymbol{\beta}^\mathsf{T}\mathbf{x})t\big).$$

Various distributions are considered.

**Exponential distribution**
For the exponential distribution, $S(t) = \exp(-\lambda t)$ and $h(t) = \lambda$. We let $\lambda$ incorporate the prognostic variables as, $\lambda\exp(\boldsymbol{\beta}^\mathsf{T}\mathbf{x})$, and so

$$h(t) = \lambda\exp(\boldsymbol{\beta}^\mathsf{T}\mathbf{x}),$$

and so the exponential distribution can be a PH model with the baseline hazard function, $h_0(t) = \lambda$.

**Weibull distribution**
For the Weibull distribution, $S(t) = \exp(-\lambda t^k)$ and $h(t) = \lambda k t^{k-1}$. Then, replacing $\lambda$ by $\lambda\exp(\boldsymbol{\beta}^\mathsf{T}\mathbf{x})$,

$$h(t) = \lambda k t^{k-1}\exp(\boldsymbol{\beta}^\mathsf{T}\mathbf{x}),$$

a PH model with the baseline hazard function, $h_0(t) = \lambda k t^{k-1}$.

**Weibull distribution revisited**
Again, for the Weibull distribution, $h(t) = \lambda k t^{k-1}$. Multiply each of $t$ and $\lambda$ by $\exp(\boldsymbol{\beta}^\mathsf{T}\mathbf{x})$, so

$$h(t) = \exp(\boldsymbol{\beta}^\mathsf{T}\mathbf{x})\lambda k\{\exp(\boldsymbol{\beta}^\mathsf{T}\mathbf{x})t\}^{k-1}.$$

This is in the form of an AFT model, with $h_0(t) = \lambda k t^{k-1}$. So the Weibull distribution can give rise to both a PH model and an AFT model – the only distribution that can do this. Fitting one model gives the same results as the other, once estimated parameters are transformed appropriately.

### Log-logistic distribution revisited

We used the log-logistic distribution as an AFT in Section 5.3. Now we consider the model in terms of the hazard function, $\lambda k t^{k-1}/(1 + \lambda t^k)$. Replace $\lambda$ by $\lambda \exp(\boldsymbol{\beta}^{\mathsf{T}}\mathbf{x})^k$,

$$h(t) = \frac{\lambda \exp(\boldsymbol{\beta}^{\mathsf{T}}\mathbf{x})^k k\, t^{k-1}}{1 + \lambda \exp(\boldsymbol{\beta}^{\mathsf{T}}\mathbf{x})^k\, t^k} = \frac{\lambda k\{\exp(\boldsymbol{\beta}^{\mathsf{T}}\mathbf{x})t\}^{k-1}}{1 + \lambda\{\exp\{(\boldsymbol{\beta}^{\mathsf{T}}\mathbf{x})t\}^k} \exp(\boldsymbol{\beta}^{\mathsf{T}}\mathbf{x}),$$

and hence when $f(t)$ is divided by $S(t)$, we see that this is an AFT model.

### Gamma distribution

The probability density function of the gamma distribution is $f(t) = \lambda^k t^{k-1} \exp(-\lambda t)/\Gamma(k)$, where $\Gamma(k) = \int_0^\infty u^{k-1} \exp(-u)du$ is the Gamma function. The survival function is $1 - \Gamma(k, \lambda t)$, where $\Gamma(k, t) = \int_0^t u^{k-1} \exp(-u)du$ is the incomplete Gamma function. The hazard function is $h(t) = f(t)/S(t)$ and does not have a "nice" formula. Replace $\lambda$ by $\lambda \exp(\boldsymbol{\beta}^{\mathsf{T}}\mathbf{x})$, and so

$$f(t) = \exp(\boldsymbol{\beta}^{\mathsf{T}}\mathbf{x})\lambda^k\{\exp(\boldsymbol{\beta}^{\mathsf{T}}\mathbf{x})t\}^{k-1} \exp\{-\exp(\boldsymbol{\beta}^{\mathsf{T}}\mathbf{x})\lambda t\}/\Gamma(k)$$
$$S(t) = 1 - \Gamma(k, \exp(\boldsymbol{\beta}^{\mathsf{T}}\mathbf{x})\lambda t),$$

and hence $h(t)$ will have the form of an AFT model.

The generalised gamma distribution has probability density function, $f(t) = p\lambda^{kp} t^{kp-1} \exp(-(\lambda t)^p)/\Gamma(k)$. For $p = 1$, the generalised gamma reduces to a gamma distribution, for $k = 1$ it reduces to a Weibull distribution, and for $p = k = 1$, to an exponential distribution.

### Gompertz distribution

For the Gompertz distribution, $S(t) = \exp(-(\lambda/a)(e^{at} - 1))$ and $h(t) = \lambda e^{at}$. Replace $\lambda$ by $\lambda \exp(\boldsymbol{\beta}^{\mathsf{T}}\mathbf{x})$,

$$h(t) = \lambda e^{at} \exp(\boldsymbol{\beta}^{\mathsf{T}}\mathbf{x}),$$

and hence the Gompertz is a PH model, with $h_0(t) = \lambda e^{at}$.

### Piecewise exponential distribution

The piecewise exponential distribution was defined in Chapter 3. It is simply a series of exponentials, one for each interval of a predefined set of contiguous intervals of the time axis. Let the time axis be split at the points, $0 = t_0, t_1, t_2, \ldots, t_J = \infty$ to produce the intervals. The PH model is

$$h(t) = \lambda_j \exp(\boldsymbol{\beta}^{\mathsf{T}}\mathbf{x}) = \exp(\alpha_j + \boldsymbol{\beta}^{\mathsf{T}}\mathbf{x}), \quad (j = 0, 1, \ldots, J).$$

The baseline hazard function is a series of horizontal lines. The intervals have to be chosen carefully. If the time axis is split at each event time, when fitted, the piecewise exponential PH model gives the same results as the Cox PH model.

## 5.6  Flexible parametric models

We now want to extend the parametric models by modelling $h_0(t)$ more flexibly. We will do this in three ways, using (i) a piecewise exponential distribution, (ii) regression splines and (iii) fractional polynomials. But first, we describe regression splines and fractional polynomials.

**Regression splines**

A cubic spline is a third degree polynomial, fitted to the points within a particular time interval. We split the time axis into a series of subintervals and fit a spline in each of them. Figure 5.18 shows a fictitious example with three splines fitted to some data, over the time interval $[0, 3]$ with *cut-points*, known as *knots*, at $t = 1$ and $t = 2$. (Note, the plot of points is just a sketch.) The equations of the three splines are,

$$
\begin{aligned}
f_1(t) &= \phantom{-}1 + \phantom{-}4t - 4.5t^2 + \phantom{-}2t^3 \quad (0 \le t \le 1) \\
f_2(t) &= \phantom{-}4.5 - 6.5t + 6t^2 - 1.5t^3 \quad (1 \le t \le 2) \\
f_3(t) &= -11.5 + 17.5t - 6t^2 + 0.5t^3 \quad (2 \le t \le 3).
\end{aligned}
$$

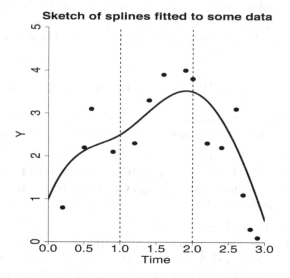

**FIGURE 5.18**
Sketch of three splines fitted to some data

You notice we end up with a smooth curve running across the whole time interval. Had we fitted the splines independently, this smoothness would not have been achieved, with the end of one spline not meeting the start of the next, let alone having a smooth join of the splines. To achieve the smoothness, we ensure the splines meet at the knots and have the same value of the first and second derivatives at the knot. For example, $f_1(1) = f_2(1) = 2.2$, $f_1'(1) = f_2'(1) = 1.0$ and $f_1''(1) = f_2''(1) = 3.0$. Each spline has four coefficients, each knot has three smoothness conditions, and so the number of degrees of freedom is $df = 3 \times 4 - 2 \times 3 = 6$, or $K + 4$ for fitting $K$ knots. The set of spline polynomials can be expressed as the single equation,

$$f(t) = \sum_{i=0}^{3} a_{0i}t^i + \sum_{j=1}^{K} a_j(t - k_j)_+^3,$$

where $k_j$ is the position of the $j$th knot and $(t - k_j)_+$ is $(t - k_j)$ for $(t - k_j)$ positive and zero for $(t - k_j)$ negative. The first term for $f(t)$ is the first spline, and the second term is a series of cubic terms, one for each knot. Basically, before the first knot the equation is the first spline function. As time progresses, upon reaching the first knot, an extra cubic polynomial, $(t - k_1)^3$, is added to the first spline, giving the second spline. At the next knot, $(t - k_2)^3$ is added, giving the third spline, etc. For our splines above,

$$f(t) = 1 + 4t - 4.5t^2 + 2t^3 - 3.5(t - 1)_+^3 + 2(t - 2)_+^3.$$

For our modelling, we will add a *boundary* knot at the minimum event time $k_{min}$, and another at the maximum death time, $k_{max}$ and make the splines linear before the first knot and after the final knot. These gives us restricted cubic splines,

$$f(t) = a_0 + a_1 t + \sum a_j \{(t - k_i)_+^3 - \xi_i(t - k_{min})_+^3 - (1 - \xi_i)(t - k_{max})_+^3\},$$

where $\xi_i = (k_{max} - k_i)/(k_{max} - k_{min})$. These splines can be used to model baseline hazard functions or baseline cumulative hazard functions.

Figure 5.19(a) shows two spline functions fitted to the estimated hazard function for the GEM arm of the ESPAC4 data, one with two knots and one with five. The five knot model shows overfitting. These are called Royston-Parmar (RP) models[52, 24].

### Fractional polynomials
A fractional polynomial, of order $m$, takes the form, $f(t) = \alpha_0 + \sum_{i=1}^{m} t^{p_i}$, where $p_i$ is one of a predefined set of powers, e.g. $\{-2, -1, -0.5, 0, 0.5, 1, 2\}$, with the convention that $x^0$ is the log function. Thus $f(x) = 6/x$ and $f(x) = 7 + 1/x + 3\log(x) - 2x^2$ are fractional polynomials of orders 1 and 3 respectively. Products of $\log(x)$ and powers of $x$ are also allowed, e.g. $f(x) = 1 - 3/x^2 + 2\log(x) + 4x^2 \log(x)$ is also a fractional polynomial.

Figure 5.19(b) shows a fractional polynomial fitted to the estimated hazard function for the GEM arm using the R function mfp() from the mfp package. The function will select the best fitting FP, the one selected for the GEM

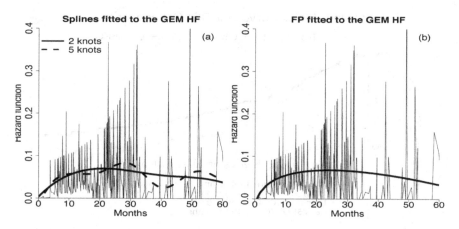

**FIGURE 5.19**
Splines (a) and a fractional polynomial (b) fitted to the GEM estimated baseline hazard function

hazard function is $-0.031 + 0.128\sqrt{t/10} - 0.041(t/10)$. It is not very different from the spline model. Note, it is best to model the log of the baseline hazard function with a fractional polynomial, rather than the baseline hazard function itself, as this will keep the baseline hazard non-negative.

### 5.6.1 Parametric models fitted to some colon cancer data

We will now fit some of these models to real data, and to make a change from the ESPAC4 data, we will use the colon dataset included in the R survival package. The data are from a clinical trial, reported in 1989, of adjuvant chemotherapy for patients with colon cancer. The treatments were Observation (Obs), Levamisole (Lev) and Levamisole combined with 5-FU (Lev+5FU). (Current chemotherapy treatments and combination treatments for colon cancer, as listed by Cancer Research UK, are: Fluorouracil (5FU); Capecitabine (Xeloda); Irinotecan (Campto); Folinic acid, Fluorouracil and Oxaliplatin (FOLFOX); Oxaliplatin and Capecitabine (XELOX).)

Details of the demographic and clinical variables are given in the survival package reference manual, or type help("colon") in R. Figure 5.20 shows Kaplan-Meier overall survival curves for the three treatments. Clearly Lev+5FU has a survival advantage. The restricted mean survival times are 66.2, 66.2 and 76.6 months for Obs, Lev and Lev+5FU respectively, where we quote this measure rather than the usual median survival time, since the median cannot be estimated for Lev+5FU, since the survival curve never falls below 0.5.

The recruited patients were stratified by invasion of the primary tumour (extent), the time interval between surgery (surg) and whether there were four

*Regression Analysis for Survival Data*

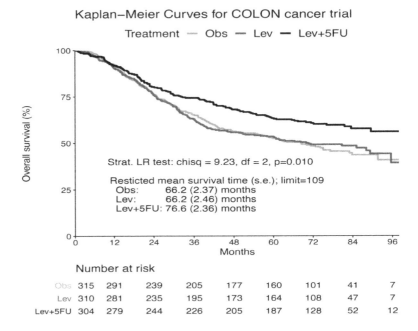

**FIGURE 5.20**
Kaplan-Meier overall survival curves for the colon cancer treatments

or less lymph nodes involved, or otherwise (nodes4), but note, in the dataset this variable does not entirely agree with the total number of nodes variable, nodes. Here we have changed nodes4 to agree with nodes. The stratified log-rank test has $X^2 = 9.23, p = 0.010$.

We now fit four different parametric models to the colon data. These are (i) a Cox PH model, (ii) a piecewise exponential model, (iii) a Gompertz model and (iv) a Royston-Parmar spline model. The prognostic variables used are: *rx* (treatment), *obstruct* (obstruction of the colon by the tumour), *nodes* (number of lymph nodes with detectable cancer), *extent* (extent of local spread, 1=sub-mucosa, 2=muscle, 3=serosa, 4=contiguous structures) and *surg* (time from surgery to registration, 0=short, 1=long). These were chosen from ten potential prognostic variables using backward selection with the Cox PH model. There were 911 patients and 441 deaths. To save space, we do not include a lot of detail, only some graphs and tables.

**Piecewise exponential**
Figure 5.21 shows the estimated baseline hazard, cumulative hazard and survival distributions from fitting a piecewise exponential model, together with the estimated cumulative hazard function from fitting a Cox PH model. The cumulative hazard functions look very similar. The cut-points for the PE were

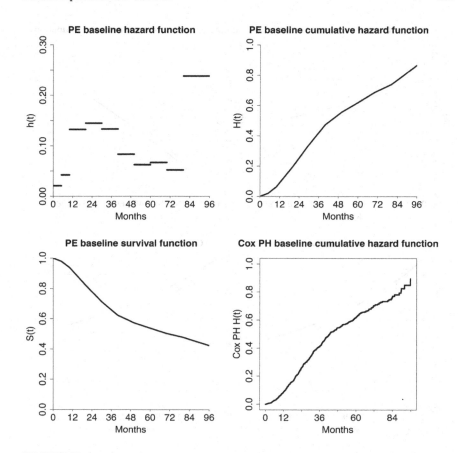

**FIGURE 5.21**
Piecewise exponetial model fitted to the colon cancer data: estimated baseline hazard, cumulative hazard and survival functions, together with the estimated cumulative hazard function from the Cox PH model for comparison

$(5, 10, 20, 30, 40, 50, 60, 70, 80)$. Note, the nodes variable, which is continuous, has been transformed to number of nodes $-$ 4, and the extent variable baseline category has been set at 3, in order to make the baseline functions look interesting. Otherwise, for example, the baseline survival function would not fall below 0.87. In practice this would not matter. The other baseline values were rx = "Obs", obstruct = "no" and surg = "short".

The piecewise exponential model was fitted using the R glm() function with a Poisson distribution, having first put the data into counting process form.

### Gompertz

Figure 5.22 shows the estimated baseline hazard function, the baseline cumulative hazard function and the baseline survival function from fitting

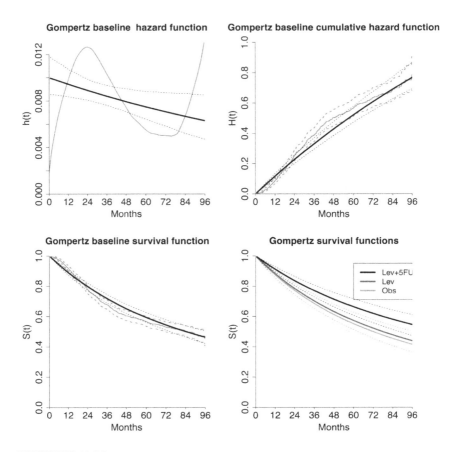

**FIGURE 5.22**
Gompertz model fitted to the colon cancer data: estimated baseline hazard, cumulative hazard and survival functions, plus the corresponding survival curves for Lev and Lev+5FU (also shown are 95% confidence intervals and Kaplan-Meier equivalents)

a Gompertz survival model, together with 95% CI's, and compares them to the KM counterparts. (The KM baseline hazard function has been smoothed.) The fitted hazard function does not display the peak expected at about 24 months, and the cumulative hazard does not capture the "kink" expected at about 40 months. The baseline survival plot does not have the "shoulder" expected at about 12 months. Also shown is the baseline survival distribution, adjusted by treatment. As expected, Lev+5FU gives a survival advantage.

The Gompertz model was fitted using the R flexsurvreg() function from the flexsurv package.

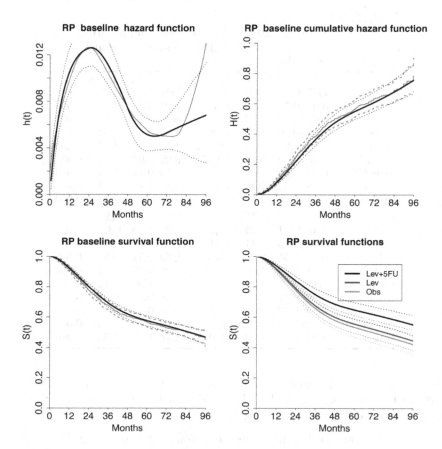

**FIGURE 5.23**

Royston-Parmar spline model fitted to the colon cancer data: estimated baseline hazard, cumulative hazard and survival functions, plus the corresponding survival curves for Lev and Lev+5FU (also shown are 95% confidence intervals and Kaplan-Meier equivalents)

**Royston-Parmar**

Figure 5.23 shows the same plots for the Royston-Parmar splines model as those we saw for the Gompertz model. Three knots were used and boundary knots were placed at 0 and 110. The fitted hazard function shows the peak expected at about 24 months, the cumulative hazard function does capture the "kink" expected at about 40 months and the baseline survival plot does have the "shoulder" expected at about 12 months.

The Royston-Parmar spline model was fitted using the R flexsurvspline() function, also from the flexsurv package. Flexsurvspline() fits splines to the cumulative hazard function, using time or log(time), the default being log(time). You can specify the locations of the knots, or specify the number of knots

**TABLE 5.9**

Estimated HRs for the models: Cox proportional hazards; Gompertz; piecewise exponential; Royston-Parmer splines, fitted to the colon cancer data

|  | **Cox PH** | | | | **Gompertz** | | | |
|---|---|---|---|---|---|---|---|---|
|  | **HR** | **LCL** | **UCL** | **p** | **HR** | **LCL** | **UCL** | **p** |
| rxLev | 0.94 | 0.75 | 1.17 | 0.558 | 0.94 | 0.75 | 1.17 | 0.556 |
| rxLev+5FU | 0.69 | 0.55 | 0.88 | 0.002 | 0.69 | 0.54 | 0.87 | 0.002 |
| obstruct | 1.24 | 0.99 | 1.56 | 0.067 | 1.22 | 0.97 | 1.54 | 0.083 |
| nodes1 | 1.10 | 1.08 | 1.12 | $< 0.001$ | 1.10 | 1.08 | 1.12 | $< 0.001$ |
| extent11 | 0.30 | 0.09 | 0.93 | 0.036 | 0.29 | 0.09 | 0.92 | 0.035 |
| extent12 | 0.59 | 0.42 | 0.84 | 0.003 | 0.60 | 0.42 | 0.84 | 0.003 |
| extent14 | 1.68 | 1.14 | 2.47 | 0.009 | 1.61 | 1.10 | 2.38 | 0.015 |
| surg | 1.29 | 1.05 | 1.59 | 0.016 | 1.28 | 1.04 | 1.57 | 0.020 |
|  | **PWE** | | | | **RP** | | | |
|  | **HR** | **LCL** | **UCL** | **p** | **HR** | **LCL** | **UCL** | **p** |
| rxLev | 0.94 | 0.75 | 1.17 | 0.553 | 0.94 | 0.75 | 1.17 | 0.552 |
| rxLev+5FU | 0.69 | 0.55 | 0.88 | 0.002 | 0.69 | 0.54 | 0.87 | 0.002 |
| obstruct | 1.24 | 0.99 | 1.56 | 0.066 | 1.24 | 0.99 | 1.56 | 0.065 |
| nodes1 | 1.10 | 1.08 | 1.11 | $< 0.001$ | 1.10 | 1.08 | 1.12 | $< 0.001$ |
| extent11 | 0.29 | 0.09 | 0.92 | 0.035 | 0.29 | 0.09 | 0.92 | 0.036 |
| extent12 | 0.59 | 0.42 | 0.84 | 0.003 | 0.59 | 0.42 | 0.84 | 0.003 |
| extent14 | 1.68 | 1.14 | 2.47 | 0.009 | 1.67 | 1.14 | 2.46 | 0.009 |
| surg | 1.29 | 1.05 | 1.58 | 0.016 | 1.29 | 1.05 | 1.58 | 0.016 |

which are then placed equally along the time axis. The number of knots to be used can be guided by the AIC of the fitted model. Here we used log(time) and had knots at $\log(24), \log(48), \log(72)$.

Table 5.9 shows the hazard ratios with 95% confidence intervals and p-values for all four fitted models. The four sets of hazard ratios are remarkably similar to each other, as are the confidence intervals and p-values. All prognostic factors are statistically significant, except rxLev and, marginally, obstruct.

As an example of calculating the survival curves from the R output, let us look at the Gompertz model. The estimated parameters are, $\hat{a} = -4.804 \times 10^{-3}$ and $\log \hat{\lambda} = -4.47$. Consider the Lev arm and the other prognostic variables set to their baseline values, then,

$$\hat{\boldsymbol{\beta}}^{\mathsf{T}} = (-0.066, -0.371, 0.202, 0.091, -1.227, -0.513, 0.479, 0.244)$$
$$\mathbf{x}_1^{\mathsf{T}} = (1, 0, 0, 0, 0, 0, 0, 0),$$
$$\hat{\lambda}_1 = \hat{\lambda} \exp(\hat{\boldsymbol{\beta}}^{\mathsf{T}} \mathbf{x}_1),$$
$$\hat{S}_1(t) = \exp\{-(\hat{\lambda}_1/\hat{a})(e^{\hat{a}t} - 1)\}, \quad \hat{h}_1(t) = (\hat{\lambda}_1/\hat{a})(e^{\hat{a}t} - 1).$$

So an estimated survival function and hazard function can be found for any patient by substituting their prognostic variable values into $\mathbf{x}$.

So, which model should we choose? It depends on what you are trying to achieve. If its just to compare HRs, then for these data, the Cox PH model is as good as any other. If you need accurate estimates of the hazard function or survival function, you would not choose the Gompertz model here. The RP spline model looks a good choice, but you do need to choose the knots, and similarly, if you choose the PWE model, you need to choose the cut-points.

# 6

## Clinical Trials: the Statistician's Role

A key person in the initial discussions of the design of a trial is the statistician. Typically, a meeting of the Chief Investigator (CI), a senior member of the Clinical Trials Centre and the statistician will take place to develop ideas. The role of the statistician is to elicit the key aspects of what is needed from the trial. Key elements are: the background to the proposed study, the main questions that the study will hopefully answer, the population of interest, the phase of the trial, the variables (outcomes) to be measured, the primary outcome, the availability of patients, the availability of centres, resources, sponsorship, etc.

An important question to be answered early on is, "How many patients do I need?" The CI would like as few as possible, the statistician as many as possible, but ethically the answer is, only as many as necessary to have enough appropriate evidence for reliable conclusions at the end of the trial. If it is a study that does not recruit patients, e.g. only historical medical records are used, then a pragmatic sample size will be chosen.

As the trial is planned, set up and run, the statistician will be involved with,

- Writing the statistical aspects of the protocol

- Writing the statistical analysis plan (SAP)

- Helping with the randomisation scheme of patients

- Data analysis and writing reports for the Independent Safety Data Monitoring Committee (ISDMC)

- Final data analysis, writing the statistical report and contributing to external publications

We cover some of this, starting with sample size calculations.

## 6.1 Sample size calculation

The sample size calculation is based on the primary outcome, but other outcomes might also be taken into account. The chosen design and type of primary

DOI: 10.1201/9781003041931-6

outcome will determine the type of hypothesis test on which to calculate the sample size. We need, (a) the level of significance, $\alpha$, (b) the power required, remembering power $= 1 - \Pr\{\text{Type II error}\} = 1 - \beta$, (c) the minimum effect size that is considered to be important and at which the power is calculated. The effect size chosen needs to be realistic and achievable in practice.

We start with the basic sample size calculations for normal and binomial random variables, progress to sample sizes for survival analysis and then take a brief look at group sequential designs. For comprehensive sample size calculations covering many cases see Chow, Shao and Wang[5]. The R package pwr can be used to calculate sample sizes for basic tests, the package gsDesign can be used for group sequential designs.

### 6.1.1   Normal random variables

**One sample test**

We wish to test whether the mean, $\mu$, of a normal random variable, $X$, is equal to a particular value $\mu_0$ with a 2-sided test. Assume the variance, $\sigma^2$, is known. Let the sample be $x_1, \ldots, x_n$. So, we are testing,

$$\text{H}_0 : \mu = \mu_0, \quad \text{H}_1 : \mu \neq \mu_0.$$

The test statistic is $z = \sqrt{n}(\bar{x} - \mu_0)/\sigma$ and we reject $\text{H}_0$ if $|\sqrt{n}(\bar{x} - \mu_0)/\sigma| > z_{\alpha/2}$, where $z_\alpha$ is the upper $\alpha$ quantile of the standard normal distribution.

Let the mean of $X$ be $\mu_0 + \epsilon$ under $\text{H}_1$, where $\epsilon$ is the effect size required, assumed positive. Then the power is $1 - \beta = 1 - \Pr(|z| \leq z_{\alpha/2}|\mu = \mu_0 + \epsilon) = \Phi(\sqrt{n}\epsilon/\sigma - z_{\alpha/2}) + \Phi(-\sqrt{n}\epsilon/\sigma - z_{\alpha/2})$, after some algebra, where $\Phi$ is the cumulative distribution function of the standard normal distribution. Now, we assume $\Phi(-\sqrt{n}\epsilon/\sigma - z_{\alpha/2})$ is small and can be ignored. Then solving, $1 - \beta = \Phi(\sqrt{n}\epsilon/\sigma - z_{\alpha/2})$, gives,

$$n = \frac{(z_{\alpha/2} + z_\beta)^2 \sigma^2}{\epsilon^2}.$$

We need an estimate of $\sigma^2$, which could come from a pilot study, similar previous studies, or by careful consideration of the nature of the random variable. A one-sided test has $z_{\alpha/2}$ replaced by $z_\alpha$.

**Example.** A single arm quality of life study is to be designed for head and neck cancer patients undergoing a new treatment, with global health status (QOL) as the primary outcome. From past studies, the typical mean value of QOL for patients on the current accepted treatment is 60, with standard deviation, 20. How many patients are needed to detect a difference of 5 in QOL, for significance level, 0.05 and power 80%, for a two-sided test? The simple calculation is,

$$n = (1.960 + 0.842)^2 \times 20^2/5^2 = 125.6,$$

and so 126 patients are required.

**Two sample test**
This time we test

$$H_0: \mu_1 = \mu_2, \quad H_1: \mu_1 \neq \mu_2,$$

where there are two groups and two samples, $x_{11}, \ldots, x_{1n_1}$ and $x_{21}, \ldots, x_{2n_2}$. We assume the variance is $\sigma^2$ for both groups. Let $\epsilon = \mu_2 - \mu_1$ and $k = n_1/n_2$. Then in a similar manner to the one-sample test, it is easily shown that,

$$n_1 = kn_2, \quad n_2 = \frac{(z_{\alpha/2} + z_\beta)^2 \sigma^2 (1 + 1/k)}{\epsilon^2}.$$

For equal sample sizes, $k = 1$, and the total sample size is four times that for the single sample case.

**Example**. Following on from the previous example, we now plan a two-sample test, this time using a control arm. A difference in QOL means of 5 is to be tested, the two arms are to be of equal size ($k = 1$) and the same significance level and power are the same as before. The calculation is,

$$n_2 = 2(1.960 + 0.842)^2 \times 20^2/5^2 = 251.2, \quad n_1 = n_2 = 251.2,$$

and so 504 patients are required.

**Paired sample test**
The paired sample test is a special case of the one-sample test. For example, suppose a measurement is made on the left kidney and on the right kidney of patients. We simply subtract the left kidney value from the right kidney value, or vice versa, and treat the differences as a single sample. Note the variance of the difference is $2\sigma^2$ and that is assuming the two measurements on each subject are uncorrelated. Hence the required sample size is,

$$n = \frac{2(z_{\alpha/2} + z_\beta)^2 \sigma^2}{\epsilon^2}.$$

### 6.1.2 Binomial random variables

**One sample test**
Let the random variable of interest be from a binomial distribution, $X \sim B(n, p)$. Our hypotheses are,

$$H_0: p = p_0, \quad H_1: p \neq p_0.$$

For a reasonably large sample, $X \sim N(np, np(1-p))$ as long as $p$ is not close to 0 or 1. Under $H_0$, $X \sim N(np_0, np_0(1-p_0))$ and under $H_1$, $X \sim N(np, np(1-p))$ for a particular value of $p$. Now, if we assume $p_0(1 - p_0)$ is approximately equal to $p(1-p)$, then the variances of X under $H_0$ and $H_1$ are approximately the same and so we can use the results for the normal distribution above to calculate sample sizes. Hence,

$$n = \frac{(z_{\alpha/2} + z_\beta)^2 p(1-p)}{\epsilon^2}, \tag{6.1}$$

where $\epsilon = p - p_0$. We need to choose a value of $p$ that makes the difference $\epsilon$ clinically meaningful and is the point at which we want to have power at the chosen level. The true value of $p$ might be different from this. We could estimate the variance, $p(1-p)$, in the sample size formula with the chosen value of $p$, or we could use the more conservative value of 0.25, as this is the maximum possible value of $p(1-p)$, and noting that for $0.25 \leq p \leq 0.75$, $p(1-p)$ only varies by 0.01.

The arcsine transformation, $\arcsin(\sqrt{x})$, can be used to obtain a better normal approximation of the binomial distribution, which is more important for $p$ near to 0 or 1. The exact binomial distribution can also be used for the calculation, but there will not be a easy formula to use for the sample size calculation.

**Example.** Suppose we wish to design a trial where the primary outcome is survival at one year after treatment. (We assume no patient leaves the trial before the end of the year.) The accepted current one-year survival rate is $p_0 = 0.5$. We believe a new treatment will increase the survival rate to 0.65. So $\epsilon = 0.15$ and using the formula above, $n = 80$ if $p = 0.65$ is used in the estimate of $p(1-p)$. If $p(1-p)$ is estimated by 0.25, $n = 88$, and if the arcsine transformation is used, $n = 85$. Note, if $p_0 = 0.1$ and $p = 0.25$, the corresponding three sample size estimates are, 65, 88 and 49, the latter using the arcsine transformation being the preferred value, as it is more accurate.

**Two sample test**

We now have two random variables, $X_1 \sim B(n_1, p_1)$ and $X_2 \sim B(n_2, p_2)$ and we test,

$$H_0: p_1 = p_2, \quad H_1: p_1 \neq p_2.$$

In similar manner, assuming normal approximations, after some algebra we have,

$$n_1 = kn_2, \quad n_2 = \frac{(z_{\alpha/2} + z_\beta)^2 \{p_1(1-p_1)/k + p_2(1-p_2)\}}{\epsilon^2}.$$

**Example.** Following on from the previous example, we want the sample size for a two arm test, with $p_1 = 0.5$ and $p_2 = 0.65$, equal arm size and the same significance level and power. From the above formula, $n_1 = n_2 = 167$, and so 334 patients are required.

## 6.1.3   Survival random variables

In the survival setting, we will calculate sample sizes based on the log-rank test. In Section 3.4.1, we saw that the log-rank test statistic for testing a difference in survival distributions between two groups, and before squaring, is,

$$Z = \frac{\sum_{i=1}^{r}(d_{1i} - e_{1i})}{\sqrt{\sum_{i=1}^{i} v_i}} = \frac{U}{\sqrt{V}}, \tag{6.2}$$

where $d_{1i}$ is the number of deaths in Group 1 at time $t_i$ and $e_{1i}$ is the expected number of deaths in Group 1 at time $t_i$ under the null hypothesis of identical survival functions. We saw that asymptotically, $U \sim N(0, V)$, or equivalently, $Z$ had a standard normal distribution, $Z \sim N(0, 1)$.

Now the sample size depends on the number of deaths observed from the start of the trial until the time of analysis and not on the total number of patients in the trial. This is obvious when you think about it – if nobody has died by the time you analyse the data, then it is impossible to find a difference between the two groups. It does not matter how many patients you have in the trial.

Let the hazard ratio be $\gamma$. The null and alternative hypotheses are,

$$H_0 : \gamma = 1, \quad H_1 : \gamma = \gamma_1,$$

where we will assume $\gamma_1 < 1$ and working with the hazard ratio being test treatment relative to control.

Let the significance level of the test be $\alpha$ and the power required, $1 - \beta$. To obtain the significance level, $\alpha$, we need to find $k$ such that,

$$\Pr(|U| > k | \gamma = 1) = \alpha,$$

and because the normal distribution is symmetric about zero, this is equivalent to

$$\Pr(U > k | \gamma = 1) = \alpha/2.$$

For power $1 - \beta$, we need,

$$\Pr(|U| > k | \gamma = \gamma_1) = 1 - \beta.$$

Thus for the significance level,

$$1 - \Phi(k/\sqrt{V}) = \alpha/2.$$

For the power, we will use the fact that under $H_1$, $U$ has an approximate normal distribution, $N\big(\log(\gamma_1)V, \ V\big)$ (not proved here) and hence,

$$\Pr(|U| > k | \gamma = \gamma_1) = \Phi\left\{ \frac{-k - \log(\gamma_1)V}{\sqrt{V}} \right\} + 1 - \Phi\left\{ \frac{k - \log(\gamma_1)V}{\sqrt{V}} \right\} = 1 - \beta.$$

As $\gamma_1 < 1$, the final $\Phi$ term will be close to 1, and so approximately,

$$\Phi\left\{ \frac{-k - \log(\gamma_1)V}{\sqrt{V}} \right\} = 1 - \beta.$$

Thus we have the two equations to solve,

$$\Phi\{k/\sqrt{V}\} = 1 - \alpha/2$$
$$\Phi\{(-k - \log(\gamma_1)V)/\sqrt{V}\} = 1 - \beta,$$

and so,

$$k/\sqrt{V} = z_{\alpha/2}$$
$$(-k - \log(\gamma_1)V)/\sqrt{V} = z_\beta.$$

Substituting $k = \sqrt{V} z_{\alpha/2}$ from the first equation into the second gives

$$V = (z_{\alpha/2} + z_\beta)^2/\{\log(\gamma_1)\}^2. \tag{6.3}$$

Now consider the formula for $V$ in Section 3.4.1,

$$V = \sum_{i=1}^{r} \frac{n_{1i} n_{2i} d_i (n_i - d_i)}{n_i^2 (n_i - 1)}.$$

Let the proportion of patients allocated to Group 1 be $p_1$ and the proportion to Group 2 be $p_2$, and assume the number of deaths at each time point relative to the number at risk is very small, and so approximately,

$$V = \sum_{i=1}^{r} p_1 p_2 d_i = p_1 p_2 d,$$

where $d$ is the total number of deaths needed. Hence, substituting for $V$ in Equation 6.3, gives the number of deaths required as

$$d = (z_{\alpha/2} + z_\beta)^2/[\{\log(\gamma_1)\}^2 p_1 p_2]. \tag{6.4}$$

For equal allocation, this becomes $d = 4(z_{\alpha/2} + z_\beta)^2/\{\log(\gamma_1)\}^2$.

So we have the number of deaths needed, but usually there will be surviving subjects and censored subjects at the time of analysis and so we have to estimate the number of subjects to be recruited so that, by the time of the analysis, there are the required number of deaths.

### 6.1.3.1  Number to recruit

The number needed to be recruited in order to obtain $d$ deaths depends on factors such as the length of the recruitment period, the length of the follow-up period, the survival function and the drop-out rate. We need to calculate the probability of death, Pr(death), for an arbitrary patient recruited into the trial. Then the number of patients to be recruited is,

$$n = \frac{d}{\text{Pr(death)}}. \tag{6.5}$$

An allowance is then made for drop-out, e.g if drop-out is assumed to be 10%, then the number to be recruited is increased to $n/0.9$.

We consider the case of exponential distributions. Let the recruitment period last $a$ units of time, the follow-up period, $f$ units of time. Assume the accrual rate is constant over the accrual period. Let the two hazard rates be $\lambda_1$ and $\lambda_2$. So the hazard ratio is HR $= \lambda_1/\lambda_2$. The sketch below shows the time line for the trial and a patient from Group 1 entering the trial at time $t$. It will help in calculating probabilities.

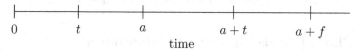

The probability that the patient dies before time $a + f$, given that they entered the trial at time $t$ is

$$1 - \exp\{-\lambda_1(a + f - t)\}.$$

The probability of the patient entering the trial at time $T$ follows a uniform distribution on $[0, a]$. Thus the unconditional probability that an arbitrary patient from Group 1 dies before time $a + f$ is given by,

$$\Pr(\text{death}) = \text{prob}_1 = \int_0^a \frac{1}{a}[1 - \exp\{-\lambda_1(a + f - t)\}] \tag{6.6}$$
$$= 1 - \lambda_1^{-1}\exp(-\lambda_1 f)\{1 - \exp(-\lambda_1 a)\}.$$

A similar probability, $\text{prob}_2$, is calculated for a patient recruited to Group 2. Then, as the group allocation proportions are $p_1$ and $p_2$, we calculate the probability of death for an arbitrary patient as $p_1\text{prob}_1 + p_2\text{prob}_2$. Hence we can now use Equation 6.5 to find $n$. In practice, of course, you need estimates of $\lambda_1$ and $\lambda_2$. Current knowledge of survival times and in particular the median survival time, $M_1$, of patients on current treatment regimes can be used to estimate $\lambda_1$, $(\hat{\lambda}_1 = \log(2)/M_1)$. Then estimate $\lambda_2$ by $\hat{\lambda}_1 \times \lambda$. (Don't confuse the exponential distribution rates $\lambda_1$ and $\lambda_2$ with the hazard ratio $\lambda$.)

Collett[6] shows a way of calculating $\Pr(\text{death})$ within a Group when the survival distribution is not assumed to be exponential, nor any other particular parametric form, but from estimated survival curves from previous studies. The probability of death for an arbitrary patient in Group 1 is estimated as

$$\Pr(\text{death}) = 1 - \frac{1}{6}\{\hat{S}_1(f) + 4\hat{S}_1(0.5a + f) + \hat{S}_1(a + f)\},$$

where $\hat{S}_1(f)$, $\hat{S}_1(0.5a + f)$ and $\hat{S}_1(a + f)$ are estimates of the value of the survival curve at times $f$, $a/2 + f$ and $a + f$. These could be read from a Kaplan-Meier curve from a past similar trial. The same is done for Group 2 patients and hence the sample size calculated.

## 6.2    Examples of sample size calculations; Phases I to III

We now briefly describe various trials and their sample size calculations, that have been published, starting with a discussion of Phase I trials.

### 6.2.1    Phase I trials: dose escalation studies

The aim of a Phase I dose escalation study is to find the most appropriate dose of a drug or combination of drugs that can be taken forward to Phase II trials. You have to find a dose that is effective against the cancer, but does not cause too many toxic effects. These trials are usually single-arm, open label, sequential trials. To find the appropriate dose, we need to specify the *dose limiting toxicity* (DLT), which is the level of side effects experienced by patients that would stop any increase is the amount of drug administered. The *maximum tolerated dose* (MTD) is the highest dose that does not cause unacceptable side effects.

A common design is the *3 + 3 design*, where groups of three subjects are tested sequentially. The sequence is basically as follows.

1.  Three subjects are giving the starting level dose and the number with DLT noted.

2.  Then,
    - if the number of DLTs = 0, proceed to the next dose level,
    - if the number of DLTs = 1 give the same dose to three more subjects and note the total number with DLT out of 6; if the number of DLTs < 2 proceed to the next dose level,
    - if the number of DLTs ≥ 2 stop and reconsider the dose starting level.

3.  If the next dose level is appropriate, repeat steps 1 and 2 using the new dose level, and continue doing this until the procedure finally stops due to DLT or other criteria.

The MTD is the dose level prior to the level that caused the process to stop, but make sure 6 subjects have been treated at the chosen MTD and that for these, there are less than 2 DLTs. The dose levels have to be chosen prior to starting the trial. Dose increases between levels can be constant, or could be reducing, e.g. 100%, 67%, 33%, 33%. What constitutes DLT also has to be decided.

There are other Phase I rule based designs as well as model based designs, such as the *Continual Reassessment Method* (CRM). The CRM has a probability model for DLT for dose level, $d$, for example, using a two-parameter logistic function,

$$\Pr(\text{DLT at dose } d) = F(d; \beta_1, \beta_2) = \frac{\exp(\beta_1 + \exp(\beta_2 d))}{1 + \exp(\beta_1 + \exp(\beta_2 d))},$$

where $\beta_1$ and $\beta_2$ are parameters to be estimated. Note, the model must be a non-decreasing function of $d$. The model is fitted using maximum likelihood or Bayesian methods, using data collected sequentially as subjects are given various doses, chosen on previous doses and DLTs. The decision on MTD is then based on the fitted model, the MTD being the dose where $F(\text{MTD}; \hat{\beta}_1, \hat{\beta}_2)$ is at the maximum acceptable probability of DLT. See [22] for a discussion on whether $3 + 3$ designs are best.

### Example: gemcitabine plus nab–paclitaxel for PDAC

To show that not all Phase I dose escalation studies follow the classic designs, we briefly describe a study to find the MTD of neoadjuvant chemotherapy using a combination of gemcitabine (GEM) plus nab–paclitaxel (nab-PTX), for resectable pancreatic ductal adenocarcinoma (PDAC) patients, see Tajima et al. [64]

The dose levels $(\text{mg/m}^2/\text{day})$ were:

Level 1:   GEM: 600       nab-PTX: 75
Level 2:   GEM: 800       nab-PTX: 100
Level 3:   GEM: 1,000     nab-PTX: 125

and the definition of DLT was based on grades 3/4 of neutropenia, thrombo-cytopenia, other non-haematologic toxicities, and whether there was delayed recovery from treatment-related toxicity lasting more than two weeks. The rule for not progressing to the next dose level was: DLTs observed in more than 50% of patients or treatment could not be completed in more than 50% of patients; otherwise progression to the next level.

Of 21 patients recruited to the level 1 dose, 3 patients were withdrawn, leaving 18, all of whom completed the treatment (100%), but with 8 patients experiencing DLTs (44%). Hence the dose was escalated to level 2. There were 18 patients recruited to the level 2 dose, but 3 of these were withdrawn. All the remaining 15 patients completed the treatment (100%), but 9 had DLTs (60%) and so progression to the next dose level did not happen and the MTD was set at dose level 1.

You might ask why so many patients were recruited to the two dose levels, given we have been talking about cautiously assigning low numbers, like 3 to the dose levels. It is because GEM and nab-PTX are already used for cancer treatment and so considered safe. This study was looking to see if GEM and nab-PTX might make a good combination treatment for resectable PDAC. The advantage of having these larger groups of patients is that efficacy data were also recorded which could help inform the design of future Phase II trials using the combination treatment.

### 6.2.2   Phase II trials

Phase II trials build on the results of a Phase I trial where a new treatment is being proposed, or they may use established drugs and other treatments in a new clinical setting. They usually consist of 30–100 subjects and are aimed at establishing the effectiveness of the treatment, to check dosing levels and assesses adverse events and reactions due to the treatment. Phase II trials can be single arm, two-arm, or possibly multi-arm. Trials may be two-stage, or multi-stage, where a set of subjects are recruited to each stage, before moving to the next stage if the results, so far, are promising. Here we just look at one Phase II trial which has a five arm, two stage design.

### Example: cixutumumab for soft-tissue sarcoma

Schöffski et al. [56] carried out an open-label, single arm Phase II trial to evaluate treatment by cixutumumab, of patients with previously treated advanced or metastatic soft-tissue sarcomas and the Erwing family of tumours. There were five groups of patients, those with: rhabdomyosarcoma, leiomyosarcoma, adipocytic sarcoma, synovial sarcoma and the Ewing family of tumours. The data for each group was analysed separately. The primary outcome was *progression-free rate* (PFR), using the RECIST criteria, measured at 12-weeks. If RECIST indicated stable disease (SD), partial response (PR) or complete response (CR), the patient was progression-free. We use the number of patients out of $n$ who were progression-free together with the binomial distribution to calculate sample size.

Let $p$ be the PFR. We use the hypotheses, $H_0 : p \leq p_0$ vs $H_1 : p \geq p_1$ and a 1-sided test. The trial design used $p_0 = 0.2$ and $p_1 = 0.4$, with $\alpha = 0.10$ and $\beta = 0.10$. For a one-sided test, using Equation 6.1 and these values, the estimated sample size is, $n = 39.4$, and so in practice, use $n = 40$. If the R function pwr.p.test() from the package pwr is used, then $n = 34$, the arcsin transformation having been used. Note, you will often find small differences in software packages that calculate sample sizes. This is usually because they are using different approximations.

The actual trial design was a two-stage Simon design. The idea is that you recruit a proportion of the total patients in the first stage and, depending on the results, either enter the second stage and recruit the remaining patients, or stop, concluding that you have enough evidence to say it is not worth continuing with the trial and the treatment will (probably) not achieve the required efficacy.

Suppose $n_1$ patients are to be recruited into the first stage and if $r_1$ or more of these are progression-free at 12-weeks, then $n - n_1$ further patients will be recruited. At the end of the trial, the treatment is rejected if less than $r$ patients are progression-free. The probability of rejecting the treatment, for PFR $= p$ is,

$$P(p) = B(r_1; n_1, p) + \sum_{i=r_1+1}^{\min(n_1,r)} b(i; n - n_1, p)\, B(r - i; n - n_1, p), \qquad (6.7)$$

where $b(i; n, p) = \binom{n}{i} p^i (1 - p)^{n-i}$ is the probability mass function of the binomial distribution and $B(r; n, p) = \sum_{i=0}^{r} b(i; n, p)$ is the cumulative distribution function. The first term in the formula for $P(p)$, is the probability that the trial stops at the first stage as $r_1$ or less patients were progression free. The second term is the sum of the probabilities that $i\,(> r_1)$ patients are progression-free at stage one, and then $r - i$ patients or less are progression-free in stage two.

The idea is to find $n$, $r$, $n_1$ and $r_1$ that gives us our chosen $\alpha$ and $\beta$ for the selected values of $p_0$ and $p_1$. To do this, we solve the equations, $P(p_0) = 1 - \alpha$ and $P(p_1) = \beta$. However, there is no unique solution and so we choose a solution that satisfies another constraint, for example a solution that will have smallest expected sample size, $n_1 \times \Pr(X_1 \leq r_1) + n \times (1 - \Pr(X_1 \leq r_1))$, where $X_1$ is the number of progression-free patients in the first group. Another criterion is a minimax one which minimises the maximum sample size, $n$.

Using the Simon design in the sarcoma trial, for $p_0 = 0.2$, $p_1 = 0.4$, $\alpha = 0.1$ and $\beta = 0.1$, gives $n_1 = 17$, $r_1 = 3$, $n = 37$ and $r = 10$ using the minimum expected sample size criteria. Because the binomial distribution is discrete, the true $\alpha$ is 0.095 and the true $\beta$ is 0.097. The expected sample size is 26.02. If the minimax criterion is $n_1 = 19$, $r_1 = 3$, $n = 36$ and $r = 10$.

Of the five sarcoma groups, only the adipocytic group progressed to the second stage and then, out of the 37 patients, 12 were progression-free, suggesting efficacy of cixutumumab for this group. Kaplan-Meier plots of PFS supported this as the plot for the adipocytic group showed superior survival.

There are other two or more stage designs. For example, Fleming's two-stage design is the same as Simon's but allows the trial to stop at the first stage if a high success rate is observed. These two-stage designs are expanded to the more comprehensive approach of group sequential designs in Section 6.3.

### 6.2.3 Phase III trials

Phase III trials are confirmatory trials that try to establish if one treatment is more efficacious than another. They can be two arm or multi-arm. They provide more information on adverse events and reactions, as well as subjects' quality of life. These are *superiority* trials, but sometimes Phase III trials are run to test whether a treatment is unacceptably less efficacious than another treatment (*non-inferiority trial*), or run to see whether a treatment has no better, nor worse efficacy than another treatment (*equivalence trial*).

**Example: ESPAC4 trial**
The ESPAC4 trial was described in Section 3.5. The sample size calculation was based on a two-year overall survival rate for GEMCAP of at least a 10% increase in the two-year survival rate for GEM; an increase from 0.475 to 0.575. The test to be a two-sided log-rank test with $\alpha = 0.05$ and power 90% and 1:1 allocation to the two groups. Assuming proportional hazards, the two survival curves satisfy, $S_1(t) = S_0(t)^\gamma$, and so $\gamma = \log\{(S_1(t)\}/\log\{S_0(t)\}$. For $t = 2$, $\gamma = \log(0.575)/\log(0.475) = 0.743$ and using Equation 6.4 above, the required number of deaths is 480 (rounded up slightly).

To find the number of patients to be recruited, we need the probability of death to be used in Equation 6.5. Using exponential distributions for the control GEM and GEMCAP arms, the average death rate is, $\lambda = (-\log(0.475) - \log(0.575))/2 = 0.027$. Using Equation 6.6 with this average death rate, $\Pr\{death\} = 0.770$ and hence, $N = 621$. We could have used Equation 6.6 separately for each arm and taken the mean of the two death rates, but the resulting $N$ does not change by very much. Lastly, an allowance of 10% is made for withdrawal of patients and 5% for lost to follow-up, giving a final number of patients to be recruited as, $N/(0.90 \times 0.95) = 726$. The ESPAC4 protocol has 722 patents to be recruited – a slight but insignificant difference.

Using the R function, nSurv(), from the gsDesign package to calculate sample size, gives the number of patients to recruit as 720 with 476 deaths. The loss of patients to the study is calculated slightly differently.

## 6.3  Group sequential designs

Following on from the two-stage designs earlier, this is a good place to consider group sequential designs. With this type of design, patients are recruited in groups, sequentially, and an interim analysis is carried out every time the data becomes available for the current group of patients. The results are used in a stopping rule that dictates (or suggests) whether another group of patients is to be recruited, or that the trial should stop. If the trial stops, then the results will have shown one treatment to be superior to the other, or there is likely to be no difference between the treatments. This continues until the final group of patients are recruited and a final analysis undertaken.

We will start with a simple design with a normally distributed primary outcome with known variance, $\sigma^2$, the same for each of two treatments, A and B. There will one interim analysis and then the final analysis, i.e. two groups in total. The hypotheses are

$$H_0 : \mu_1 = \mu_2, \quad H_1 : \mu_1 \neq \mu_2.$$

Let $n$ patients be recruited into each group, with equal allocation to treatments, ready for the interim analysis. Let the data for treatment A be $x_1, \ldots, x_n$ and for treatment B, $y_1, \ldots, y_n$. If the trial is to continue, let a further $n$ patients be recruited for each group, with data, $x_{n+1}, \ldots, x_{2n}$, and $y_{n+1}, \ldots, y_{2n}$. The two-sample $Z$-test will be used as the test statistic, first at the interim analysis and then at the final analysis. Let these two test statistics be denoted by $Z_1$ and $Z_2$ respectively, where

$$Z_1 = (1/n)\left\{ \sum_{i=1}^{n} x_i - \sum_{i=1}^{n} y_i \right\}/(\sigma/\sqrt{2n}) = \frac{\bar{x}_1 - \bar{y}_1}{\sigma\sqrt{2/n}}, \qquad (6.8)$$

$$Z_2 = (1/2n)\left\{ \sum_{i=1}^{2n} x_i - \sum_{i=1}^{2n} y_i \right\}/(\sigma/\sqrt{n}) = \frac{\bar{x}_2 - \bar{y}_2}{\sigma\sqrt{1/n}},$$

and so the distributions of $Z_1$ and $Z_2$ are both standard normal $N(0,1)$ under $H_0$.

Our stopping rule, at the interim analysis, is to stop and declare superiority of one treatment if the value of $Z_1$ is in the critical region of the test, otherwise carry on recruitment with a final test at the end of the trial. Let us use a significance level of 0.05 and so the critical region for the test is $|z_1| >$ 1.96. Suppose the value of $Z_1$ is not in the critical region and so we carry on recruiting and then we use $Z_2$ at the end of recruitment. We check the value of $Z_2$ to see if it is in the critical region, $|Z_2| > 1.96$ and declare our results at the 5% significance level. But this is wrong! For a start, the two tests are not independent because they both contain the first $n$ $x$ and $y$ values. Results of the first test affect the results of the second, and as a consequence, the significance level is not 5% as we thought.

The actual significance level is calculated as follows,

$$\alpha = 1 - \text{Pr(interim test and final test do not reject } H_0 \,|\, H_0 \text{ true})}.$$

We have to find the joint distribution of $Z_1$ and $Z_2$. Each has a normal distribution and so their joint distribution is a bivariate normal distribution (see Appendix for notes on the multivariate normal distribution). Now $E(Z_1) = E(Z_2) = 0$ and $\text{Var}(Z_1) = \text{Var}(Z_2) = 1$. The covariance of $Z_1$ and $Z_2$ is found as follows. Let $U_1 = (x_1 + \ldots + x_n) - (y_1 + \ldots + y_n)$ and $U_2 = (x_{n+1} + \ldots + x_{2n}) - (y_{n+1} + \ldots + y_{2n})$. Then from Equations 6.8,

$$Z_1 = \frac{1}{\sqrt{2n}\sigma} U_1 \quad Z_2 = \frac{1}{\sqrt{n}\sigma}(U_1 + U_2)$$

Hence

$$\text{cov}(Z_1, Z_2) = \frac{1}{\sqrt{2n}\sigma^2} \text{cov}(U_1, U_1 + U_2).$$

Now $U_1$ and $U_2$ are independent since they do not contain common $x$ or $y$ values and so,

$$\text{cov}(Z_1, Z_2) = \frac{1}{\sqrt{2n}\sigma^2} \text{var}(U_1) = \frac{1}{\sqrt{2n}\sigma^2} \times n\sigma^2 = \frac{1}{\sqrt{2}}.$$

As $Z_1$ and $Z_2$ have unit variance, their correlation is equal to this covariance. Hence the joint probability density function of $Z_1$ and $Z_2$ is,

$$f(z_1, z_2) = \frac{1}{2\pi} \exp\left\{ -(z_1^2 - \sqrt{2}z_1 z_2 + z_2^2)/2 \right\},$$

and so the probability of a Type I error is,

$$\alpha = 1 - \int_{-1.96}^{1.96} \int_{-1.96}^{1.96} \frac{1}{\pi} \exp\left\{ - (z_1^2 - \sqrt{2}z_1 z_2 + z_2^2)/2 \right\} dz_1 dz_2.$$

Numerical integration is needed for this integral. Its value is 0.083, rather larger than our chosen 0.05! The above can be extended to $K$ analyses, $K - 1$ interim analyses and a final analysis. There will be $K$ $U_i$'s and their joint distribution is multivariate normal with mean vector and covariance matrix

$$\boldsymbol{\mu} = (0, \ldots, 0), \quad \boldsymbol{\Sigma} = [\sigma_{ij}], \text{ where } \sigma_{ij} = \sqrt{i/j},$$

and then $\alpha$ is calculated using a $K$-fold integral of the multivariate normal pdf. For $K = 3, 4$ and 5, the corresponding values of $\alpha$ are 0.107, 0.126 and 0.142, if always using a critical region, $\pm 1, 96$. As $K$ approaches $\infty$, $\alpha$ approaches 1.

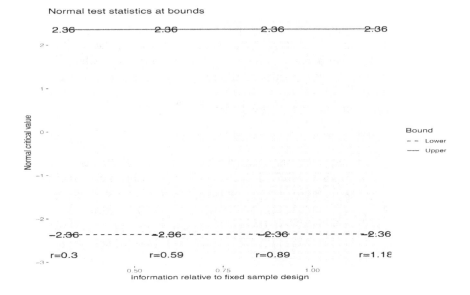

**FIGURE 6.1**
Pocock boundaries for a group sequential design

This is not a desirable situation having the Type I error increase with the number of interim analyses you carry out. To keep the stated Type I error, the critical regions have to be changed. If, for $K = 2$, the critical points are changed from $\pm 1.96$ to $\pm 2.18$, then the Type I error is back to 0.05. For $K = 4$, critical points of $\pm 2.36$ will bring the Type I error back to 0.05. Figure 6.1 shows the four critical regions for this case and is called a *Pocock design*. The design has nominal p-value of 0.018 at each of the four analyses, but the actual p-values are 0.018, 0.032, 0.042 and 0.05, as we go through the interim analyses and end with the final analysis.

Note, the design does not give the estimated sample size for the chosen significance level and power, but the factor, $r$, by which the sample size from a standard design needs to be multiplied in order to obtain the required sample size. This applies to the interim analyses and the final analysis. From Figure 6.1 we see than the total sample size has to be 1.18 times larger.

The R function gsDesign() in the R package gsDesign was used to draw the Pocock design in Figure 6.1. Note, care is needed when setting p-values in the gsDesign() function. For a two sided design, with p-value 0.05 say, the p-value to be entered is 0.025, referring to one side of the two-sided test.

Now we let the critical regions vary over the four analyses. We might want to make it harder to reject $H_0$ at the first interim analysis so that we ensure at least a reasonable sample size before we might stop the trial. Then make it slightly easier to reject at the second interim analysis and so forth. One could use any set of critical points as long as the overall Type I error is 0.05.

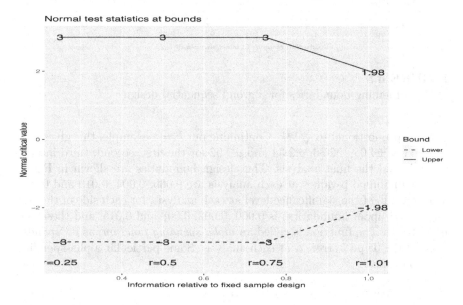

**FIGURE 6.2**
Haybittle-Peto boundaries for a group sequential design

The *Haybittle-Peto design* uses the critical region, $|Z| > 3$ at each interim analysis, and then a critical region that makes the overall significance level $\alpha$. Figure 6.2 shows the design. The nominal p-values are 0.0026 at the three interim analyses. The critical region for the final analysis is very close to that for a standard test, $\pm 1.96$, and so as an approximation the standard test could be used at the final analysis. A criticism of this design is that it is very conservative and trials are rarely stopped at the interim analyses.

We can use a function to specify the critical points at the interim analyses, for example the *O'Brien and Fleming design* has critical points at the $k$th

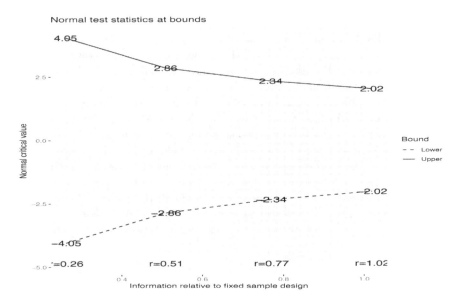

**FIGURE 6.3**
O'Brien-Fleming boundaries for a group sequential design

analysis proportional to $\sqrt{1/k}$. Continuing our $k = 4$ example, this gives critical points $\pm4.05$, $\pm2.86$, $\pm2.34$ and $\pm2.02$ for the first, second, third interim analysis and the final analysis. This design boundaries are shown in Figure 6.3. The nominal p-values at each analysis are 0.000, 0.004, 0.019 and 0.043. The "spend" of the significance level at each analysis, for each side of the test (lower and upper boundaries) is 0.000, 0.002, 0.008 and 0.015, and these add up to 0.025. The function is called an *alpha-spending function*, as it "spends" some of the Type I error, $\alpha$ at each analysis. Some particular alpha-spending functions are:

$$\begin{aligned} \text{Linear}: \quad & f(t) = \alpha t, \\ \text{O'Brien/Fleming}: \quad & f(t) = 2 - 2\Phi(z_{\alpha/2}/\sqrt{t}), \\ \text{Pocock}: \quad & f(t) = \alpha \log(1 + (e - 1)t), \\ \text{Hwang/Shih/DeCani}: \quad & f(t) = \alpha(1 - e^{-\gamma t})/(1 - e^{-\gamma}) \quad (\gamma \neq 0), \\ & = \alpha t \quad (\gamma = 0), \end{aligned}$$

for $t = n_i/n$, the relative time of the $i$th interim analysis that has sample size $n_i$. The $\gamma$ in the Hwang-Shih-DeCani alpha–spending function has a constant, $\gamma$, to be chosen. The Wang-Tsiatis alpha-spending function is similar to that of the O'Brien-Flemimg, but with $\sqrt{t}$ replaced by $t^\delta$, where $\delta$ is a chosen constant.

### 6.3.0.1 Stopping for futility

As a trial progresses, it may become obvious that our new treatment, say, will probably never be shown to be superior to the control treatment. At a particular interim analysis, the probability of finally showing the new treatment is superior to the control might be very small. If this is so, then we should stop the trial. However, under some circumstances we actually might want to continue, as long as no patient is at a disadvantage by being on the new treatment. For instance, we might wish to continue collecting adverse event and quality of life data, because the new treatment might be better tolerated than the control.

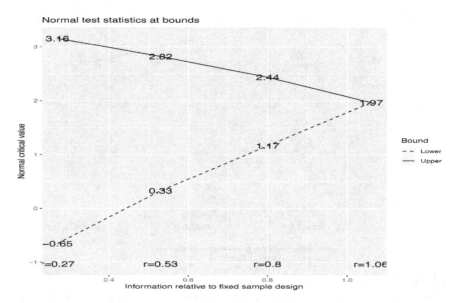

**FIGURE 6.4**

Group sequential design with Hwang–Shih–DeCani binding superiority and futility boundaries

Analogous to the alpha-spending function, we can use a *beta-spending function* that spends the Type II error, $\beta$. The default alpha-spending and beta-spending functions for gsDesign() are Hwang-Shih-DeCani spending functions with $\gamma = -4$ for alpha-spending and $\gamma = -2$ for beta-spending. For our example, the boundaries for these default values are shown in Figure 6.4. At each of the interim analyses, the trial is stopped for superiority if $z_i$ is greater than the upper bound and stopped for futility if $Z_i$ is less than the lower bound, noting it is a 1-sided test, with $\alpha = 0.025$ and 90% power ($\beta = 0.1$). The $\alpha$ spending is 0.001, 0.002, 0.006 and 0.016, at the four analyses, totalling 0.025. The $\beta$ spending is 0.010, 0.017, 0.027 and 0.046, totalling 0.1.

The critical values for the boundaries in Figure 6.4 have been calculated assuming that if $z_i$ is below the lower boundary, the trial will definitely stop for futility – no discussion! If we allowed discretion on whether to stop or not for futility, then we have not allowed for the inflation of the Type I error probability that will happen. A more pragmatic approach is to make the lower boundary non-binding and calculate the critical values that stops this inflation. This is shown in Figure 6.5, where you will notice the critical value at the final analysis has now increased from 1.97 to 2.01.

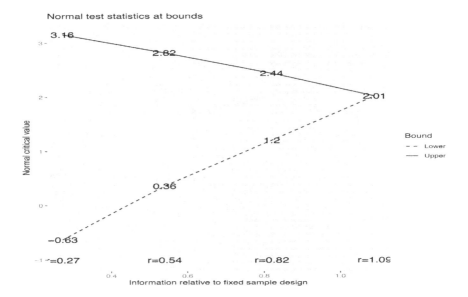

**FIGURE 6.5**
Group sequential design with Hwang-Shih-DeCani binding superiority and non-binding futility boundaries

Lastly, we might want to have interim analyses at particular time points, rather than being equally spaced. Also, we might wish to have sample sizes calculated rather than proportions of the fixed sample size. This is easily achieved within gsDesign(). Also, another method for testing for futility is to use conditional power, calculating the probability of rejecting $H_0$, given the current value of $Z_i$. If this is low, we might want to stop the trial. We will not take this any further here.

**6.3.0.2   Group sequential designs for survival endpoints**

The log-rank test statistic, $Z$, of Equation 6.2 has an approximate standard normal distribution. Hence we can use the group sequential design theory in the survival setting. Things are slightly more complicated than in the previous designs because of the censoring and random occurrence of events. Basically, we have to work with the information, $I_k = 1/\text{var}(U_k)$, at the $k$th analysis. For our previous designs, $I_k$ was proportional to $k$ which made the theory easier. However, we will not take this any further here.

The R function gsSurv() allows us to produce group sequential designs, that takes account of total study duration, amount of follow-up time and dropout rate, as well as the spending functions. An example is given re-designing the ESPAC4 trial.

### Group sequential designs for the ESPAC4 trial
The ESPAC4 trial used a Haybittle-Peto design with four interim analyses spaced at intervals when 100, 200, 300 and 400 deaths had been reached. As stated before, the Haybittle-Peto design is very conservative and a trial will only be stopped in extreme circumstances.

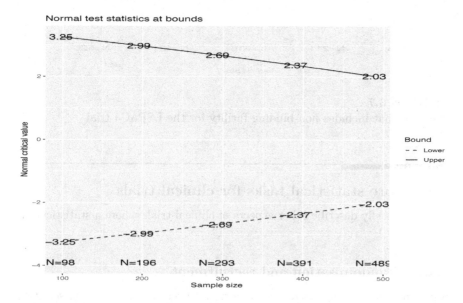

**FIGURE 6.6**
A symmetric design for the ESPAC4 trial

Let us look at a less conservative approach that could have been used. Figure 6.6 shows a symmetric design with a Hwang-Shih-DeCani alpha-spending function, but no beta-spending function. A total of 489 deaths are required, needing 736 patients to be recruited, compared to 476 deaths and 720 patients for a standard design. Figure 6.7 shows an asymmetric design with Hwang-Shih-DeCani alpha- and beta-spending functions. A total of 526 deaths are required, with 792 patients to be recruited.

Further details on group sequential designs can be found in Jennison and Turnbull[26] and Wassmer and Brannath[73].

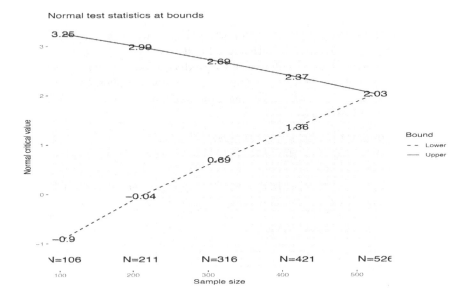

**FIGURE 6.7**
A design that includes non-binding futility for the ESPAC4 trial

## 6.4    More statistical tasks for clinical trials

Here, we briefly describe other aspects of clinical trials where a statistician is involved.

### 6.4.1    Randomisation and recruitment

Randomisation of patients to arms of a trial is necessary to eliminate selection and accidental bias. It aims to produce comparable groups of patients. There are various methods of randomisation which we discuss briefly, working with two arms, labelled A and B, e.g. a control arm and a treatment arm. For most trials, patients enter a trial sequentially – they are not all recruited at the start of the trial.

#### 6.4.1.1    Simple randomisation

Simple randomisation allocates each patient to either A or B according to a randomly generated value, $p$, from a uniform distribution on $[0, 1]$. For equal allocation, if $p \leq 0.5$ the patient is allocated to A, and otherwise to B. This will not produce exactly equal numbers in the two groups (occasionally it will) and could lead to a substantial imbalance between the groups. For this reason, simple randomisation is not recommended.

### 6.4.1.2 Block randomisation

To ensure equal randomisation, at least approximately, patients are allocated to arm in blocks. For example, for two arms, and a block size of four, the six possible allocation orderings, two patients to each arm, are: AABB, ABAB, ABBA, BAAB, BABA and BBAA. Blocks patterns are randomly chosen and joined together to form the allocation order of the patients, for example ABBA|BBAA|BAAB|BAAB|.... One problem is that it is possible to predict some of the allocations – if the first two patients are allocated to A, then we know the next two will be allocated to B. To overcome this, different block sizes can be used, for example, blocks of two, four or six, and block sizes are randomly chosen as well as the allocation pattern within a block.

### 6.4.1.3 Stratified randomisation

We would like our patient allocation to arms to be balanced within various covariates or factors, for example sex, age, resection margin and country. This can be done using stratified randomisation, with randomisation of patients carried out separately within each combination of the chosen stratification factors. For example, the ESPAC4 trial had country (England, Wales, Scotland, Germany, France, Sweden) and resection margin (positive, negative) as stratification factors. The twelve strata are England/positive, England/negative, Wales/positive, Wales/negative, etc. Each patient recruited into the trial is placed in the appropriate stratum and is allocated to the next arm in the allocation list for that particular stratum. Although you might like to have many stratification factors, practically you can only have a few, dictated by the total number of strata.

### 6.4.1.4 Minimisation

One way to have more factors for balancing the two arms of the trial, is to use the method of minimisation. We show how the method works with a simple example. Suppose there are three factors to be used, F, G and H, each with two categories, "–" and "+". As patients are recruited, they are allocated to arm A or arm B according to the numbers already in factor categories. Suppose five patients have been allocated to each of A and B and the numbers in the categories are as follows,

| Arm | F– | F+ | G– | G+ | H– | H+ |
|-----|----|----|----|----|----|----|
| A | 2 | 3 | 3 | 2 | 4 | 1 |
| B | 1 | 4 | 2 | 3 | 3 | 2. |

Now suppose a new patient is recruited and has factor values, F–, G+ and H+. If the patient were allocated to A, then the six counts in the A row would become, 3, 3, 3, 3, 4, 2. The discrepancy with the B row is $|3 - 1| + |3 - 4| + |3 - 2| + |3 - 3| + |4 - 3| + |2 - 2| = 5$. If the patient were allocated to B, then the six counts in the B row would become, 2, 4, 2, 4, 3, 3, and the discrepancy with the A row is $|2 - 2| + |3 - 4| + |3 - 2| + |2 - 4| + |4 - 3| + |1 - 3| = 7$. The patient is allocated to A since this gives the lowest discrepancy. This shows the basic idea of minimisation, but more sophisticated allocation rules have been suggested. One criticism of minimisation is that only the marginal distributions of the factors are used and interactions are ignored.

The R packages, randomizeR and blockrand can be used for generating randomisation lists and for other aspects of randomisation in clinical trials.

### 6.4.2   Statistical contribution to the protocol

The International Council for Harmonisation of Technical Requirements for Pharmaceuticals for Human Use (ICH) has sets of guidelines on quality, efficacy, safety and other multidisciplinary aspects of clinical trials. The efficacy guideline, E6, *Good Clinical Practice*, states that the following should be included in the Statistics Section of the protocol:

1. A description of the statistical methods to be employed, including timing of any planned interim analysis(ses).

2. The number of subjects planned to be enrolled. In multicentre trials, the numbers of enrolled subjects projected for each trial site should be specified. Reason for choice of sample size, including reflections on (or calculations of) the power of the trial and clinical justification.

3. The level of significance to be used.

4. Criteria for the termination of the trial.

5. Procedure for accounting for missing, unused, and spurious data.

6. Procedures for reporting any deviation(s) from the original statistical plan should be described and justified in the protocol and/or in the final report, as appropriate.

7. The selection of subjects to be included in the analyses (e.g. all randomised subjects, all dosed subjects, all eligible subjects, evaluable subjects).

The ICH guideline, E9, *Statistical Principles for Clinical Trials*, gives detailed advice for the statistician. Particular consideration needs to be given to the following for inclusion in the protocol:

• Variables: Identify the primary variable and the hypothesis associated with it. If multiple primary variables are needed to assess efficacy, then an appropriate set of hypotheses need to be stated and the method for controlling the Type I error described. Secondary variables need to be described along with associated hypotheses .

• Blinding: Ideally the trial is *double blinded*, where the investigator is blind to which subjects receive which treatment, and the subjects are also unaware of which treatment they are receiving. A *single blinded* trial has either the investigator or the subjects blind to the treatment, for example a trial of two types of radiotherapy can have the subjects blinded but not the investigator. Often, neither the investigator nor the subjects can be blinded, for example surgery versus chemotherapy.

- Analysis sets: During the course of a clinical trial, some subjects may comply with the protocol, for instance, some might not complete the full course of treatment, some are lost to follow-up, errors may have been made in checking eligibility criteria, etc. Those subjects that have been deemed to comply, e.g. those subjects that completed 75% of the treatment regime and had no other protocol violations, form the *per protocol* (PP) analysis set. The *intention to treat* (ITT) analysis set comprises all randomised subjects. However, sometimes it may be desirable to omit a few patients from the ITT group, for example those who were randomised, but there are no post-randomisation data recorded for them, or a patient for whom it is subsequently found they were misdiagnosed and do not have the cancer under investigation. This gives us the *full analysis* set of subjects. The PP analysis set might give a clearer comparison of the difference in treatments than the ITT analysis set, however the ITT analysis set represents the situation of treating patients in real life.

- Missing data: How missing data are to be handled should be discussed in the protocol. There is no universal method for dealing with missing data. Missing data can be classed as *missing completely at random* (MCAR), *missing at random* (MAR) and *missing not at random* (MNAR). Data are MCAR if the likelihood of it being missing is not related to any observed or unobserved variables, for example, loss of a subject's blood sample. Data are MAR if the likelihood of it being missing is related to observed variables, but not to unobserved variables, for example, subjects with a high ECOG Scale of Performance Status might be more likely to drop out than patients with a lower score. Data are MNAR if the likelihood of being missing depends on unobserved data, for example, in a quality of life study, whether a subject completes a questionnaire or not, may depend on their quality of life at the time the questionnaire is administered. Methods for dealing with missing data include, likelihood based approaches, multiple imputation, inverse probability modelling, see O'Kelly[45].

- Sensitivity analyses: It is good practice to carry out some sensitivity analyses on the clinical trial data, to see if the results and conclusions are robust. For example: compare the treatment differences obtained using the ITT and the PP datasets; establish whether outlying observations affect the estimated treatment differences significantly; check whether allowances for missing data make a difference; check whether the incorporation of a baseline covariate affects results.

### 6.4.3  The Statistical Analysis Plan (SAP)

The Statistical analysis plan (SAP) is basically an expanded version of the statistics section of the protocol. Gamble et al. [15] have written a guideline on the writing of SAPs that aims to support transparency and reproducibility of the handling of trial data. The recommended items to be included in the SAP are:

- *Administrative Information*: title and trial registration, SAP version, protocol version, roles and responsibilities, signatures

- *Introduction*: background and rationale, objectives

- *Study methods*: trial design, randomisation, sample size, framework, statistical interim analyses and stopping guidance, timing of final analysis, timing of outcome assessment

- *Statistical principals*: confidence intervals and p-values, adherence and protocol deviations, analysis populations

- *Trial Population*: screening data, eligibility, recruitment, withdrawal/follow-up, baseline patient characteristics

- *Analysis*: outcome definition, analysis methods, missing data, additional analyses, harms, statistical software, references

See the reference for further information.

The SAP should be written soon after the protocol has been agreed, but it may change as the trial progresses. Suppose a trial starts with three arms, but after the first ISDMC meeting, it is decided to stop recruitment into one arm of the trial for efficacy or safety reasons, then the SAP will have to be revised.

### 6.4.4  Recruitment

As part of the design of a clinical trial, recruitment capability and recruitment rates need to be considered. Potential centres need to be contacted to see if there is interest in the trial that is being proposed, whether they will take an active part in recruiting their patients onto the trial, and how many patients they think they can contribute. If a centre is interested, and offer $x$ patients per month (on average), it may be wise to discount this by 30% say, to temper their enthusiasm. A large trial will probably have many centres, in the home country and abroad, and so the trial is an "international" trial which can be more influential in changing standard practice. Some trials have more than 100 centres.

A trial with a large number of centres cannot open with all centres recruiting patients on day one; centres have to be opened sequentially. Each centre needs the paperwork in place. The statistician needs to show how recruitment

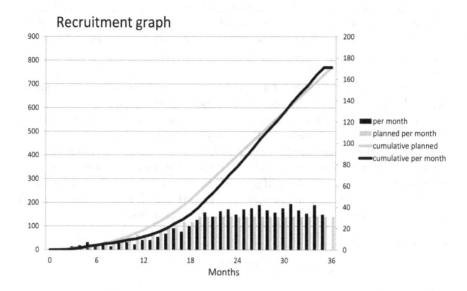

**FIGURE 6.8**

Predicted and actual recruitment into a clinical trial

will happen over the recruitment period. This is best done using a recruitment graph. The following is a simple fictitious example where 770 patients are to be recruited within a recruitment period of 36 months. There are ten centres that are to be opened, in order, at a rate of one every two months, opening assumed to be at the start of a month and with the first centre to be opened in month 1. The recruitment rates for the centres are:

Centre No.                         1  2  3  4  5  6  7  8  9 10
Recruitment rate (no. per month)   1  2  2  3  3  3  3  4  5  5

Figure 6.8 shows the predicted recruitment per month (right hand axis) and the cumulative recruitment per month (left hand axis), as well as the actual monthly recruitment and cumulative recruitment. Recruitment starts slowly at first. Recruitment picks up and the last patient is recruited at month 35. Note, had the higher recruiting centres been opened first, the recruitment period would have been about 4 months shorter.

## 6.4.5   Statistical reports

During the course of a trial, statistical reports have to be written for the ISDMC. At the end of the trial, a final statistical report is required. The ICH has a guideline, E3, *Structure and content of clinical study reports*.

The structure and content of reports for the ISDMC have to be agreed by the committee. Part of a ISDMC meeting may be with other study groups or

trial staff, e.g with the chief investigator and trial coordinator in attendance –
this is the *open meeting*, where no efficacy data are discussed. The rest of the
meeting is for the members of the ISDMC only – this is the *closed meeting*,
where efficacy data are discussed. Separate reports are written for the open
and closed meetings.

The final statistical report for a clinical trial should be relatively straightforward, following the SAP, but often a time consuming task.

The statistician will usually play a major role in writing publications of
clinical trials. It is recommended that the Consolidated Standards of Reporting Trials, 2010, guideline (CONSORT) is followed when reporting parallel-
group randomised controlled trials. The guideline helps authors to write a
clear and transparent account of a trial and its results. Details can be found
at http://www.consort-statement.org/, where the CONSORT checklist of 25
items can be downloaded that covers the following areas:

- *Title and abstract*

- *Introduction*: background and objectives

- *Methods*: trial design, participants, interventions, outcomes, sample size,
randomisation, blinding, statistical methods

- *Results*: participant flow, recruitment baseline data, numbers analysed, outcomes and estimation, ancillary analyses, harms

- *Discussion*: limitations, generalisability, interpretation

- *Other information*: registration, protocol, funding

Also, a CONSORT participant flow diagram template can be downloaded
from the website so that a clear illustration can be given of the path of the
participants through the course of the trial, in terms of numbers, as they are
enrolled, randomised, followed up, leave the trial and, at the end, when their
data are analysed.

### 6.4.6   Post study analyses

What do you have after a trial has finished? A lot of data! There is usually a
lot of recorded clinical data that was not used in the trial publications. There
may also be tumour, normal tissue and blood samples stored in freezers. This
resource can and should be used for further research using secondary analyses.
For example, based on the ESPAC4 trial, the publication,

Patterns of Recurrence After Resection of Pancreatic Ductal Adenocarcinoma A Secondary Analysis of the ESPAC-4 Randomized Adjuvant
Chemotherapy Trial [28],

reports on a study that showed treatment with gemcitabine plus capecitabine (GEMCAP) was found to be associated with reduced rate of local recurrence and improved overall survival, compared with treatment by gemcitabine alone.

The translational medicine publication,

GATA6 regulates EMT and tumour dissemination, and is a marker of response to adjuvant chemotherapy in pancreatic cancer [40],

reports on the use of ESPAC3 data to show that the transcription factor GATA6 is associated with outcome and response to chemotherapy for pancreatic patients.

Similarly, the publication,

Pancreatic cancer hENT1 expression and survival from gemcitabine in patients from the ESPAC-3 Trial [19],

reports on the prognostic value of the hENT1 protein for pancreatic cancer patients. This is explored further in Chapter 9 that discusses cancer biomarkers.

# 7

---

# Cancer Epidemiology

The word "epidemiology" conjures up the word "epidemic", where a disease moves through a population and for which mathematicians, statisticians and epidemiologists have their epidemic models, such as the SIR model that models the spread of a disease within a population. The "S" stands for the number of individuals susceptible to the disease, the "I" stands for the number of individuals infected and capable of passing it on, and the "R"stands for the number of individuals removed from the population, as recovered, immune or dead. At time, $t$, $N = S(t) + I(t) + R(t)$, where $N$ is the total population size. The equations for the spread of the disease are

$$\frac{dS}{dt} = -\frac{\beta SI}{N}, \quad \frac{dI}{dt} = \frac{\beta SI}{N} - \gamma I, \quad \frac{dR}{dt} = \gamma I.$$

These can be solved, but let us not digress.

The word epidemiology was first used to describe the study of epidemics in the early 1800's, but it is meaning is much wider now and covers all, non-clinical, aspects of diseases within a population. Epidemiology is the study of the incidence and distribution of diseases in a population. The population might be patients in a particular hospital where a disease, such as MRSA (methicillin-resistant staphylococcus aureus) is spreading, or the population within a city, for example, the Great Plague of London (bubonic plague) in 1665/1666, or within a country such as Nigeria that has the highest number of cases of malaria, or, ultimately, the whole world population with coronavirus disease (COVID-19), that as I write this book, is affecting every country on earth – a pandemic. Epidemiology is also about control of diseases, health-related conditions such as high blood pressure and obesity, and how diseases affect a country's and the world's economy.

There are three key components that characterise a particular disease:

### Person    Place    Time

- *Person* will have relevant features, such as age, sex, ethnic group, education and susceptibility.

- *Place* will involve entities like country, urban/rural, healthcare provision, economic factors and spatial coordinates.

- *Time* will cover calendar time, seasonality, time since onset of disease and others.

DOI: 10.1201/9781003041931-7

- *Patterns* of diseases are sought by descriptions and modelling of Person, Place and Time, relating to the disease.

- *Determinants* are causes and factors that influence the occurrence and development of a disease or health condition.

Epidemiologists talk about the *Epidemiologic Triangle* composed of:

- *Agent* – the cause of the disease

- *Host* – who or what gets the disease

- *Environment* – factors causing or related to the disease

For example, malaria is a mosquito-borne disease affecting humans and other animals with factors such as high temperature, rainfall and presence of stagnant water affecting infection rates.

Here we are only concerned with cancer epidemiology, but all the methods and discussion would be applicable to other diseases. We do not need to cover epidemics because cancer cannot generally be passed on to other people, apart from genetic mutations affecting a person's offspring, and where passing on a virus could cause the risk of certain cancers being increased in the recipient. For example, the Epstein-Barr virus is associated with Burkitt's lymphoma and the hepatitis virus is associated with liver cancer. First, we look at how we measure cancer incidence and death rates and proceed from there.

## 7.1   Measuring cancer

Cancer is measured in various ways and we concentrate in this chapter on epidemiological aspects rather than clinical aspects, such as tumour size, median survival time, etc., which were covered in Chapter 2. The key measurements are:

**cancer prevalence, incidence, mortality and risk,**

which we will describe in turn.

### Cancer prevalence

Cancer prevalence is the proportion of people, in a population, with cancer generally, or a particular cancer (or some other disease or a condition). Those with cancer, a particular cancer, etc. are called *cases*. To be more precise, there are three types of prevalence:

- *Point prevalence*: the proportion of existing cases, old or new, at a single point in time.

- *Period prevalence*: the proportion of people who were cases in a specific period of time.

- *Lifetime prevalence*: the proportion of people who are, or were ever, a case in their lifetime, up to the time of measurement.

Point prevalence is what we usually mean when we talk about prevalence. It will tell us the proportion of the population that has cancer at present, or at a particular time point in the past. Lifetime prevalence will tell us the proportion that have or have had cancer. Period prevalence is not particularly useful for cancer; it is good for seasonal diseases such as influenza.

## Cancer incidence
Cancer incidence is the number of new cancer registrations (cases) during a particular time period for a particular population. Cancer *incidence rate* (IR) is the number of cases relative to the population size and it is usually quoted per 100,000 people.

## Cancer mortality
Cancer mortality is measured as a mortality rate of which there are several types. These are:

- *Crude mortality rate* (MR): the mortality rate from all causes of death within a population and often quoted per 100,000 people.

- *Cause-specific mortality rate*: the mortality rate from a specific cause for a population, e.g. all cancers, leukaemia.

- *Age-specific mortality rate* (ASMR): the mortality rate for various age groups.

- *Case fatality rate* (CFR): the ratio of the number of deaths from a cancer, to the number of cases of the cancer.

## Cancer risk
Interest is in *risk factors* associated with cancer. CRUK lists the main risk factors for cancer to be, in order: smoking, obesity and weight, sun and UV radiation, diet and healthy eating, physical activity, alcohol, infections (such as HPV), air pollution and radon gas, hormones, workplace causes of cancer, inherited genes, age, and they suggest four out of every ten cancers can be prevented. A factor might be a risk factor for one cancer but not another, for example, smoking is linked to several cancers: lung, mouth, pharynx, nose and sinuses, larynx, oesophagus, liver, pancreas, stomach, kidney, bowel, ovary, bladder, cervix, and some types of leukaemia, while UV radiation appears to be associated only with skin cancers. The risk factor *age* – the older you are, the higher your risk of cancer, but however much you try, you cannot avoid this risk factor. We talk about *exposure* to risk factors and we might work with exposed groups of individuals and non-exposed groups of individuals in order to measure risk.

**Cancer risk measurements**

The word *risk* means different things to different people. To a statistician it is usually associated with the word probability, but to others it can mean other things, like the likelihood of an event happening, together with a measure of loss if it does. For example, the likelihood of downturn in a company's profit and the subsequent loss in share value, or the likelihood of developing cancer, with the loss being possible death. Here we measure cancer risk just by probability.

Let there be two groups of individuals, for instance, one group might be smokers and the other non-smokers. Let the probability of an individual in the smoking group developing lung cancer be $p_1$, and probability of an individual in the non-smoking group developing lung cancer be $p_0$. These are the risks for the two groups. There are several ways of comparing the two risks/probabilities, these are:

- The *risk difference* (RD) for the two groups is $p_1 - p_0$.

- The *risk ratio* or *relative risk* (RR) for the two groups is $\text{RR} = p_1/p_0$.

- The *odds* of developing the cancer for an individual in a group are $p_i/(1 - p_i)$, $(i = 0, 1)$.

- The *odds ratio* (OR) is the ratio of the odds for the two groups, $\text{OR} = \frac{p_1}{(1-p_1)} / \frac{p_0}{(1-p_0)} = \frac{p_1(1-p_0)}{(1-p_1)p_0}$.

- The *population attributable fraction* (PAF) is the proportion of individuals with lung cancer that would not have developed lung cancer, had they not been exposed to tobacco. Let $P_1$ be the proportion of individuals with lung cancer that have been exposed to tobacco in the population, and then, $\text{PAF} = P_1 \times (p_1 - p_0)/p_1$. This can be expressed in terms of RR, as $\text{PAF} = P_1 \times (\text{RR} - 1)/\text{RR}$.

## 7.1.1   Study designs for measuring cancer

Chapter 4, Figure 4.1 showed the types of cancer studies that are carried out. Epidemiological studies are observational studies with the three main designs being:

**cohort studies, case-control studies, cross-sectional studies**

We will use data from the seminal studies undertaken by Sir Richard Doll and Sir Austin Bradford Hill on smoking and lung cancer, the first was a case-control study and the second, known as the British Doctors Study, was a cohort study.

**Cohort studies**

Cohort studies are forward looking, where individuals are selected to be members of one or more cohorts, the selection being based on risk exposure and demographic variables. The cohort(s) are followed up for a certain period of time, with data collected on cancer outcomes and other variables. Data is often collected at various timepoints during the course of the study. This is a *prospective cohort study* which may take years to complete. Another type of cohort study is an *historical cohort study* or sometimes called a *retrospective cohort study*, where the data are obtained from historical records, appropriate individuals being selected for the cohort(s), again based on risk and demographic variables. Their records are then traced forward to see how the cancer outcome variables performed up to a particular time point. It is as if the researcher travels back in time and then conducts a prospective cohort study, eventually arriving back to the present.

*Example*: In the British Doctors Study[10], doctors in Great Britain were invited to join in the study. Those participating completed a relatively short questionnaire about smoking habits and were then followed up for twenty years, deaths and their causes being recorded. Further questionnaires were sent out during the course of the study and interim reports were published. Breslow and Day[2] gives a good description of the study and its results.

**Case-control studies**

A case-control study is a retrospective study with two groups of individuals being selected. The individuals in one group have the cancer in question, or have experienced a certain outcome – these are the *cases*. The individuals in the other group have not – these are the *controls*. The cases and controls have to be selected objectively, so that the only difference between the groups is their disease/outcome status. Various variables/factors/exposures are recorded for each individual which can be assessed as to whether they are risk factors for the disease/outcome.

There are two types of case-control study – *matched* and *unmatched*. For a matched study, a control is matched to each case, or very often, several controls to each case. The matching takes account of relevant factors, such as age, sex, BMI, so that a case is very similar to its controls, apart from the disease/outcome status. In an unmatched study, the only matching is to make the two groups similar overall.

*Example*: The Doll and Hill[8] case-control study used patients with lung carcinoma as cases and non-cancer patients as controls, all patients being selected from London hospitals between April 1948 and October 1949. This study, not only established a link between smoking and lung cancer, but helped medical statistics to become firmly embedded in public health and medical care. The cases and controls were matched on age and sex, but only at the group level, not one-to-one. Questionnaires on demographics and smoking habits were administered by almoners (we would call them social workers now).

**Cross-sectional studies**

Cross-sectional studies collect data at one time point. They are inexpensive to run compared to cohort studies, can collect data on multiple outcomes and exposures and can estimate prevalence. On the downside, incidence cannot be measured and causality cannot be established, only association. You cannot show whether a risk factor caused the cancer, or that having cancer caused individuals to become more exposed to the risk factor. The latter is unlikely in most cases – I am not going to take up smoking because I have been diagnosed with lung cancer! To carry out a cross-sectional study, a sample of individuals needs to be chosen from the population of interest. Data collected can come from questionnaires, interviews and medical examinations. Bias can be a problem because of non-responders, participant recall of past events and selection bias.

*Example*: The first questionnaire sent out in the British Doctors Study can be considered to be a cross-sectional study.

### Advantages and disadvantages of the three types of study

**Cohort studies**: *Advantages*: as the main study outcomes do not influence the selection of study individuals, incidence, risk and relative risk can be estimated; can assess exposures that are rare. *Disadvantages*: possibly long follow-up time and high cost for prospective cohort studies, which is not such a problem for historical studies; no control over variables in historical cohort studies; no randomisation and so imbalances can occur; possible selection bias; rare cancers would need a very large cohort

**Case-control studies**: *Advantages*: faster and cheaper to carry out compared to a prospective cohort study; many potential risk factors can be studied; they are useful studies for rare diseases. *Disadvantages*: as the main study outcomes do influence the selection of study individuals, incidence, risk and relative risk cannot be estimated, but the odds ratio can; studies can exhibit bias.

**Cross-sectional studies**: *Advantages*: can measure prevalence and association between exposure risk factors and cancer, but cannot establish causality; fast results; not so costly as other designs. *Disadvantages*: cannot measure incidence; biases, such as recall and interviewer bias can be present.

Before discussing these designs further and statistical models for analysing their data, we start by looking at how countries generate their cancer statistics.

## 7.2   Cancer statistics for countries

Data on cancer are collected by cancer registries. The National Cancer Registration and Analysis Service (NCRAS), part of Public Health England (PHE), collect cancer data for England. The equivalent registries for Scotland, Wales and Northern Ireland are: Scottish Cancer Registry and Intelligence Service (SCRIS), the Welsh Cancer Intelligence and Surveillance Unit (WCISU) and the Northern Ireland Cancer Registry (NICR). In the USA there is the

National Program of Cancer Registries (NPCR). The European Network of Cancer Registries (ENCR) has the European countries cancer registries as members.

Concentrating on NCRAS, when somebody is diagnosed and treated for cancer, data on that person are generated in GP (General Practitioner) surgeries, hospitals, screening services, death certificates and other. Relevant parts of the data are sent to the cancer registry from one or more of these sources. There, the data are merged and then put on the cancer database. The collected data includes personal information, name, age, sex etc. as well as data relating to the person's cancer, type, stage, treatment, etc. Access to the database has to be sought and permission to use the data is only given for medical purposes: surveillance, clinical audit, service evaluation, ethically approved research and genetic counselling. A person whose data is on the database can ask for, and will be given, a copy of the data held on them.

The Office of National Statistics (ONS) publishes a yearly bulletin, "Cancer registration statistics, England", which gives pertinent summaries of the NCRAS data. Cancer Research UK (CRUK) also publishes a lot of summary cancer data. The United Kingdom and Ireland Association of Cancer Registries (UKIACR) has the four UK cancer registries and the National Cancer Registry Ireland (NCRI) as members. It too publishes reports, books, and scientific papers about cancer.

The International Agency for Research on Cancer (IARC) is the World Health Organisation's (WHO) cancer agency that promotes international collaboration on cancer research. It has accessible databases of cancer incidence from many locations in the world and also publishes reports and research monographs.

From these registry databases cancer incidence, mortality and other statistics are produced. We will use publicly available data to illustrate cancer incidence and cancer mortality.

## 7.2.1 Cancer incidence

Table 7.1 shows the incidence of breast cancer and prostate cancer in England and Iceland for the years 2003–2007. The data were downloaded from IACR. Comparing the incidence rates (IR) for prostate cancer in England and Iceland, the rate for Iceland is 20.2 higher than that for England. The female breast cancer rate for England is 26.0 higher than than for Iceland. Male breast cancer is rare in both populations, and when we discuss breast cancer we will only be dealing with female breast cancer.

Why the difference in incidence rates of these two cancers for the two countries? Could it be due to the population structure? We know the risk of cancer increases with age; are there relatively more older people in one country than the other? Figure 7.1 shows the population age distribution for both countries, in 5-year intervals, split by sex. The Icelandic population has relatively more younger people than the English population and less older people. The ratios

**TABLE 7.1**

Incidence of prostate and breast cancer in England and Iceland, 2003–2007

| | | England | | | Iceland | | |
|---|---|---|---|---|---|---|---|
| | | Cases* | Popn.† | IR‡ | Cases* | Popn.† | IR‡ |
| Breast | Female | 181,414 | 25,706,980 | 141.1 | 852 | 148,087 | 115.1 |
| | Male | 1297 | 24,755,010 | 1.0 | 10 | 150,602 | 1.3 |
| Prostate | Male | 150,941 | 24,755,010 | 121.9 | 1,070 | 150,602 | 142.1 |

\* Number of cases in 5 year period; † Population averaged over 5 years; ‡ per 100,000

of the proportion of the Icelandic population to the proportion of the English population in each age group are:

| age gp.: | 0- | 5- | 10- | 15- | 20- | 25- | 30- | 35- | 40- | 45- | 50- | 55- | 60- | 65- | 70- | 75- | 80- | 80+ |
|---|---|---|---|---|---|---|---|---|---|---|---|---|---|---|---|---|---|
| size ratio: | 1.2 | 1.2 | 1.2 | 1.1 | 1.1 | 1.2 | 1.0 | 0.9 | 1.0 | 1.1 | 1.0 | 0.8 | 0.8 | 0.7 | 0.8 | 0.8 | 0.7 | 0.7 |

In crude numbers, relatively, Iceland has 20% more younger people than England and 20% less older people and this will affect the cancer incidence rates, making them hard to compare directly.

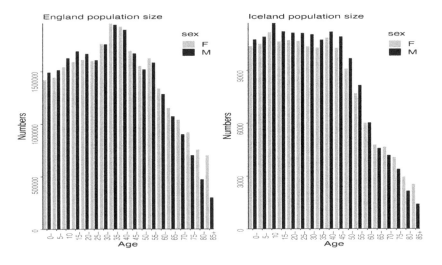

**FIGURE 7.1**

Population distribution for England and Iceland by age

To overcome this problem, we can calculate the cancer incidence rate for each age group individually. Figure 7.2 shows these plotted for breast and prostate cancers for the two countries. There is a big "wobble" in the incidence rate of breast cancer for Iceland in the older age groups – this will be due to the low group sizes. The same happens for prostate cancer but with a smaller wobble. It appears from the graphs that, overall, the breast cancer incidence rates for the two countries are not very dissimilar from each other,

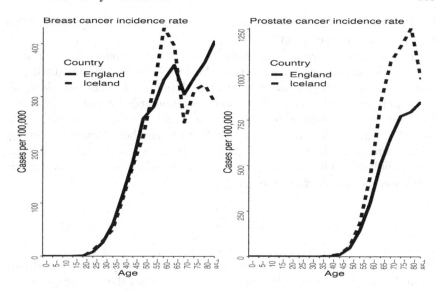

**FIGURE 7.2**
Breast and prostate cancer incidence rate by age group for England and
Iceland

**TABLE 7.2**
Age distribution standards

| Age group | World standard | European standard |
|---|---|---|
| 0–4 | 12,000 | 5,000 |
| 5–9 | 10,000 | 5,500 |
| 10–14 | 9,000 | 5,500 |
| 15–19 | 9,000 | 5,500 |
| 20–24 | 8,000 | 6,000 |
| 25–29 | 8,000 | 6,000 |
| 30–34 | 6,000 | 6,000 |
| 35–39 | 6,000 | 7,000 |
| 40–44 | 6,000 | 7,000 |
| 45–49 | 6,000 | 7,000 |
| 50–54 | 5,000 | 7,000 |
| 55–59 | 4,000 | 6,500 |
| 60–64 | 4,000 | 6,000 |
| 65–69 | 3,000 | 5,500 |
| 70–74 | 2,000 | 5,000 |
| 75–79 | 1,000 | 4,000 |
| 80–84 | 500 | 2,500 |
| 85–89 | 500 | 1,500 |
| 90+ | * | 1,000 |
| Total | 100,000 | 100,000 |

* The World standard does not have a 90+ category;
  the 85–89 category will be 85+

but for prostate cancer, the Icelandic rate starts to become higher than that for England after the age of 50.

To compare the overall cancer incidence rates for countries, we have to standardise each country's population. We do this using a standardised population. Table 7.2 shows two standard populations, the World standard and the European standard, which are accepted standards for comparing cancer incidence between various countries in the world, or specifically within Europe. For a particular country, we calculate the number of cases in each age group there would be, if the population, per 100,000 people followed the standard population. Then we add the cases and calculate the population *cancer incidence, age-standardised rate* (ASR), as,

$$\text{ASR} = \sum_{i=1}^{I} \frac{c_i}{n_i} \frac{N_i}{N},$$

where $I$ is the number of age groups, $n_i$ is the number of people in the $i$th age group for the country, $c_i$ is the number of cases within the $i$th age group , $N_i$ is the number of people in the $i$th age group of the standard population and $N$ is the total number in the standard population, here 100,000.

The population crude (non-standardised) IR and the ASR using the World standard population and the European standard, for the two cancers for the two countries are:

|         | Breast cancer | | | Prostate cancer | | |
|---------|-------|-------|----------|-------|-------|----------|
|         | Crude | World | European | Crude | World | European |
| England | 141.1 | 85.4  | 151.5    | 121.9 | 65.0  | 164.0    |
| Iceland | 115.1 | 86.7  | 150.2    | 142.1 | 98.9  | 246.5    |

The World standard reduces the crude rates, while the European standard increases the crude rate. This is a reflection of the pattern of the World and European standard populations. Comparing England with Iceland, we see that breast cancer IR is very similar in the two countries, while Iceland has a much higher prostate cancer IR.

In some situations it may be that in trying to calculate incidence within a particular time period, not everybody under consideration is at risk for the whole of the time period. For example, suppose a study is carried out on a group of $n$ women aged 65 at the start of the study and all of whom are followed up for 10 years. The number of women developing breast cancer in this time period is counted. Let $X_i = 1$, if the $i$th women develops breast cancer and $X_i = 0$ if she does not. Then the incidence rate is $IR = \sum_{i=1}^{n} X_i/n$. Suppose $n = 1000$ and 52 women develop breast cancer. Then IR $= 52/1000 = 0.052$ for the 10-year period, or 520 per 100,000, per year. In practice, some of the women will be lost to the study, because of death or withdrawal from the study. Let $T_i$ be the time the $i$th female spends on the study or the time to which they develop the cancer, if they do so. Then IR $= \sum_{i=1}^{n} X_i / \sum_{i=1}^{n} T_i$ per person-year. For our example, suppose $\sum_{i=1}^{n} T_i = 8125$, then IR $= 52/8125 = 0.0064$ per person-year, or 640 per 100,000, per year.

**Cumulative incidence rate**

The cumulative incidence rate (CIR) is the sum of age-specific incidence rates over a particular age range, for example, 0–85+ years would give the lifetime risk. Probably you would use 0–84, or 0–74 years rather than 0–85+. Figure 7.3 shows the cumulative incidence rates for breast and prostate cancer for England and Iceland, accumulating the age specific incidence rates at every year. Note, if we add the incidence rates given at each 5-year interval, we have to multiply these rates by five, since the rate applies for each year in the 5-year interval. From the graphs, we can read off the cumulative incidences for various ages, for example, by age of 55 the cumulative incidence is 3256 per 100,000 women in England and 2979 per 100,000 in Iceland. For prostate cancer, the cumulative incidences at 55 years, are 246 and 378 for England and Iceland respectively.

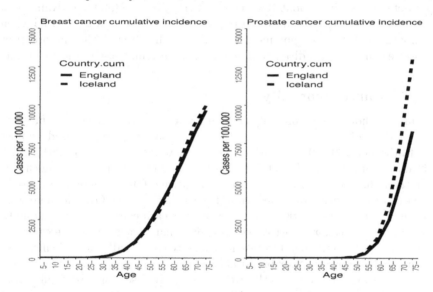

**FIGURE 7.3**
Breast and prostate cancer cumulative incidence for England and Iceland

What does the cumulative incidence rate measure? For cancer incidence rate, we essentially have a cross-sectional study, where the new registrations within a 5-year period in a particular population, have been placed in a frequency table of 5-year age groups. Each age group has its number of cases normalised to give cases per year, per 100,000 people. So for a particular age group, the number of normalised cases, divided by 100,000, can be interpreted as the probability of you becoming a case in the next year, given you lived long enough to have reached this age group. For example, the English breast cancer 55–59 age group in Figure 7.2 has 259 cases per year and so the chance of somebody aged between 55 and 59 would have probability of 0.00259 of becoming a case within the next year, given they have reached their current age. This is essentially the hazard rate at that age. Thus IR estimates the hazard function, $\lambda(t)$, and so the cumulative incidence rate estimates the cumulative hazard function, $\Lambda(t) = \int_0^t \lambda(u)du$ (see Chapter 3). Then, $S(t) = \exp(-\Lambda(t))$, and so,

$$\mathrm{Pr}(\text{becoming a case by time } t) = 1 - \exp(\mathrm{CIR}(t)).$$

So, for an English female, the chance of having breast cancer by the age of 55 is $100(1 - \exp(-3256/100000)) = 3.2\%$. For Icelandic women the risk is very similar to that for English women, at 2.9%. For English men, the chance of having prostate cancer by age 55 is 0.3% and for Icelandic men, the chance is 0.4%. The chance of having breast cancer by the age of 75 is about 9% for both English and Icelandic women, but the chance men of having prostate cancer by age of 75 is about 8% for English men and 12% for Icelandic men. For percentages less than about 10%, $1 - \exp(\mathrm{CIR}(t)) \approx \mathrm{CIR}(t)$. One advantage of using CIR's is that a standardised population does not need to be used when comparing risk between populations. But note, these risks of having breast or prostate cancer are conditional on not dying of anything else in the meantime.

## 7.2.2   Cancer mortality

Figure 7.4 shows the number of deaths per year for breast cancer and prostate cancer in the UK by age group and the corresponding standardised number of deaths per 100,000 population. The data were for the years 2015–2017, obtained from the CRUK website. Deaths from breast cancer start much earlier than those from prostate cancer, but by age 60, the number of deaths from prostate cancer overtakes those from breast cancer. Overall there were between 11,000 and 12,000 deaths per year for each cancer. Correspondingly, the mortality rate for breast cancer is higher than that for prostate cancer in the earlier years, but then the rate for prostate cancer, after age 65, increases dramatically above that for breast cancer.

What is also of interest, is the ratio of deaths to cases, the case fatality rate (CFR). This is plotted in Figure 7.5 over the age range 20–89; the number of cases before age 20 are rare, the data for 90+ are unreliable. Up to age 75, there is a higher CFR for breast cancer than for prostate cancer, but then the rate for prostate cancer overtakes. Note, if we had followed up large cohorts of cancer patients, for many years, and then calculated CFRs, the results would have probably been different. The CRUK data is a cross-sectional study, where deaths do not necessarily correspond to cases. Somebody who was a case in 2011 could become a death in 2016 and so they are not counted as a case in 2016, but are as a death. For instance, the number of cases per year of prostate cancer in 2015–2017, in the 90+ age group was 1192, and the number of deaths

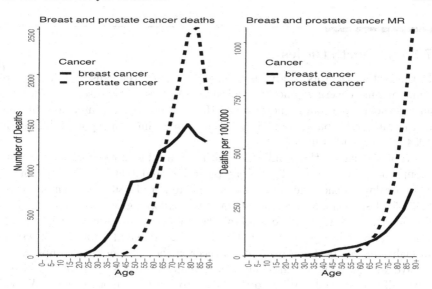

**FIGURE 7.4**
Breast and prostate deaths and mortality rates by age group for the UK

was 1832! However, overall the CFR's in Figure 7.5 give a good overview. Finally, we note that the overall CFR for breast cancer is 21% and that for prostate cancer, 24%.

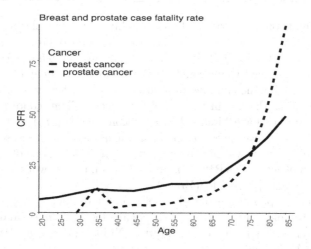

**FIGURE 7.5**
Case fatality rate for breast and prostate cancer by age group for the UK

## 7.3 Cohort studies

In the British Doctors Study, as you would expect, the causes of death were wide ranging, covering cancers, respiratory diseases, heart diseases, other diseases, suicide, poisoning and trauma. Here we are mainly interested in lung cancer and first look at some data taken from an interim report published in 1954 by Doll and Hill[9].

Table 7.3 shows the number of deaths from lung cancer for two exposure groups, heavy smokers ($> 1$ gm tobacco per day) and light smokers ($> 0$, but $\leq$ 1 gm per day) at the time of the analysis. There were 12,865 heavy smokers and 8,431 light smokers. Heavy smokers form the exposed group; the light smokers the non-exposed group. (We wouldn't say that now for light smokers!) Table 7.4 shows the labels for the counts in the $2 \times 2$ table and Table 7.5 shows the probabilities, $p_i$, of dying from lung cancer in the two smoking groups.

We are now going to carry out a small analysis on this $2 \times 2$ table, which is not really appropriate. Why do this? Just to illustrate methodology. We will then move on to more appropriate analyses.

**TABLE 7.3**
Doctors Study: $n_{ij}$'s

|  | Dead | Alive |
|---|---|---|
| Heavy | 24 | 12, 841 |
| Light | 12 | 8, 419 |

**TABLE 7.4**
Doctors study: a,b,c,d

|  | Dead | Alive | $n_i$ |
|---|---|---|---|
| Heavy | $a$ | $b$ | $n_1$ |
| Light | $c$ | $d$ | $n_0$ |

**TABLE 7.5**
Doctors study: $p$'s

|  | Dead | Alive |
|---|---|---|
| Heavy | $p_1$ | $1 - p_1$ |
| Light | $p_0$ | $1 - p_0$ |

The estimate of $p_1$, the risk of dying from lung cancer for the heavy smokers is $\hat{p}_1 = 24/12865 = 1.87 \times 10^{-3}$, i.e. 1.87 in 1000 doctors. For the light smokers, $\hat{p}_0 = 12/8431 = 1.42 \times 10^{-3}$. The *risk ratio* or *relative risk* is RR $= \hat{p}_1/\hat{p}_0 = 1.31$. We need the variance of RR in order to calculate a confidence interval. Let $Y_1$ be the number of deaths out of $n_1$ heavy smokers and $y_0$ the number of deaths out of the $n_0$ light smokers. Then, conditioned on $n_1$ and $n_0$, $Y_1$ and $Y_0$ have binomial distributions, Bin$(n_1, p_1)$ and Bin$(n_0, p_0)$ respectively. (Note, we have to condition on $n_1$ and $n_0$, because the study only has one cohort; it was not designed with two cohorts, one of smokers and one of non-smokers.) Thus RR $= (Y_1/n_1)/(Y_0/n_0)$, but we work with log(RR). So log(RR) $= \log(Y_1/n_1) - \log(Y_0/n_0)$.

From the Appendix, using the delta method, $\log(Y_1/n_1)$ has an approximate normal distribution with mean $\log(E[Y_1/n_1])$ and variance, $\text{var}(Y_1/n_1)/(E[Y_1/n_1])^2$. Thus, $\log(Y_1/n_1) \sim N\big(\log(p_1), (1 - p_1)/\{n_1 p_1\}\big)$ and similarly for $\log(Y_0/n_0)$. Hence,

$$E[\log(\text{RR})] = \log(p_1/p_0), \quad \text{var}[\log(\text{RR})] = (1 - p_1)/\{n_1 p_1\} + (1 - p_0)/\{n_0 p_0\},$$

estimated by $\log\left\{ \frac{a(c+d)}{c(a+b)} \right\} = 0.270$, and $\frac{b}{a(a+b)} + \frac{d}{c(c+d)} = 0.125$ respectively. Hence, a 95% confidence interval for log(RR) is $271 \pm 1.96 \times 0.353$. Upon exponentiating, RR $= 1.31$ with 95% confidence interval, $[0.66, 2.62]$, giving no evidence of a difference in the lung cancer death rates for light and heavy smokers.

Other comparisons between the risks of the exposure and non-exposure groups are risk difference, odds ratio and population attributable fraction. These are estimated as,

$$\text{RD} = \hat{p}_1 - \hat{p}_0 = a/(a+b) - c/(c+d) = 4.42 \times 10^{-4}$$

$$\text{OR} = \frac{\hat{p}_1/(1-\hat{p}_1)}{\hat{p}_0/(1-\hat{p}_0)} = \frac{ad}{bc} = 1.31$$

$$\hat{P}_1 = \frac{a}{a+c} = 0.67; \quad \text{PAF} = \hat{P}_1 \times \frac{(RR-1)}{RR} = \frac{ad - bc}{(a+c)(c+d)} = 0.16.$$

For PAF, the proportion of individuals with lung cancer exposed to heavy smoking, $P_1$, can be estimated from these data since the cohort study did not select doctors based on their smoking habits. In contrast, a study that had 10,000 doctors selected for the heavy smoking group and 10,000 for the light smoking group, you could not estimate $P_1$. Note, PAF $= 0.16$, i.e. only 16% of lung cancer cases could be avoided by changing from heavy smoking to light smoking. Also, we did not use non-smokers for the unexposed group for illustration here, since there were no deaths from lung cancer within the non-smoking group.

The variance of RD is $p_1(1-p_1)/n_1 + p_0(1-p_0)/n_0$, estimated by $\frac{ab}{(a+b)^3} + \frac{cd}{(c+d)^3} = 3.13 \times 10^{-7}$, and so an approximate a 95% CI for RD is, $[-6.5 \times 10^{-4}, 1.54 \times 10^{-3}]$. The variance of $\log(\text{OR})$ will be shown to be $1/a + 1/b + 1/c + 1/d$ in Section 7.4.1. This has value, $0.125$, and so a 95% CI for OR is, $[0.66, 2.62]$.

Note that OR is equal to RR to two decimal places. This is because if $a$ is small relative to $n_1$ and $b$ is small relative to $n_0$, then OR $\approx$ RR.

Now let us see why this analysis is not appropriate. You can of course use these methods on cohort and other studies where the analysis is appropriate. Over the twenty years of follow-up, some doctors will have withdrawn from the study for various reasons, some will have died of lung cancer or other causes. A more appropriate analysis is to use methods like those discussed for mortality rates at the start of the chapter, allowing for the time spent in the study by the doctors.

Doll and Peto[10] published a report on the British Doctors study in 1976. We will highlight a small part of the results they obtained. The time spent in a cohort study by each individual has to be ascertained. Suppose a doctor, aged 43, is recruited at the start of the study in 1951, and dies in 1962. He has contributed eleven *person-years* to the study. But it is not as simple as that when age groups are to be considered. Without birth and death dates, we assume each individual is born and dies mid-year. Suppose the age groups are 35–39, 40–44, 45–59, 60–64, etc. Then a crude split of our doctor's eleven person-years into the age groups is: 0 for the 35–39 age group, 1.5 for the

40–44 age group, 5 for the 45–49 age group, 4.5 for the 50–54 age group. With birth dates, the split can be made more accurately.

For the Doctors Study, person-years were calculated for each of the subgroups of smoking by age group, the age groups being five-year age groups, from 20 to 84, plus 85 years and older. Table 7.6 shows the deaths from twelve cancers and other causes, as well as seven smoking categories, from non-smoker to > 25 grms of tobacco per day.

## TABLE 7.6
Doctors Study: death rates and smoking

| Cause | Deaths | Non | Cur/Ex | Ex | Cur 1-14 | 15-24 | $\geq$ 25 | Non v Rest | Trend |
|-------|--------|-----|--------|-----|----------|-------|-----------|------------|-------|
| Lung | 441 | 10 | 83 | 43 | 104 | 52 | 106 | 224 | 41.98 | 197.04 |
| Oesophagus | 65 | 3 | 12 | 5 | 16 | 12 | 13 | 30 | 3.94 | 14.94 |
| Other resp. sites | 46 | 1 | 9 | 4 | 11 | 6 | 9 | 27 | 3.31 | 21.68 |
| Stomach | 163 | 23 | 28 | 21 | 32 | 28 | 38 | 32 | - | - |
| Colon | 195 | 27 | 34 | 34 | 34 | 35 | 33 | 31 | - | - |
| Rectum | 78 | 6 | 14 | 14 | 14 | 10 | 14 | 27 | 2.81 | 10.76 |
| Pancreas | 92 | 14 | 16 | 12 | 18 | 14 | 18 | 27 | - | 3.98 |
| Prostate | 186 | 39 | 30 | 31 | 30 | 28 | 31 | 38 | - | - |
| Kidney | 46 | 3 | 8 | 9 | 8 | 8 | 9 | 9 | - | - |
| Bladder | 80 | 9 | 14 | 11 | 16 | 16 | 16 | 12 | - | - |
| Marrow * | 152 | 33 | 24 | 26 | 24 | 27 | 22 | 19 | - | (3.51) |
| Unknown site | 64 | 12 | 11 | 9 | 12 | 10 | 13 | 14 | - | - |
| Other | 151 | 25 | 26 | 29 | 24 | 19 | 24 | 35 | - | - |

St. annual death rate per 100,000 / $\chi^2$

Categories: Non - non-smokers; Cur - current smokers; Ex - Ex-smokers;
1-14 - Cur 1-14g/day; 15-24 - Cur 15-24g/day; $\geq$ 25 - Cur $\geq$ 25g/day;
* Marrow and reticuloendothelial systems

## TABLE 7.7
Notation for the Doctors Study

| Smoking cat. | | 1 | 2 | ... | J | Total |
|--------------|---------------|-----------|-----------|-----|-----------|----------|
| 1 | Deaths | $d_{11}$ | $d_{12}$ | ... | $d_{1J}$ | $d_{1.}$ |
| | Person-years | $n_{11}$ | $n_{12}$ | ... | $n_{1J}$ | $n_{1.}$ |
| | Mortality rate | $\lambda_{11}$ | $\lambda_{12}$ | ... | $\lambda_{1J}$ | |
| | Exp. deaths | $E_{11}$ | $E_{12}$ | ... | $E_{1J}$ | $E_{1.}$ |
| 2 | Deaths | $d_{21}$ | $d_{22}$ | ... | $d_{2J}$ | $d_{2.}$ |
| | Person-years | $n_{21}$ | $n_{22}$ | ... | $n_{2J}$ | $n_{2.}$ |
| $\vdots$ | $\vdots$ | $\vdots$ | $\vdots$ | | $\vdots$ | $\vdots$ |
| I | Deaths | $d_{I1}$ | $d_{I2}$ | ... | $d_{IJ}$ | $d_{I.}$ |
| | Person-years | $n_{I1}$ | $n_{I2}$ | ... | $n_{IJ}$ | $n_{I.}$ |
| | Mortality rate | $\lambda_{I1}$ | $\lambda_{I2}$ | ... | $\lambda_{IJ}$ | |
| | Exp. deaths | $E_{I1}$ | $E_{I2}$ | ... | $E_{IJ}$ | $E_{I.}$ |
| | Ref. | $\lambda_1^*$ | $\lambda_2^*$ | ... | $\lambda_J^*$ | |
| Total | Deaths | $d_{.1}$ | $d_{.2}$ | ... | $d_{.J}$ | $D$ |
| | Person-years | $n_{.1}$ | $n_{.2}$ | ... | $n_{.J}$ | $N$ |
| | Expected deaths | $E_{.1}$ | $E_{.2}$ | ... | $E_{.J}$ | $E$ |

Age group

We follow the theory given in the text on Cohort studies by Breslow and Day[2] and the paper by Armstrong[1]. We also use similar notation as theirs, which is summarised in Table 7.7. In summary,

There are $I$ smoking categories and $J$ age groups.

There is a reference mortality rate, $\lambda_j^*$, for the $j$th age group, which needs to be known.

For the $i$th smoking/$j$th age group: there are $d_{ij}$ deaths; $n_{ij}$ person years; the population mortality rate parameter is $\lambda_{ij}$ and its estimate will be $\hat{\lambda}_{ij}$; the expected number of deaths, $E_{ij}$, is based on the reference mortality rate, so $E_{ij} = n_{ij} \times \lambda_j^*$.

Summing over $j$, $d_{i.} = \sum_j d_{ij}$ is the number of deaths in the $i$th age group; summing over, $i$, $d_{.j} = \sum_i d_{ij}$ is the number of deaths in the $i$th smoking category; summing over $i$ and $j$ gives the total number of deaths. Similar summations apply to $n_{ij}$ and $E_{ij}$.

Table 7.8 shows the actual numbers of deaths and person years for each smoking/age category.

**TABLE 7.8**
Doctors Study: lung cancr deaths and person-years by age and cigarette consumption

| No. cigs | Av. no. | | 40-44 | 45-49 | 50-54 | 55-59 | 60-64 | 65-69 | 70-74 | 75-79 |
|---|---|---|---|---|---|---|---|---|---|---|
| 0 | 0 | deaths | 0 | 0 | 1 | 2 | 0 | 0 | 1 | 2 |
| | | PY | 17864.5 | 15832.5 | 12226.0 | 8905.5 | 6248.0 | 4351.0 | 2723.5 | 1772.0 |
| 1-4 | 2.7 | deaths | 0 | 0 | 0 | 1 | 1 | 0 | 1 | 0 |
| | | PY | 1216.0 | 1000.5 | 853.5 | 625.0 | 509.5 | 392.5 | 242.0 | 208.5 |
| 5-9 | 6.6 | deaths | 0 | 0 | 0 | 0 | 1 | 1 | 2 | 0 |
| | | PY | 2041.5 | 1745.0 | 1562.5 | 1355.0 | 1068.0 | 843.5 | 696.5 | 517.5 |
| 10-14 | 11.3 | deaths | 1 | 1 | 2 | 1 | 1 | 2 | 4 | 4 |
| | | PY | 3795.5 | 3205.o | 2727.0 | 2288.0 | 1714.0 | 1214.0 | 862.0 | 547.0 |
| 15-19 | 16.0 | deaths | 0 | 1 | 4 | 0 | 2 | 2 | 4 | 5 |
| | | PY | 4824.0 | 3995.0 | 3278.5 | 2466.5 | 1829.5 | 1237.0 | 683.5 | 370.5 |
| 20-24 | 20.4 | deaths | 1 | 1 | 6 | 8 | 13 | 12 | 10 | 7 |
| | | PY | 7046.0 | 6460.5 | 5583.0 | 4357.5 | 2863.5 | 1930.0 | 1055.0 | 512.0 |
| 25-29 | 25.4 | deaths | 0 | 2 | 3 | 5 | 4 | 5 | 7 | 4 |
| | | PY | 2523.0 | 2565.5 | 2620.0 | 2108.5 | 1508.5 | 974.5 | 527.0 | 209.5 |
| 30-34 | 30.2 | deaths | 1 | 2 | 3 | 6 | 11 | 9 | 2 | 2 |
| | | PY | 1715.5 | 2123.0 | 2226.5 | 1923.0 | 1362.0 | 763.5 | 317.5 | 130.0 |
| 35-40 | 38.0 | deaths | 0 | 0 | 3 | 4 | 7 | 9 | 5 | 2 |
| | | PY | 892.5 | 1150.0 | 1281.0 | 1063.0 | 826.0 | 515.0 | 233.0 | 88.5 |

(From Doll and Peto and Breslow and Day)

We assume $d_{ij}$ follows a Poisson distribution, with mean $n_{ij}d_{ij}$. The likelihood for $\lambda_{ij}$ is, $L = (n_{ij}\lambda_{ij})^{d_{ij}} \exp(-n_{ij}\lambda_{ij})/d_{ij}!$, and hence, after a small amount of algebra, the maximum likelihood estimate of $\lambda_{ij}$ is $\hat{\lambda}_{ij} = d_{ij}/n_{ij}$, not surprisingly.

First, let us consider just two smoking categories, non-smokers and smokers, summing over age group and number of cigarettes per day. The resulting four data values are: non-smokers – deaths $d_0 = 6$, person-years 69,923; smokers – deaths $d_1 = 195$, person-years 109,368. Then the risk ratio, smokers to non-smokers, is, RR $= 195 \times 69923/6 \times 109368 = 20.8$. Now, log(RR) is approximately normally distributed, with variance, $1/d_1 + 1/d_0 = 0.172$. Hence, a 95% confidence interval for RR is [9.22, 46.82].

Next, we look at the *mortality ratios* (MR) for the smoking categories. The crude mortality ratio for the $i$th smoking category/$j$th age group is $\hat{\lambda}_{ij} = d_{ij}/n_{ij}$. So for example, smoking category, "20–24"/age group, "55–59", $d_{64} = 8/4357.5 = 0.001836$, or 184 per 100,000. We would like to standardise this by comparing it to the mortality rate for lung cancer for non-smokers, $\lambda_4^*$, in the same age group. We do not have the data though, and we cannot use the non-smokers in the study as there are only 6 deaths, making the standardisation unreliable. Instead, as a proxy, we use estimates of mortality for lung cancer, for non-smokers by age, made by Lariscy[34] for the US population. The rates are $(0.001, 0.01, 0.11, 0.18, 0.31, 0.43, 0.63, 0.77)/1000$, for the eight age groups. Note, we have had to make up the first two rates since these were not included in their analysis. (We could have left these two age groups out of our analysis.) So, $\lambda_4^* = 0.18 \times 10^{-3}$, and the RR for smoking category, "20–24"/age group, "55–59" is RR$_{64} = 0.001836/0.00018 = 10.2$. The expected number of deaths, based on the non-smoker rates, is $E_{64} = 4357.5 \times 0.00018 = 0.78$, much less than the observed number of 8.

We now combine the data over age group. The *standardised mortality ratio* (SMR) for smoking category, $i$, is

$$\text{SMR}_i = \sum_{j=1}^{J} d_{ij} / \sum_{j=1}^{J} n_{ij}\lambda_{ij} = d_{i.}/E_{i.},$$

i.e. the total number of deaths in the category divided by the sum over the age groups of the expected number of deaths.

**TABLE 7.9**

Doctors Study: SMRs for the smoking categories

| Category: | 0 | 1-4 | 5-9 | 10-14 | 15-19 | 20-24 | 25-29 | 30-34 | 35-40 |
|---|---|---|---|---|---|---|---|---|---|
| SMR: | 0.60 | 3.50 | 2.03 | 5.79 | 6.76 | 13.66 | 14.45 | 21.63 | 28.93 |

Table 7.9 shows the SMRs for all the smoking categories. Clearly, the more the exposure to tobacco, the higher the risk of death from lung cancer.

The SMR for smokers overall is given by summing the deaths and expected deaths over all rows in Table 7.8 apart from the first and all columns,

$$\text{SMR} = \sum_{i=2}^{I} \sum_{j=1}^{J} d_{ij} / \sum_{i=2}^{I} \sum_{j=1}^{J} n_{ij}\lambda_{ij} = 11.29,$$

compared to the (unreliable) SMR $= 0.60$ for non-smokers.

We could go on to consider confidence intervals and significance tests for the SMRs, based on Poisson distributions, but instead, we will show a neat analysis using Poisson regression.

### 7.3.1 Poisson regression models

First, we consider only two exposure categories, smokers and non-smokers, and ignore the age groups. Assume the number of deaths, $d_1$ and $d_2$, are Poisson random variables, independent of each other. The Poisson generalised linear model is,

$$d_i \sim \text{Poisson}(\mu_i), \ \mu_i/n_i = \exp(\beta_0 + \beta_1 x), \ \text{or} \ \log(\mu_i) = \log(n_i) + \beta_0 + \beta_1 x,$$

where $x = 1$ for the smoker category and $x = 0$ for the non-smoker category. Note, the rate $(\mu/n)$ for a non-smoker, in any age group, is $\exp(\beta_0)$, and the rate $(\mu/n)$ for a smoker, in any age group, is $\exp(\beta_0 + \beta_1)$. Taking the ratio of these, RR $= \exp(\beta_1)$.

Using the R function glm() to analyse the data of Table 7.8, the parameters (standard errors) are estimated as, $\beta_0 = -9.363\,(0.408)$ and $\beta_1 = 3.034\,(0.415)$. The deviance is 263.6 on 14 df and AIC=315.1. Exponentiating $\beta_1$ gives the crude risk ratio as, RR $= 20.8$, with 95% CI, [10.1, 52.8].

Not surprisingly, when the deaths and person-years are added together over age group for smokers and non-smokers, and the Poisson model fitted with just four data values, RR is still estimated by 20.8 and with the same confidence interval. (The sum of independent Poisson distributions is also a Poisson distribution.)

Next, we bring in the confounding age group variable using the same model, extending $\mu_i$ to,

$$\log(\mu_i) = \log(n_i) + \beta_0 + \beta_1 x + \sum_{j=2}^{J} \beta_j z_j,$$

where the $z_j$'s are indicator variables for the age groups, the reference category being the first. So, the first age group has $z$-values, $(0,0,0,0,0,0,0)$, the second, $(1,0,0,0,0,0,0)$, etc.

Table 7.10 show the fitted parameters and their standard errors for the regression model, together with the estimated risk ratios. The risk ratio for smokers/non-smokers is 19.07, very similar to the previous value where the confounding variable, age, was not taken into account. The new confidence interval is [9.27, 48.4], slightly more narrow than the previous. The risk ratios for age increase with age group, each being compared to the 40–44 age group. If we wanted to change the reference age group, say, comparing the group 60–64 with group 55–59 as the reference, it is easily seen from the model, that $\log(\text{RR}_{45}) = \beta_5 - \beta_4$, and is estimated by 0.720, Hence, $RR_{45} = 2.05$. The variance of $\log(\text{RR}_{45})$ is $\text{var}(\text{RR}_{44}) + \text{var}(\text{RR}_{55}) - 2\text{cov}(\text{RR}_{45})$. The R

**TABLE 7.10**
Doctors Study: Poisson regression with smoking and age

| Category | Ref. group | parameter | $\hat{\beta}$ (s.e.) | RR (95% CI) |
|---|---|---|---|---|
| Intercept | | $\beta_0$ | $-11.98\,(0.702)$ | |
| Smokers | Non-smokers | $\beta_1$ | $2.948\,(0.415)$ | $19.07\,[9.27,\,48.4]$ |
| Age gp. 45-49 | 40-44 | $\beta_2$ | $0.927\,(0.690)$ | $2.53\,[0.70,\,114.71]$ |
| Age gp. 50-54 | .. | $\beta_3$ | $2.177\,(0.616)$ | $8.82\,[3.06,\,37.27]$ |
| Age gp. 55-59 | .. | $\beta_4$ | $2.603\,(0.609)$ | $13.51\,[4.77,\,56.56]$ |
| Age gp. 60-64 | .. | $\beta_5$ | $3.323\,(0.599)$ | $27.75\,[10.09,\,11.73]$ |
| Age gp. 65-69 | .. | $\beta_6$ | $3.717\,(0.603)$ | $41.15\,[14.96,\,170.10]$ |
| Age gp. 70-74 | .. | $\beta_7$ | $4.143\,(0.601)$ | $63.01\,[22.75,\,261.3]$ |
| Age gp. 75-79 | .. | $\beta_8$ | $4.394\,(0.610)$ | $80.92\,[28.50,\,339.39]$ |
| Deviance=7.88, df=7, AIC=73.3 | | | | |

function, vcov(), will return the covariance matrix of the parameter estimates of the model from which we calculate $\log(RR_{45})$ as $0.358 + 0.370 - 2 \times 0.333 = 0.062$, and hence the 95% CI for $\log(RR_{45})$ is $[1.26, 3.35]$. Alternatively, the R function, relevel() with arguments, (*vbl*, ref = *"ref category"*), will change the reference category and then proceed as usual, thus eliminating the need to find the covariance matrix.

**TABLE 7.11**
Doctors Study: Poisson regression with age continuous

| Model | AIC | Smoker yes/no RR (95% CI) | Age RR (95% CI) |
|---|---|---|---|
| 1. smoker + age/10 | 73.3 | $19.53\,[9.49, 49.58]$ | $2.94\,[2.56, 3.39]$ |
| 2. smoker $*$ age/10 | 75.2 | | |
| 3. smoker + age/10 + $(age/10)^2$ | 65.0 | $19.1\,[9.27, 48.4]$ | |
| 4. smoker + $\log(age/10 - 4)$ | 65.8 | $18.99\,[9.23, 48.23]$ | $7.44\,[5.45, 10.39]$ |
| | | No. cigarettes | Age |
| 5. avcigs + age/10 | 236.1 | $1.09\,[1.08, 1.10]$ | $3.19\,[2.76, 3.70]$ |
| 6. avcigs + $\log(age/10-3)$ | 231.5 | $1.09\,[1.08, 1.10]$ | $5.09\,[5.09, 3.70]$ |
| 7. $\log(avcigs+1) + \log(age/10)$ | 219.6 | $3.32\,[2.65, 4.29]$ | $5.09\,[5.09, 3.70]$ |
| | | Exposure | |
| 8. avcigs $\times$ (age $- 20)/100$ | 355.2 | $1.28\,[1.25, 1.31]$ | |
| 9. $\log(avcigs + 1) \times$ (age $- 30)/20$ | 231.7 | $2.28\,[1.96, 2.28]$ | |

We could treat age as a continuous variable, with ages taken as mid-points of the categories. Table 7.11 shows the results of fitting various Poisson models. The first model has smoker (yes/no) and age divided by ten as the explanatory variables. The risk ratio for smoking is 19.5, very similar to that for age as a categorical variable, and with similar confidence intervals. Note also, that the AICs are equal to 1 dp. The RR for age is 2.94, so that the risk of death from lung cancer in your 60's, is about three times higher than the risk in your 50's, and similarly for the other adjoining decades of your life. We divided age by 10 for ease of interpretation, as otherwise we would have statements like, the risk of dying from lung cancer is 1.11 higher at age 57 than at age 56. (Note, compounded: $1.11^{10} = 2.84$.)

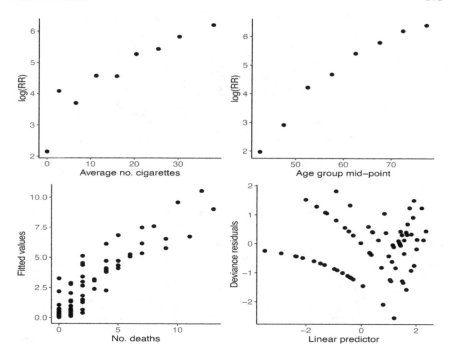

**FIGURE 7.6**
Poisson regression: plots of log(RR) against average no. of cigarettes and age; plot of fitted values against no. of deaths and a residuals plot

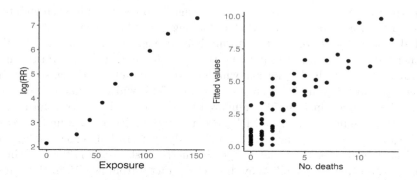

**FIGURE 7.7**
Poisson regression: plots of log(RR) against exposure and fitted values against no. of deaths

Can we find a better fitting model? Model 2 in Table 7.11 shows an interaction term between smoker and age included; its deviance is higher than for the first model and so is excluded. In one way this is good, because the

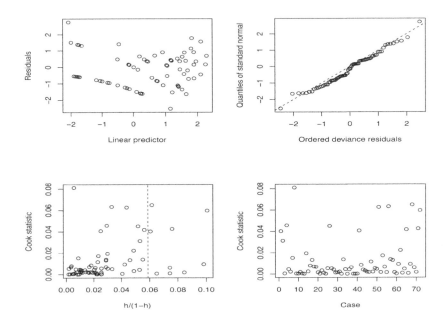

**FIGURE 7.8**
Poisson regression: model fit statistics

interpretation of the RRs for smoking and age becomes difficult when an inter-action term is involved. The RR for smoking depends on age and vice versa. Models 3 and 4 show two transformations of age, one a quadratic and the other log, giving models that fit better than the first. The quadratic trans-formation means the RR for age will depend on the reference age, so the RR when you are in your 60's compared to your 50's, will be different from the RR when you are in your 70's compared to your 60's. The log transformation can make the interpretation of RR harder, since the RR is for a unit increase in $\log(\text{age}/10 - 4)$, not age. So the risk of dying from lung cancer at age 67 is 7.44 times higher than that at age 50, and the risk at age 113 is 7.44 times higher than that at age 67! Why did we subtract 4 from age/10 in the trans-formation? Answer: the model fitted better than no subtraction. But can we justify it? Yes, I think so – allow 20 years (2 decades) due to the fact that people do not start smoking at birth, and possibly 20 years of smoking before lung cancer develops.

Next, we look at the effect of how much people smoke on the risk of dying from lung cancer, by fitting Poisson models using average number of cigarettes smoked per day and age. We use all the 72 data points in Table 7.8. Model 5 in Table 7.11 shows the results using average number of cigarettes and age/10 as risk factors. The RR for number of cigarettes, is 1.09. so smoking one extra cigarette per day increases the risk of you dying from lung cancer by 1.09. Compounded, the RR for smoking 10 extra cigarette per day increases the risk by 2.38.

To find appropriate transformations of average number of cigarettes and age, the top left graph in Figure 7.6 shows log(RR) plotted against average number of cigarettes, where average number of cigarettes and corresponding person-years have been summed over age. Similarly, the top right graph shows log(RR) plotted against age, the summations being over average number of cigarettes. The plots use log(RR) since the link-function is log. The graphs suggest that a log transformation might be appropriate in both cases. Models 6 and 7 show two of several models that were tried. Model 7 has the lowest AIC. The bottom two graphs in Figure 7.6 shows the fitted values plotted against the observed number of deaths, and the deviance residuals plotted against the linear predictors for model 7. The model fits well. The deviance residual plot is interesting! Again, the interpretation of RR for log-transformed cigarettes and age variables is not straightforward.

Lastly, we take a different approach and work with exposure based on average number of cigarettes smoked per day times age minus 20, or transformations thereof. Model 8 shows the results without any transformation – the AIC is 355.2. Various transformations were tried, the one with the lowest AIC of 231,7, is shown as model 9. The left hand graph of Figure 7.7 shows log(RR) plotted against exposure for the definition of exposure in model 9. Exposure has been sorted into ascending order and then summed over the first nine values, the second nine values, etc., with the corresponding number of deaths also summed. The points are almost linear. The right hand graph shows fitted values plotted against number of deaths. More graphs of model fit statistics can be easily obtained using the R function glm.diag.plots(), as shown in Figure 7.8.

Did we need to search for an appropriate transformation of age and average number of cigarettes smoked per day to give us a well fitting model? The answer is probably not in this case. It is clear that the risk of lung cancer increases with age and is linked to smoking. The risk is higher for higher exposure and it is not as if we are trying to find a safe level of tobacco use. However, in other situations we might need to consider carefully the models we fit in order to obtain robust information regarding levels of risk.

### 7.3.2  Two more examples of cohort studies

Since a cohort study, prospective or historical, follows a group of individuals into their future, any appropriate statistical model can be used on the data, depending on context. We give brief accounts of two cohort studies published in the literature, one uses logistic regression, the other uses survival analysis techniques.

**Cancer-associated ischaemic stroke**

The first is a retrospective or historical cohort study looking at cancer-associated ischaemic stroke which uses logistic regression as its main statistical model (see Graziolia et al. [17]). There is a known association between cancer and cerebrovascular disease. Of the patients admitted with ischaemic stroke between March 2005 and March 2015 to Perugia and Varese hospitals in Italy, 2209 were included in the study. Their mean age was 73 years, 55% were male, and 98(4.4%) had active cancer. Data on twenty nine demographic and risk factor variables were obtained from hospital records.

Univariate logistic regression was used on each variable. The model is

$$\Pr(\text{cancer}) = \exp(\beta_0 + \beta_1 x)/(1 + \exp(\beta_0 + \beta_1 x)),$$

where $x$ is the prognostic variable, and $\beta_0$ and $\beta_1$ are parameters to be estimated. The log odds is, $\log(\text{ODS}) = \beta_0 + \beta_1 z$. If $x$ is an indicator variable with values, 0 and 1, then the two log odds are $\beta_0$ and $\beta_0 + \beta_1$. The difference is the log odds ratio (OR), equal to $\beta_1$. So, if $\hat{\beta}_1$ is the estimate of $\beta_1$, then $\exp(\hat{\beta}_1)$ is the estimate of the OR. The model was fitted separately using various prognostic variables, and those with p-values less than 0.2 were included in a multivariate logistic regression model. These results are shown in Table 7.12 together with estimated ORs and 95% confidence intervals. Age, diabetes, VTE (venous thromboembolic events), LDL (low-density lipoprotein) and cryptogenic subtype (a stroke of undetermined aetiology) are significantly associated with cancer.

**TABLE 7.12**

Cancer and ischaemic stroke: multivariate logistic regression analysis

|                    | OR   | 95% CI        |                          | OR   | 95% CI        |
| ------------------ | ---- | ------------- | ------------------------ | ---- | ------------- |
| Age>65 years       | 3.34 | [1.64, 6.81]  | Platelet>450,000/mm3     | 1.96 | [0.53, 7.28]  |
| Sex male           | 0.95 | [0.62, 1.44]  | LDL cholesterol<70 mg/dL | 1.92 | [1.06, 3.47]  |
| Diabetes mellitus  | 0.56 | [0.31, 1.00]  | Triglycerides>175 mg/dL  | 0.63 | [0.28, 1.40]  |
| Obesity            | 0.49 | [0.17, 1.38]  | Cryptogenic subtype      | 1.93 | [1.22, 3.04]  |
| VTE                | 2.84 | [1.12, 7.19]  |                          |      |               |

Before discussing the second example, we need now to briefly describe *propensity score matching*. It is a useful technique to be aware of.

**Propensity score matching**

Historical cohort studies can suffer from confounding due to inconsistency in baseline variables. For example, those exposed to the risk factor might be older on average than those not exposed. There are three common methods to allow for this problem: divide the individuals into subgroups homogeneous in the baseline variables, and analyse these separately; use regression analysis allowing for the confounders, as we have previously seen; use propensity scoring.

Propensity scoring is a method of balancing groups by their baseline variables. Each individual in the cohort is given a score, which is the probability the individual would be in the exposed group, given their baseline data. Let the baseline covariates be $\mathbf{X}$ and let $Z = 1$ if an individual is in the exposed group and $Z = 0$ if in the non-exposed group. The propensity score, $e(\mathbf{x})$, is defined as

$$e(\mathbf{x}) = \Pr(Z = 1|\mathbf{x}),$$

which is calculated for each individual. Logistic regression is a common method for calculating the propensity scores. Simply carry out the logistic regression with the $z_i$'s as the dependent observations and the $\mathbf{x}_i$'s are independent observations. Then $e(\mathbf{x}_i) = \exp(\hat{\beta}_0 + \hat{\beta}_1 x_{1i} + \hat{\beta}_2 x_{2i} + \ldots)/(1 + \exp(\hat{\beta}_0 + \hat{\beta}_1 x_{1i} + \hat{\beta}_2 x_{2i} + \ldots))$, where the $\hat{\beta}$'s are the estimated parameters.

Next, check on the overlap of propensity scores for the two groups to see how much they overlap. If there is very little overlap, then it is pointless going further. Typically, there will be much overlap, but some extreme values might need to be discarded.

Then non-exposed individuals are matched to exposed individuals, based on their propensity score. There are several ways this can be done using a variety of algorithms. A common one is greedy nearest neighbour matching, where, in turn, exposed individuals each have a non-exposed individual matched to them based on the closest $e(\mathbf{x})$ value. If there is a tie in the $e(\mathbf{x})$ values, one of them is chosen at random. The chosen non-exposed individual drops out of the pool of non-exposed individuals. This will give 1:1 matching. If multiple non-exposed individuals are to be matched to each exposed individual, you just go round the loop several times. How do you choose the order in which exposed individuals are found a match? It can be in random order, by best match possible, or by some other means. A more computer intensive algorithm is optimal matching, where matched pairs are chosen so that the average within-pair difference in propensity scores is minimised.

Lastly, once the matching is complete, the balance in baseline variables is examined. If this is acceptable, we now have a case-control study born out of the cohort study.

### Another lung cancer example

The second example takes us back to lung cancer with much more up to date data than the Doll and Hill Doctors Study data. Gao and Zhou[16] used 30,864 lung cancer cases taken from the SEER database (https://seer.cancer.gov) as their cohort and looked at prognostic factors of survival, with the recorded number of lymph nodes (LN) infiltrated by the cancer (positive lymph nodes) in particular. The number of positive lymph nodes was dichotomised to "$\leq 3$", and "$> 3$". There were 29,012 patients with LN $\leq 3$ and 1852 patients with LN $> 3$. Figure 7.9 (a) shows a sketch of the Kaplan-Meier curves for overall survival for the two groups based on a figure in the paper by Gao and Zhou. The hazard ratio, from a univariate analysis, is 2.07 with 95% CI, [1.94, 2.20], $p < 0.001$, with a clear survival advantage for the LN $\leq 3$ group.

The authors thought it a good idea to choose a subset of the LN $\leq 3$ group of patients using propensity scoring, so that the baseline factors are matched.

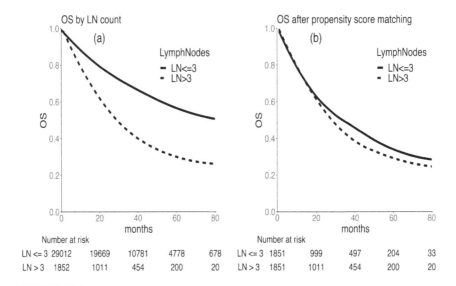

**FIGURE 7.9**

Overall survival for lung cancer patients by lymph node count: (a) raw data, (b) data after propensity score matching

They used 1:1 propensity score matching on the baseline factors using logistic regression and a greedy nearest neighbour matching algorithm, resulting in 1851 patients in both groups. Figure 7.9 (b) shows a sketch of the new Kaplan-Meier curve for the LN ≤ 3 group. What a difference! The hazard ratio is now 1.09, with 95% CI, [1.00, 1.19], $p = 0.047$ – not such a large survival advantage now! In a multivariate analysis of demographic and clinical prognostic factors, before matching, the hazard ratio for the LN > 3 group is 1.07, relative to the LN ≤ 3 group. However, one problem here is that the lymph node stage (N within TNM staging) has also been included in the model. This is highly correlated with the lymph node count. In a sense, the model wants to share out the hazard ratio between two very similar prognostic factors. After matching, the number of patients with no lymph node infiltration (N0), in the reduced LN ≤ 3 group is zero! Thus, the results post-matching are conditioned on patients having some lymph node infiltration. The first analysis is best!

## 7.4 Case-control studies

We use the data from the Doll and Hill case-control study[8] as an example of an unmatched study and to illustrate the statistical theory used. Then an oesophageal cancer example is used for a matched case-control study.

## 7.4.1 Unmatched case-control study

For the Doll and Hill study, we concentrate on females first. There were 60 lung cancer cases, of whom 68.3% smoked, and 60 controls, of whom 46.7% smoked. The data are shown in Table 7.13.

| TABLE 7.13 Lung cancer: data, $n_{ij}$ | | | TABLE 7.14 Lung cancer: $p$'s | | | TABLE 7.15 Lung cancer: $q$'s | | |
|---|---|---|---|---|---|---|---|---|
| | Case | Control | | Case | Control | | Case | Control |
| Smoker | 41 | 28 | Smoker | $p_1$ | $1 - p_1$ | Smoker | $q_1$ | $q_0$ |
| Non-smoker | 19 | 32 | Non-smoker | $p_0$ | $1 - p_0$ | Non-smoker | $1 - q_1$ | $1 - q_0$ |

Doll and Hill used a chi-squared test to assess the association of smoking with lung cancer; the results were $X^2 = 5.76$, df $= 1$, $p = 0.0164$, showing an association. They did not use a Yates' continuity correction, which would haven given $X^2 = 4.91$, $p = 0.0267$.

We now model the situation. Let cases be called "Disease" (D), and controls, "Healthy" (H), the change because cases and controls start with the same letter! (The controls in this example were not healthy at the time, as they were hospital patients, but it does not matter.) Let smoking use be denoted by "S" and no smoking use be denoted by "NS". Then

$$\Pr(D|S) = p_1, \quad \Pr(H|S) = 1 - p_1,$$
$$\Pr(D|NS) = p_0, \quad \Pr(H|NS) = 1 - p_0,$$

as shown in Table 7.14. Table 7.15 gives the alternative conditional probabilities as,

$$\Pr(S|D) = q_1, \quad \Pr(NS|D) = 1 - q_1,$$
$$\Pr(S|H) = q_0, \quad \Pr(NS|H) = 1 - q_0.$$

If we want to use relative risk (RR), we have to estimate $p_1/p_0$. But we cannot from these data, because we have chosen a fixed number of cases and a fixed number of controls and ascertained whether they smoked or not. We do not have the total number of females at risk in order to calculate risk. To calculate RR, we would need a sample of females who had smoked and a sample of women had not. We would follow them up for several years and see how many in each group developed lung cancer, i.e. a cohort study. Alternatively we could use a large sample of females and see who had, and who had not developed lung cancer, and who had, and who had not smoked, i.e. a cross-sectional study.

We will choose a different measure from RR, the odds ratio, $OR_p = \frac{p_1(1-p_0)}{p_0(1-p_1)}$, and use Bayes's theorem to express this in terms of the $q$'s. Let

$\Pr(D) = p$, and so $\Pr(H) = 1 - p$. Using Bayes' theorem,

$$p_1 = \Pr(D|S) = \frac{\Pr(S|D)\Pr(D)}{\Pr(S|D)\Pr(D) + \Pr(S|H)\Pr(H)} = \frac{q_1 p}{q_1 p + q_0(1 - p)},$$

$$p_0 = \Pr(D|NS) = \frac{\Pr(NS|D)\Pr(D)}{\Pr(NS|D)\Pr(D) + \Pr(NS|H)\Pr(H)} = \frac{(1 - q_1)p}{(1 - q_1)p + (1 - q_0)(1 - p)}.$$

Hence $\frac{p_1}{(1-p_1)} = \frac{q_1 p}{q_0(1-p)}$ and $\frac{p_0}{(1-p_0)} = \frac{(1-q_1)p}{(1-q_0)(1-p)}$. Substituting these into $OR_p$ gives,

$$OR_p = \frac{q_1(1 - q_0)}{q_0(1 - q_1)} = OR_q. \tag{7.1}$$

Thus the disease odds ratio for disease ($OR_p$) is equal to the odds ratio for exposure ($OR_q$) and so we have a way of comparing the risk of developing lung cancer for exposed and unexposed females using this connection.

Let the observed data values be denoted by the labels in Table 7.16, so the number of cases is $n_1$, the number of cases who are smokers is $a$, etc. Later we will use alternative labels shown in Table 7.17 when we wish to think of the counts as random variables.

**TABLE 7.16**
Case-Control: count labels

|  | Case | Control |  |
|---|---|---|---|
| Smoker | $a$ | $c$ | $m_1$ |
| Non-smoker | $b$ | $d$ | $m_0$ |
| Total | $n_1$ | $n_0$ | $n$ |

**TABLE 7.17**
Alternative count labels

|  | Case | Control |  |
|---|---|---|---|
| Smoker | $x_1$ | $x_0$ | $m_1$ |
| Non-smoker | $n_1 - x_1$ | $n_0 - x_0$ | $m_0$ |
| Total | $n_1$ | $n_0$ | $n$ |

The likelihood, $L$, of the data is,

$$L = q_1^a(1 - q_1)^c q_0^b(1 - q_0)^d,$$

and then from the log-likelihood, it is easy to show $\hat{q}_1 = a/(a + c)$ and $\hat{q}_0 = b/(b + d)$, Hence the maximum likelihood of $OR_q$, and equivalently $OR_p$ is $\widehat{OR}_p = \widehat{OR}_q = ad/bc$.

From the data, $OR = 2.47$, dropping the subscript $p$ and the hat.

**Variance and confidence intervals for OR**

We work with the log odds ratio, $\log(OR)$ and so,

$$\begin{aligned} \log(OR) &= \log(p_1) - \log(1 - p_1) - \log(p_0) - \log(1 - p_0) \\ &= \log(q_1) - \log(1 - q_1) - \log(q_0) - \log(1 - q_0), \end{aligned}$$

using Equation 7.1 to go from $OR_p$ to $OR_p$. Let the number of females with lung cancer be $n_1$, of whom $X_1$ smoked, and so $X_1 \sim \text{Bin}(n_1, q_1)$. Similarly for the controls, $X_0 \sim \text{Bin}(n_0, q_0)$. From the Appendix, we know that

$\log(X_1) - \log(1 - X_1)$ has an approximate normal distribution, $N(\log(q_1/(1 - q_1)), 1/\{n_1 q_1 (1 - q_1)\})$, and similarly for $X_0$. As $X_1$ and $X_0$ are independent,

$$\log(\mathrm{OR}) \dot\sim N\left( \log\left( \frac{q_1(1-q_0)}{q_0(1-q_1)} \right), \frac{1}{n_1 q_1 (1 - q_1)} + \frac{1}{n_0 q_0 (1 - q_0)} \right).$$

The observed number of females with lung cancer who smoke is $x_1$ and the observed number of controls who smoke is $x_0$, and hence, after a little algebra,

$$\log(\mathrm{OR}) \dot\sim N\left( \log\left( \frac{x_1(n_0 - x_0)}{x_0(n_1 - x_1)} \right), \frac{1}{x_1} + \frac{1}{(n_1 - x_1)} + \frac{1}{x_0} + \frac{1}{(n_0 - x_0)} \right),$$

$$\text{or, } \log(\mathrm{OR}) \dot\sim N\left( \log\left( \frac{ad}{cb} \right), \frac{1}{a} + \frac{1}{b} + \frac{1}{c} + \frac{1}{d} \right),$$

noting in the calculation we have used $1/\{q(1-q)\} = 1/q + 1/(1-q)$. A 95% confidence interval for $\log(\mathrm{OR})$ is $[0.16, 1.15]$, and for OR, by exponentiating, $[1.17, 5.19]$, showing evidence of an association between smoking and lung cancer in females.

Table 7.18 shows the lung cancer and smoking numbers for males. The data illustrate how popular smoking was in those days. As the smallest cell number is 2, Fisher's exact test has to be used, rather than the chi-squared test. The odds ratio is, OR $= 14.03$, with 95% confidence interval, $[3.50, 122.3]$, very strong evidence of the association of smoking and lung cancer.

**TABLE 7.18**

Lung cancer and smoking: males

|  | Case | Control | Total |
|---|---|---|---|
| Smoker | 647 | 622 | 1269 |
| Non-smoker | 2 | 27 | 29 |
| Total | 649 | 649 | 1298 |

## 7.4.2 Confounding in case-control studies

Our second example is taken from Breslow and Day's book on Case-Control studies, accessed from the IARC website. The data are from the "Ijje-et-Viliane' study[68] where cases are 200 males diagnosed with oesophageal cancer, and controls are 775 males selected from electoral lists. Each participant completed a questionnaire on diet, the book showing the data for tobacco and alcohol consumption. The book can be accessed from https://publications.iarc.fr.

Table 7.19 shows the number of cases and controls with low alcohol consumption (0-79 g/day) and high consumption (80+ g/day), stratified by age-group. First, let us ignore age-group and calculate the odds ratio for cases and controls for alcohol consumption. There are 96 cases with high consumption and 104 with low consumption, with corresponding numbers, 109 and

**TABLE 7.19**

Oesophageal cancer study: age-groups

| | Alcohol | Age groups 25-44 | 45-54 | 55-64 | 65+ |
|---|---|---|---|---|---|
| Cases | 80+g/day | 5 | 25 | 42 | 24 |
| | 0-79g/day | 5 | 21 | 34 | 44 |
| Controls | 80+g/day | 35 | 29 | 27 | 18 |
| | 0-79g/day | 270 | 138 | 139 | 119 |

666 for controls. The odds ratio is OR = 5.64, with 95% confidence interval, [4.00, 7.95], a high association of alcohol consumption with oesophageal cancer. However, could the high odds ratio be due, at least in part, to difference in the age profiles of the cases and controls? The frequencies in the age-groups for cases and controls are {10(5%), 46(23%), 76(38%), 68(34%)} and {305(39%), 167(22%), 166(21%), 137(18%)} respectively, and so there is a difference with the cases having more older participants than the controls. A chi-squared test gives $X^2 = 96.0$, df = 3 and $p < 0.0001$ for a difference between the groups. Age could be a confounding variable. It could be influencing the value of the odds ratio. To allow for this confounding, we can use the pooled odds ratio.

The pooled odds ratio can be estimated by the Cochran-Mantel-Haenszel odds ratio, $OR_{CMH}$, defined as

$$OR_{CMH} = \frac{\sum_{i=1}^{K} x_1^{(i)} \left(n_0^{(i)} - x_0^{(i)}\right)/n^{(i)}}{\sum_{i=1}^{K} x_0^{(i)} \left(n_1^{(i)} - x_1^{(i)}\right)/n^{(i)}} = \frac{\sum_{i=1}^{K} a^{(i)} d^{(i)}/n^{(i)}}{\sum_{i=1}^{K} b^{(i)} c^{(i)}/n^{(i)}},$$

where there are $K$ strata, $x_1^{(i)}$, $x_0^{(i)}$, $n_1^{(i)}$, $n_0^{(i)}$ and $a^{(i)}$, etc. are the $x_1$, $x_0$, $n_1$, $n_0$ and $a$ etc. for the $i$th strata. The Mantel-Haenszel odds ratio is not the same as the maximum likelihood estimator, but is an approximation to it.

**TABLE 7.20**

Cochran-Mantel-Haenszel odds ratio

| Age-Group | $OR_{MH}$ | 95% CI |
|---|---|---|
| 25-44 | 3.61 | [1.68, 7.75] |
| 45-54 | 6.36 | [3.30, 12.28] |
| 55-64 | 5.67 | [2.63, 12.16] |
| 65+ | 7.71 | [1.67, 34.96] |
| MH pooled | 5.29 | [3.65, 7.66] |

The variance of $OR_{CMH}$ can be derived exactly but is complicated. There have been several approximations for the variance, or the variance of $\log(OR_{CMH})$, the Robins-Breslow-Greenland variance estimator being a commonly used. We will not quote the formula for it, but it can be found in Silcocks[59] who gives a good discussion about the variance of $OR_{CMH}$.

For our data, the function, CCInter(), from the R library, EpiStats, was used to calculate $OR_{CMH}$, shown in Table 7.20, along with the odds ratios for the four age groups. It is not too different from the "crude" OR, of 5.64. Also shown are the 95% confidence intervals.

We can test for homogeneity of the OR's in Table 7.20, using the test statistic,

$$X^2_{CMH} = \frac{\sum_{i=1}^{K} \left( a^{(i)} - n_1^{(i)} m_1^{(i)} / n^{(i)} \right)^2}{\sum_{i=1}^{K} n_1^{(i)} n_0^{(i)} m_1^{(i)} m_0^{(i)} / \{ [n^{(i)}]^2 (n^{(i)} - 1) \}},$$

The statistic, $X^2_{CMH}$, follows a chi-squared distribution, with df=1, under the null hypothesis of homogeneity. For our data $X^2_{CMH} = 86.5$, with $p < 0.0001$, showing heterogeneity among the strata.

### 7.4.3 Logistic regression for unmatched case-control studies

Let us take another approach to analysing unmatched case-control study data and use logistic regression. The model is,

$$\text{logit}(q) = \log(q/(1-q)) = \beta_0 + \beta_1 x, \tag{7.2}$$

where $x = 1$ if an individual is exposed to the potential risk factor and $x = 0$ if they are not. For the exposed group, $\log(q_1/(1-q_1)) = \beta_0 + \beta_1$ and for the unexposed group, $\log(q_0/(1-q_0)) = \beta_0$. The difference of these two equations is,

$$\log(q_1/(1-q_1)) - \log(q_0/(1-q_0)) = \log\left( \frac{q_1/(1-q_0)}{q_0/(1-q_1)} \right) = \beta_1 = \log(OR),$$

and hence $OR = \exp(\beta_1)$.

The logistic regression model of Equation 7.2 was used on the lung cancer data in Table 7.13, fitted with the R function glm(). The estimates of $\beta_0$ and $\beta_1$, with standard errors, were $\hat{\beta}_0 = -0.521\,(0.290)$ and $\hat{\beta}_1 = 0.903\,(0.380)$, giving the estimate of $OR = 2.47$, with 95% confidence interval, $[1.17, 5.19]$. You notice that this estimate of OR and its confidence interval are exactly the same those we obtained initially using $OR = ad/bc$ from Table 7.16.

The advantage of logistic regression is that it is easy to include a wide range of explanatory variables, continuous, discrete, categorical and ordinal, in the regression. For example, let us model the Ille-et-Vilaine oesophageal cancer data, incorporating alcohol and age-group as explanatory variables. The model is

$$\text{logit}(q) = \beta_0 + \beta_1 z_1 + \beta_2 z_2 + \beta_3 z_3 + \beta_4 z_4,$$

where $z_1, \ldots, z_4$ are indicator variables with values 0 and 1 as follows,

|       | Alc:80+ | Alc:0-79 | Age:25-44 | Age:45-54 | Age:55-64 | Age:65+ |
|-------|---------|----------|-----------|-----------|-----------|---------|
| $z_1$ | 1       | 0        |           |           |           |         |
| $z_2$ |         |          | 0         | 1         | 0         | 0       |
| $z_3$ |         |          | 0         | 0         | 1         | 0       |
| $z_4$ |         |          | 0         | 0         | 0         | 1.      |

**TABLE 7.21**

Logistic regression: oesophageal cancer study

|            | Estimate | Std. error | p-value    | OR [95%CI]          |
| ---------- | -------- | ---------- | ---------- | ------------------- |
| Intercept  | $-3.83$  | 0.334      | $< 0.0001$ |                     |
| Age:45-54  | 1.97     | 0.371      | $< 0.0001$ | $7.18\ [3.48, 14.85]$  |
| Age:55-64  | 2.49     | 0.358      | $< 0.0001$ | $12.03\ [5.96\ 24.25]$ |
| Age:65+    | 2.74     | 0.363      | $< 0.0001$ | $15.50\ [7.61, 31.56]$ |
| Alc:80+    | 1.68     | 0.189      | $< 0.0001$ | $5.37\ [3.70, 7.78]$   |

Deviance: 791.96, df=970, AIC = 801.96

Table 7.21 shows the fitted model results, together with ORs and their confidence intervals. The ORs for the three age groups are the odds ratios relative to the age-group, 45-54, independent of alcohol consumption. The OR for Alc:80+ and its confidence interval is close to the Mantel-Haenszel odds ratio. The logistic regression model has "allowed for age" in estimating the OR for alcohol.

If you get confused in interpreting the ORs, just put the values of the indicator variables into the model. For example, for individuals in alcohol category 80+ and age category 55-64 (group A say), $z_1 = 1$, $z_2 = 0$, $z_3 = 1$, $z_4 = 0$, and so $\mathrm{logit}(q_A) = \beta_0 + \beta_1 + \beta_3$. For individuals in alcohol category 0-79 and age category 55-64 (group B), $z_1 = 0$, $z_2 = 0$, $z_3 = 1$, $z_4 = 0$, and so $\mathrm{logit}(q_B) = \beta_0 + \beta_3$. The difference of the logits is $\mathrm{logit}(q_A) - \mathrm{logit}(q_B) = \log(\mathrm{OR}) = \beta_1$, and hence $q_A(1 - q_B)/q_B(1 - q_A) = \exp(\beta_1)$, giving the OR for alcohol consumption for this age-group. Note, we would obtain this value of OR for each of the age-groups. If we want the OR comparing age group 65+ with age-group 45-54, we find $\mathrm{logit}(q_A) - \mathrm{logit}(q_B) = \log(\mathrm{OR}) = \beta_4 - \beta_2$, for both categories of alcohol consumption. Hence, $\mathrm{OR} = \exp(\beta_4 - \beta_2) = \exp(\beta_4)/\exp(\beta_2)$, and for our data, this gives $\mathrm{OR} = 15.50/7.18 = 2.16$.

For this model with alcohol consumption and age-group as explanatory variables, Akaike's Information Criterion (AIC) is AIC = 801.96. We could refit the model and include an interaction term of alcohol consumption and age-group. Interpretation of the resulting ORs is then not straightforward. When it is fitted though, the three extra indicator variables for the interaction term are not significant, and also, AIC = 806.12, showing the interaction term is not needed. In similar vein, if age-group is deleted from the model, giving us our first logistic regression model with only alcohol consumption as an explanatory variable, AIC = 897.06, and so age-group should be included in the model.

Finally we might like to try age as a continuous variable, by replacing age-group membership for each individual by the mid-point of the age-range of the age-group. In Table 7.19, the first and last categories are amalgamations of two categories, because of some very low numbers of cases or controls. We will revert to the original categories to include the Age:25-34 category and the Age:65-74 category, but exclude the last category, Age:75+. So the mid-points are, 30, 40, 50, 60 and 70. When fitted, AIC = 764.98. Transformations

**TABLE 7.22**

Logistic regression: oesophageal cancer study, using log(age)

|  | Estimate | Std. error | p-value | OR [95% CI] |
|---|---|---|---|---|
| Intercept | −17.48 | 1.929 | < 0.0001 | |
| log(age) | 3.92 | 0.475 | < 0.0001 | 50.21 [19.81, 127.3] |
| Alc:80+ | 1.67 | 0.192 | < 0.0001 | 5.22 [3.61, 7.66] |
| Deviance: 752.95, df=928, AIC = 758.95 | | | | |

of age were tried, $age^2$, $\sqrt{age}$ and log(age), the final one having the lowest AIC of 758.95. The fitted model coefficients, standard errors and odds ratios are shown in Table 7.22, where it can be seen that the odds ratio for alcohol consumption is again, not very different from that obtained from the other models. The odds ratio for log(age) is extremely high, and represents the OR for an increase in log(age) by unity. To put this in perspective, $\log(50) = 3.91$ and $\log(136) = 4.91$, and so we are comparing an age of 56 with an age of 136, as a unit increase in log(age). If we had fitted age in units of decades, i.e. $3, 4, \ldots, 7$, the estimated coefficient is 0.717, with standard error, 0.0856, giving OR = 2.05 and 95% confidence interval, $[1.173, 2.42]$. So, for every 10-years increase in age, the odds ratio is 2.05.

### 7.4.4 Matched case-control studies

We now look at matched case-control studies. For every case, there is at least one matched control, the matching done on demographic and other appropriate variables. Again, logistic regression is a good way of analysing this type of data. We start with just one control for each case, with pairs of individuals – one case and one control.

First, let us build an unconditional logistic regression model, ignoring the fact that there has to be one case and one control in each pair. Let $Y_{ij} = 1$ if the $j$th individual in the $i$th pair is a case, and $Y_{ij} = 0$ if a control, for $j = 1, 2$. Let $x_{ij}$ be an indicator variable, with $x_{ij} = 1$ if the $j$th individual of the pair is exposed and $x_{ij} = 0$ if not exposed. For the $i$th pair, model $p_{ij} = \Pr(Y_{ij} = 1)$ as,

$$p_{ij} = \exp(\beta_{0i} + \beta_1 x_{ij})/\{1 + \exp(\beta_{0i} + \beta_1 x_{ij})\},$$

each pair having their own constant parameter, $\beta_{0i}$. The likelihood is,

$$L = \prod_{\substack{i=1 \\ j=1,2}}^{N} \left\{ \frac{\exp(\beta_{0i} + \beta_1 x_{ij})}{1 + \exp(\beta_{0i} + \beta_1 x_{ij})} \right\}^{y_{ij}} \left\{ \frac{1}{1 + \exp(\beta_{0i} + \beta_1 x_{ij})} \right\}^{(1-y_{ij})}.$$

There are $N + 1$ parameters to be estimated with $2N$ data points. That is usually too many for a sensible model. It would be even worse if each pair had its own $\beta_0$ parameter – $2N$ data values, $2N$ parameters, and the model would fit perfectly, but would be nonsense.

The unconditional logistic model is actually conditioned on the exposure variable, $x_{ij}$. But now we also condition on there being one case and one control. Tables 7.23–7.26 show the four possible $2 \times 2$ tables, where E is "Exposed" and $\bar{\text{E}}$ is "Not Exposed". Given all marginal totals are fixed, once one table entry has been set at either 0 or 1, the other three table entries are determined. Hence we have only $N$ data points.

| **TABLE 7.23** | | | **TABLE 7.24** | | | **TABLE 7.25** | | | **TABLE 7.26** | | |
|---|---|---|---|---|---|---|---|---|---|---|---|
| Table 1 | | | Table 2 | | | Table 3 | | | Table 4 | | |
| | Cas | Con | Tot | | Cas | Con | Tot | | Cas | Con | Tot | | Cas | Con | Tot |

| | Cas | Con | Tot |
|---|---|---|---|
| E | 1 | 1 | 2 |
| $\bar{\text{E}}$ | 0 | 0 | 0 |
| Tot | 1 | 1 | 2 |

| | Cas | Con | Tot |
|---|---|---|---|
| E | 1 | 0 | 1 |
| $\bar{\text{E}}$ | 0 | 1 | 1 |
| Tot | 1 | 1 | 2 |

| | Cas | Con | Tot |
|---|---|---|---|
| E | 0 | 1 | 1 |
| $\bar{\text{E}}$ | 1 | o | 1 |
| Tot | 1 | 1 | 2 |

| | Cas | Con | Tot |
|---|---|---|---|
| E | 0 | 0 | 0 |
| $\bar{\text{E}}$ | 1 | 1 | 2 |
| Tot | 1 | 1 | 2 |

For one case and one control, we have,

$$\Pr(Y_{i1} = 1, Y_{i2} = 0) = p_{10} = \frac{\exp(\beta_{0i} + \beta_1 x_{i1})}{1 + \exp(\beta_{0i} + \beta_1 x_{i1})} \times \frac{1}{1 + \exp(\beta_{0i} + \beta_1 x_{i2})},$$

$$\Pr(Y_{i1} = 0, Y_{i2} = 1) = p_{01} = \frac{1}{1 + \exp(\beta_{0i} + \beta_1 x_{i1})} \times \frac{\exp(\beta_{0i} + \beta_1 x_{i2})}{1 + \exp(\beta_{0i} + \beta_1 x_{i2})}.$$

Then $\Pr(Y_{i1} = 1, Y_{i2} = 0 | Y_{i1} + Y_{i2} = 1) = p_{10}/(p_{10} + p_{01})$, and $\Pr(Y_{i1} = 0, Y_{i2} = 1 | Y_{i1} + Y_{i2} = 1) = p_{01}/(p_{10} + p_{01})$, and so after a little algebra,

$$\Pr(Y_{i1} = 1, Y_{i2} = 0 | Y_{i1} + Y_{i2} = 1) = \frac{\exp(\beta_1 x_{i1})}{\exp(\beta_1 x_{i1}) + \exp(\beta_1 x_{i2})},$$

$$\Pr(Y_{i1} = 0, Y_{i2} = 1 | Y_{i1} + Y_{i2} = 1) = \frac{\exp(\beta_1 x_{i2})}{\exp(\beta_1 x_{i1}) + \exp(\beta_1 x_{i2})},$$

and the $\beta_{0i}$'s have been conditioned out.

Now, without loss of generality, let the first individual be the case within each pair, and so the conditioned likelihood is,

$$L = \prod_{i=1}^{N} \left\{ \frac{\exp(\beta_1 x_{i1})}{\exp(\beta_1 x_{i1}) + \exp(\beta_1 x_{i2})} \right\}. \tag{7.3}$$

Differentiating the log-likelihood with respect to $\beta_1$ and equating to zero,

$$\sum_{i=1}^{N} x_{i1} - \sum_{i=1}^{N} \frac{x_{i1} \exp(\beta_1 x_{i1}) + x_{i2} \exp(\beta_1 x_{i2})}{\exp(\beta_1 x_{i1}) + \exp(\beta_1 x_{i2})} = 0.$$

Now, $x_{i1}$ and $x_{i2}$ only take the values 0 and 1. We can count the number of pairs that follow the patterns of the $2 \times 2$ tables in Tables 7.23, ..., 7.26. Let: $n_{11}$ be the number of pairs following the first table, i.e. both case and control are exposed; $n_{10}$ be the number of pairs following the second table, i.e. case exposed, control not; $n_{01}$ be the number of pairs following the third table, i.e.

case not exposed, control exposed; $n_{00}$ be the number of pairs following the fourth table, i.e. both case and control not exposed. Substituting 0/1 for the $x_{ij}$'s, the likelihood equation reduces to,

$$(n_{11} + n_{10}) - \left\{ n_{11} + \frac{(n_{10} + n_{01}) \exp(\hat{\beta}_1)}{1 + \exp(\hat{\beta}_1)} \right\} = 0,$$

and hence, $\exp(\hat{\beta}_1) = n_{10}/n_{01}$, and this is an estimate of the odds ratio, OR, for exposed and non-exposed individuals. It can be shown that the standard error of $\log(\text{OR})$ is $\sqrt{1/n_{10} + 1/n_{01}}$.

For a simpler approach, not resorting to conditional logistic regression, the McNemar chi-square test can be used to test the hypothesis, $H_0 : \text{OR} = 1$. The test statistic is $x^2 = (n_{10} - n_{01})^2/(n_{10} + n_{01})$, with $X^2 \sim \chi_1^2$ under $H_0$. The main point of using conditional logistic regression is that, as before, various baseline and risk factor variables can be used as we will see in the following example.

Equation 7.3 can be adjusted for 1:M matching.

### Oesophageal cancer example

A case-control study was carried out on oesophageal cancer in southern Thailand, by V. Chongsuvivatwong[3]. The question to be answered was whether the risk factors of *alcohol consumption, smoking, rubber processing* and *nutritional deficiency* were risk factors for oesophageal cancer. Forty cases were recruited from a hospital and each matched to a number of controls, ranging from one to six controls. There were 129 controls in total. A subset of the data can be accessed from the R package, epiDisplay and makes a good example to illustrate the theory behind a matched case-control study. Table 7.27 shows a summary of the data. Also included in the package is a subset of this subset, that has just one control for each case, i.e. 1-1 matching.

**TABLE 7.27**
Oesophageal case-control study summary data

|  |  | Smoking | | Rubber | | Alcohol | |
|---|---|---|---|---|---|---|---|
|  | No. | yes | no | yes | no | yes | no |
| Case | 26 | 21 | 5 | 8 | 18 | 17 | 9 |
| Control | 93 | 62 | 31 | 36 | 57 | 33 | 60 |

We start with the one-to-one subset of the data. Tables 7.28 – 7.30 show the number of pairs out of 26 that had both case and control as exposed individuals, the number where the case was exposed but the control was not, etc. for the three risk factors. The OR's were estimated using the simple estimator $n_{10}/n_{01}$ with s.e. $\sqrt{1/n_{10} + 1/n_{01}}$, together with the McNemar test statistic. The results were:

| **TABLE 7.28** Oesophageal cancer: smoking | | | | **TABLE 7.29** Oesophageal cancer: rubber | | | | **TABLE 7.30** Oesophageal cancer: alcohol | | | |
|---|---|---|---|---|---|---|---|---|---|---|---|
| | Controls | | Total | | Controls | | Total | | Controls | | Total |
| Cases | E | Ē | | Cases | E | Ē | | Cases | E | Ē | |
| E | 16 | 5 | 21 | E | 4 | 4 | 8 | E | 8 | 9 | 17 |
| Ē | 5 | 0 | 5 | Ē | 5 | 13 | 5 | Ē | 2 | 7 | 9 |
| Total | 21 | 5 | 26 | Total | 9 | 17 | 26 | Total | 10 | 16 | 26 |

| Risk factor | OR | 95% CI | $X^2$ | p |
|---|---|---|---|---|
| Smoking | 1.00 | [0.29, 3.45] | 0 | 1.0 |
| Rubber | 0.80 | [0.21, 2.98] | 0.11 | 0.739 |
| Alcohol | 4.50 | [0.97, 20.83] | 4.45 | 0.035, |

showing alcohol consumption appears to be related to oesophageal cancer. The
R function clogit() was also used to analyse the data using exact conditional
logistic regression for each of the three risk factors separately. The results
obtained were exactly the same as those above.

Now we turn to the full data set supplied, with multiple controls matched
to each case, and use clogit() to produce a multivariate model using all three
risk factors. The results are

| Risk factor | OR | 95% CI | p |
|---|---|---|---|
| Smoking | 1.55 | [0.44, 5.51] | 0.496 |
| Rubber | 0.63 | [0.18, 2.25] | 0.480 |
| Alcohol | 5.30 | [1.65, 17.00] | 0.005 |

Again, the only significant risk factor is alcohol.

## 7.5  Cross-sectional studies

Cross-sectional studies collect data at one time point. They are inexpensive
to run compared to cohort studies, can collect data on multiple outcomes and
exposures and can estimate prevalence. On the downside, incidence cannot be
measured and causality cannot be established, only association. You cannot
show whether a risk factor caused the cancer, or that having cancer caused
individuals to become more exposed to the risk factor. The latter is unlikely
in most cases.

To carry out a cross-sectional study, a sample of individuals needs to be
chosen from the population of interest. Data collected can come from questionnaires, interviews and medical examinations. Bias can be a problem because
of non-responders, participant recall of past events and selection bias.

## Measuring prevalence in cross-sectional studies

We will use data from a cross-sectional study that compared the prevalence of prostate cancer among Asian men compared to Caucasian men, Zlotta et al. [77] Prostates were collected from 220 Caucasian men from Russia, and 100 Asian men from Japan, all of whom had died from causes other than prostate cancer. Only men who had not been screened for prostate cancer were selected. An autopsy on each prostate established whether cancer was present, with other measurements also being made.

In general, let $X_i = 1$ if and individual has the cancer and zero otherwise and let $N$ be the total population size. Then the prevalence, P, of cancer is P $= \sum_{i=1}^{N} X_i/N = \bar{X}$. Now $X_i$ has a Bernoulli distribution (i.e. Bin(1, $p$)), and so $\sum_{i=1}^{N} X_i$ has a Binomial distribution, Bin($Np, Np(1-p)$). Thus P is estimated by $\bar{x}$ with approximate 95% confidence interval, $[\bar{x} - 1.96\sqrt{\bar{x}(1 - \bar{x})/N}, \bar{x} + 1.96\sqrt{\bar{x}(1 - \bar{x})/N}]$. Table 7.31 shows the numbers of prostate cancers found for the Caucasian and Asian men, together with the estimated prevalence and confidence interval.

**TABLE 7.31**
Prevalence of prostate cancer for Caucasian and Asian men

|  | N | No. cancers | Prevalence | 95%CI |
|---|---|---|---|---|
| Caucasian men | 220 | 82 | 0.37 | [0.31, 0.44] |
| Asian men | 100 | 35 | 0.35 | [0.26, 0.44] |

The prevalences are very similar. Forming a $2 \times 2$ table from the first three columns of Table 7.31, a chi-squared test gives, $X^2 = 0.153$, $p = 0.696$, giving no evidence for a difference in the prevalence rates. Treating the prevalences as probabilities, the odds ratio is 1.10.

The mean ages of the Caucasian and Asian men were 62.5 and 68.5 years respectively. We need to assess the affect of age on the probability of having prostate cancer. Table 7.32 shows the number of men without cancer, the number of men with cancer and the number of men with cancer, whose Gleason score (GS) was seven or above, for each 10-year age group from 21 to 90. (The Gleason score ranges from 0 to 10, based on the appearance of cancer cells.)

**TABLE 7.32**
Prevalence of prostate cancer for Caucasian and Asian men

| | Caucasian men | | | Asian men | | |
|---|---|---|---|---|---|---|
| Age Gp. | Non-cancer | Cancer | GS$\geq$ 7 | Non-cancer | Cancer | GS$\geq$ 7 |
| 21-30 | 2 | 0 | - | 1 | 0 | - |
| 31-40 | 9 | 1 | 0 | 5 | 0 | - |
| 41-50 | 22 | 9 | 1 | 3 | 0 | - |
| 51-60 | 32 | 13 | 4 | 11 | 1 | 0 |
| 61-70 | 29 | 25 | 4 | 18 | 8 | 2 |
| 71-80 | 44 | 34 | 10 | 20 | 16 | 9 |
| 81-90 | 0 | 0 | - | 7 | 10 | 7 |

**TABLE 7.33**

Logistic regressions of prostate cancer with group and age as regressors

| Model | AIC | coefficients | value | s.e. | p-value | OR |
|---|---|---|---|---|---|---|
| Group | 424.1 | Intercept | −0.619 | 0.210 | 0.003 | |
| | | Group | 0.099 | 0.252 | 0.696 | 1.10 |
| Group + Age | 405.4 | Intercept | −3.749 | 0.780 | <0.001 | |
| | | Group | 0.398 | 0.270 | 0.141 | 1.49 |
| | | Age | 0.449 | 0.106 | <0.001 | 1.57 |
| Group*Age | 403.6 | Intercept | −6.444 | 1.770 | <0.001 | |
| | | Group | 3.872 | 1.932 | 0.045 | |
| | | Age | 0.825 | 0.242 | <0.001 | |
| | | Age×Group | −0.498 | 0.270 | 0.066 | |

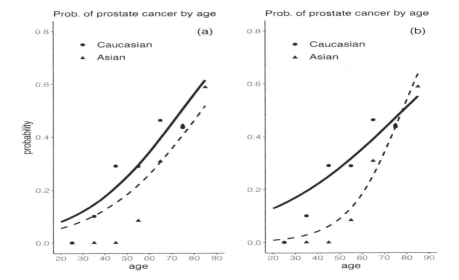

**FIGURE 7.10**

Crude prostate cancer probabilities by age, together with superimposed probabilities from logistic regression models: (a) group + age, (b) group + age + group×age

We carried out logistic regressions with cancer (yes/no) regressed on group, and then group and age with and without an interaction. Age was treated as continuous using the mid-points of the age intervals, divided by ten so we can work with decades. Table 7.33 shows the results. For the model with group alone, the results are the same as previously. Adding age to the model decreases the AIC significantly, with age a significant risk factor. When an interaction term is included, the AIC reduces again, and now both group and age are significant, and their interaction nearly significant. As before,

OR=exp(coefficient) and is only easily interpretable for additive regressors. Figure 7.10 (a) shows the crude cancer rates, calculated as the proportion of cancers in each age group for each ethnic group, plotted against by age. Superimposed are the fitted probability curves for each group, i.e

$$P(\text{cancer}) = \frac{\exp(\hat{\beta}_0 + \hat{\beta}_1\text{group} + \hat{\beta}_2\text{age}/10)}{\{1 + \exp(\hat{\beta}_0 + \hat{\beta}_1\text{group} + \hat{\beta}_2\text{age}/10)\}}, \quad \text{for group} = 0 \text{ and } 1,$$

noting we have divided age by 10 as we are working in decades.

We can see that the model does not fit well to lower age cancer rates for the Asian men. Figure 7.10 (b) shows the same but with an interaction term between group and age in the model. The fits are better now, but as we said, interpretation of odds ratios are awkward. Overall, we have found a good model but it would have been nice to have had more data; statisticians always say that!

Lastly, we look at the Gleason score, modelling the probability of a Gleason score of seven or more on age and ethnic group. We used models using: group; age; group + age; group + age + group×age, as before, resulting in AIC's of 141.3, 140.7, 139.3 and 139.4.

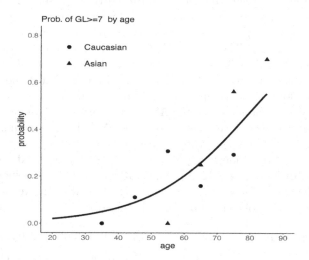

**FIGURE 7.11**
Crude probabilities of Gleason score of 7 or more by age, together with su-perimposed probabilities from a logistic regression model

Figure 7.11 shows the crude probabilities of of a Gleason score of seven or more, calculated as the proportion of cancer patients with GL ≥ 7 in each age group, for each ethnic group, plotted against by age. There are only nine data points, and in consideration of the AIC's, its sensible to use the model that just uses age as a regressor. For this model, age was significant with

$p = 0.005$. Superimposed on the plot is the fitted probability curve. Clearly, the probability of a Gleason score of seven or above (medium to high grade cancer) increases with age.

## 7.6   Spatial epidemiology

We end this chapter touching on the topic of spatial epidemiology that brings together spatial statistics and epidemiology. The book edited by Lawson[35] might be of interest to the reader. We will only give one very simple example.

It is said that there appears to be a link of low levels of vitamin D to prostate cancer. The best source of vitamin D is sun exposure. But, sun exposure increases the risk of melanoma skin cancer. So, possibly, low levels of sun exposure, means less vitamin D and a higher risk of prostate cancer, but lower risk of melanoma. Conversely, high levels of sun exposure, means a higher risk of melanoma, but lower risk of of prostate cancer. Does geography play a part in this?

To investigate further, the age-standardised rates (ASR) for prostate and melanoma skin cancers were obtained for nineteen European countries, the countries chosen to give a range of latitudes, recorded at the latitude of a country's capital city. The idea is that, as you move north away from the equator, sunlight is less intense, so vitamin D decreases, possibly giving rise to more prostate cancer, but less melanoma. Figure 7.12(a) shows prostate cancer ASR plotted against melanoma ASR for the countries. After looking at the scatterplot, the idea that the ASRs are inversely related is abandoned! As the prostate cancer ASR rises, so does the ASR for melanoma.

Figure 7.12(b) shows prostate cancer ASR plotted against latitude, together with a simple linear regression ($y = \beta_0 + \beta_1 x$), and similarly for melanoma ASR in Figure 7.12 (c). The ASR for both cancers increases with latitude. The simple linear regressions of prostate ASR and melanoma ASR on latitude had the following fitted coefficients (s.e):

$$\text{Prostate}: \ \hat{\beta}_0 = -35.0\,(49.2) \quad \hat{\beta}_1 = 2.11\,(0.95) \ p = 0.040$$

$$\text{Melanoma}: \ \hat{\beta}_0 = -3.43\,(6.17) \quad \hat{\beta}_1 = 0.26\,(0.19), \ p = 0.040.$$

The explanation for this dependence on latitude? I do not have one.

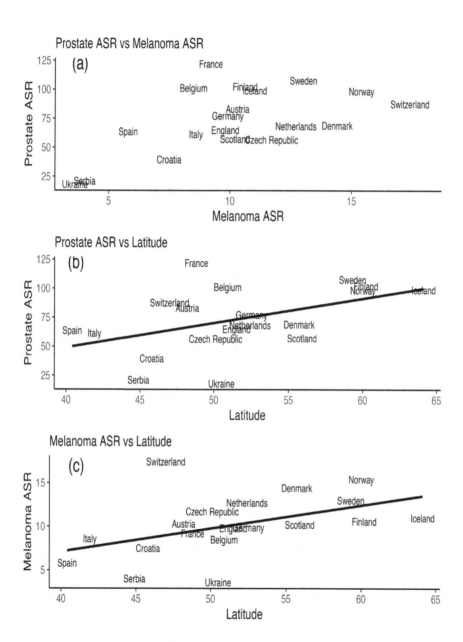

**FIGURE 7.12**
Prostate ASR plotted against melanoma ASR; prostate and melanoma ASR plotted against latitude

# 8

## Meta-Analysis

It is good news when a clinical trial has a positive outcome, perhaps showing a new chemotherapy treatment for a certain cancer has increased survival compared to the current recommended treatment. However, how do we know that the trial result is not a Type I error, (i.e. rejecting the null hypothesis of no treatment difference in favour of the new drug, when there is actually no treatment difference), and that there is, in reality, no advantage in using the new drug over the current treatment in terms of survival. If the new treatment had worse side effects than the current treatment, then patients would be worse off having the new treatment. We have taken a backward step in the treatment of this particular cancer. Before rushing to change practice, it would have been better to have had at least one more trial using the same treatments, perhaps carried out in other countries, in order to help confirm whether or not the new treatment is indeed superior to the current treatment.

Occasionally, a trial gives overwhelmingly results and there needs to be serious consideration of how current practice might be affected. At the time of writing, AstraZeneca recently gave the press release regarding a Phase III trial of Tagrisso versus placebo for non-small cell lung cancer,

"The ADAURA Phase III trial for Tagrisso (osimertinib) in the adjuvant treatment of patients with Stage IB, II and IIIA epidermal growth factor receptor-mutated (EGFRm) non-small cell lung cancer (NSCLC) with complete tumour resection will be unblinded early following a recommendation from an Independent Data Monitoring Committee (IDMC) based on its determination of overwhelming efficacy".

So, ethically, there can never be another trial for this population of patients, using a placebo again.

When there is not this overwhelming efficacy, but there are two or more trials, each comparing the same pair of treatments, then the results from the individual trials can be amalgamated to give an overall assessment of the comparison of the treatments. This amalgamation of trial results sounds straightforward, but there are always aspects that need to be taken into account, for instance what if two trials compare a new treatment, drug A, with drug B and two other trials compare the new treatment, drug A, with drug C, where both drug B and drug C are acceptable as current practice?

In the hierarchy of evidence of efficacy, as discussed in Chapter 4, the best method to assess efficacy or other outcome, is to carry out a systematic review

DOI: 10.1201/9781003041931-8

which synthesises information from several trials, often carrying out a *meta-analysis*. Simply put, a meta-analysis is a statistical analysis that combines the results of two or more trials, or other studies, that address the same medical question. For example, the mean of several estimates of a hazard ratio will more accurately estimate the population hazard ratio, than any single estimate.

How do you carry out a systematic review? Where do you start? This chapter will hopefully give you a good starting point. It may be that there are only two trials that compare a certain treatment A with a certain treatment B, using the same population of patients, with results published in the same journal. Then carrying out a systematic review and meta-analysis of the results will be straightforward. In general this will not be the case and there will be many potential publications that are candidates for inclusion in the review. Systematic reviews must be comprehensive, unbiased and repeatable, so that anybody else conducting the same review will arrive at the same conclusions. An important point to remember if you are going to undertake a systematic review, is not to underestimate the amount of work and time involved.

## 8.1   How to carry out a systematic review

The starting point is to establish the review question to be answered. This has to be clear and such that it is worth the investment in time and money to find an answer to the question. Examples of some systematic reviews that have answered questions involving cancer are:

1. "Surgical resection and whole brain radiation therapy versus whole brain radiation therapy alone for single brain metastases" (24 January 2005)

2. "Single agent versus combination chemotherapy for metastatic breast cancer" (15 April 2009)

3. "Chemotherapy for advanced gastric cancer" (29 August 2017)

These are all systematic reviews published in the Cochrane Database of Systematic Reviews (CDSR) on the dates shown in parentheses. The Cochrane Database of Systematic Reviews is one of the databases in the Cochrane Library (https://www.cochranelibrary.com/). Each Cochrane review is a peer-reviewed systematic review that has been prepared according to the Cochrane Handbook for Systematic Reviews of Interventions or the Cochrane Handbook for Diagnostic Test Accuracy Reviews. This website is a good place to start before attempting a systematic review and meta-analysis. It has a lot of advice, downloadable material and a training hub. The Cochrane Handbook for Systematic Reviews of Interventions can be accessed at https://training.cochrane.org/handbook after registering with the Cochrane Library. A systematic review does not have to be developed with Cochrane and published by Cochrane, but it is advisable to follow their advice for any systematic review undertaken.

There are implied questions in the titles of the three Cochrane reviews above. The first is the most precise in the sense that it is a simple treatment A versus treatment B question, "surgical resection and whole brain radiation therapy" versus "whole brain radiation therapy". The second and third review questions are clear, but not so precise. The "single agent" of question two will cover all single drug treatments for metastatic breast cancer, and "combination chemotherapy" will cover all drugs used in combination. For question three "chemotherapy" will cover all chemotherapy drugs used alone or in combination for advanced gastric cancer.

Before any work on a systematic review starts, a systematic review team has to be established. There needs to be at least one expert in the medical field and at least one methodology expert. It is a good idea to involve a PPI group (Patient and Public Involvement). The median number of authors for a Cochrane review is five and Cochrane will not publish a single authored systematic review. To emphasise the amount of work involved in carrying out a systematic review, the average time spent on a review for Cochrane is 30 months.

Arriving at the review question will involve decisions on,

- **P**articipants: what is the population? Would you use all patients with a certain cancer, or will age, tumour stage and other factors be taken into account?

- Interventions: what is the intervention? Is it just two chemotherapy drugs, or a subgroup of drugs, or any chemotherapy for the cancer of interest?

- **C**omparisons: what comparisons are going to be made? Is it just median survival times for drug A versus drug B, or will adverse events be compared too?

- **O**utcomes: what are the outcomes to be considered? Survival? Time to recurrence? Quality of life?

These four items are the Cochrane PICO components needed in deciding on the question and the planning of your review. For a Cochrane review, the review question is registered in their library.

The next task is to write a protocol. This is essential as it describes every step of the execution of the review. The protocol should cover:

- *Background information*: description of the disease and the intervention(s)

- *Objectives*: objectives of the systematic review

- *Inclusion/Exclusion criteria*: what studies will be included/excluded from the review. Criteria might include: date – e.g. studies from the date radical new treatments appeared; stage – e.g. stages 1–3 of the cancer of interest; location – e.g. Europe; age – e.g. patients between 18 and 65 years of age; peer review – e.g. all included studies to have been peer reviewed; reported outcomes – e.g. studies must report median survival time; study design – e.g. must be a randomised controlled trial.

- *Search methods*: how will reported studies be found? Databases that can be searched include: MEDLINE via Pubmed, Cochrane Library, EMBASE and the Web-of-Science. A comprehensive search strategy has to be established so that candidate publications for the review can be identified. For example, if you type "esophageal cancer" into the Pubmed search engine, 64,470 publications are identified, which would be an enormous amount to wade through. If you type "oesophageal cancer" (the English spelling) into the search engine, the same 64,470 publications are identified. If you type "(oesophageal cancer) AND trial", 4,725 publications are identified. You can combine "(oesophageal cancer) and trial" by using the advanced search builder, entering the two terms into separate lines and using the boolean operator "AND". The other operators are "OR" and "NOT". For example, building "((oesophageal cancer) AND trial) NOT chemotherapy", produced 1993 publications. You can restrict your search to author, title, year, or by one of the many other filters available.

- *Methods*: describe what information will be taken from each publication, e.g. sample sizes, median survival with standard errors, as a minimum. Describe how data will be stored. Describe how data will be double checked. Describe how selected publications will be assessed for bias, for example: randomisation bias, blinding of outcome measures, incomplete reporting – e.g. no standard error reported, selective reporting of outcomes.

- *Analysis and reporting*: describe how the data gleaned from the publications will be analysed and reported; e.g. tables, forest plots, assessment of heterogeneity, pooling of data via meta-analysis, funnel plot for publication bias, sensitivity analyses. (These will be described later.)

The finished protocol is then registered with Cochrane or other organisation.

### 8.1.1   An oesophageal cancer review

We are not going to carry out an original systematic review as an example, but use one from the Cochrane Library, entitled,

"Preoperative chemotherapy for resectable thoracic oesophageal cancer" [31],

You can look at this review online for full details.

The title is reasonably clear, from which we can infer the research question, which is, in layman's terms, is it best to have chemotherapy first before surgery (resection) for (localised) oesophageal cancer, or go straight for surgery? Does the chemotherapy increase survival rates? The word "localised" cannot be gleaned from the title, but surgery is only possible in the early stages of the disease.

The authors' stated objective is to determine the role of preoperative chemotherapy in this group of patients, which is wider than just looking at survival rates. They searched the databases Cochrane Central Register of Controlled Trials (CENTRAL), MEDLINE, EMBASE, CANCERLIT plus other resources. Their selection criteria were all trials of patients with potentially resectable carcinoma of the oesophagus, of any histologic type, who were randomly assigned to chemotherapy or no chemotherapy before surgery. Their primary outcome was survival, assessed by hazard ratios (HR). Secondary outcomes were rates of resection, response to chemotherapy, plus others. Teamwork was used to assess titles and abstracts, to choose studies for full review for potential inclusion and to extract data.

The quality of each trial was assessed using the Jadad method[25], which is a three-point questionnaire, with questions, "Was the study described as randomised?", "Was the study described as double blind?" and "Was there a description of withdrawals and dropouts?" Each answer "yes" scores 1, each answer "no" scores 0, with bonus points added/subtracted if the methods of randomisation and blinding are described and are appropriate/inappropriate. So a trial can score between 0 and 5. They also used the Cochrane tool, RoB 2, that gives a framework for assessing risk of bias within a trial, with domains: (1) bias arising from the randomisation process; (2) bias due to deviations from intended interventions; (3) bias due to missing outcome data; (4) bias in measurement of the outcome; (5) bias in selection of the reported result. It does not produce a score but a table with "low risk of bias", "unclear risk of bias" and "high risk of bias".

Searching the databases identified 371 records. After screening, 13 studies were left for inclusion in a qualitative and quantitative synthesis analysis. We will not summarise the qualitative analysis but concentrate on the quantitative analysis – the meta-analysis.

We now look at some statistical methodology for meta-analyses.

## 8.2   Fixed effects model

Suppose our literature search has identified $n$ studies of treatment A versus treatment B. Let $\theta$ be the outcome of interest, for example, the log-hazard ratio, the difference in treatment means or the log-odds ratio. These have normal or approximately normal distributions and here, we only consider outcomes with a normal, or approximately normal distribution.

Let $\hat{\theta}_i$ be the estimate of $\theta$ for the $i$th study. The meta-analysis fixed effects model is,

$$\hat{\theta}_i = \theta + \epsilon_i,$$

where $\epsilon_i$ has a normal distribution, $N(0, \sigma_i^2)$. Thus $\hat{\theta}_i \sim N(\theta, \sigma_i^2)$, $(i = 1, \ldots, n)$ and the studies are assumed to be pairwise independent. Note that we are assuming the studies have a common mean, $\theta$, but we do allow their variances, $\sigma_i^2$, to be different. This allows for differences in the variation encountered in the individual studies. For example, there might be ethnic, or age differences in the patients from hospitals, where each hospital has run its own study.

The likelihood for $\theta$ is,

$$\text{L} = \prod_{i=1}^{n} \frac{1}{\sqrt{2\pi}\sigma_i} \exp\{-\tfrac{1}{2}(\hat{\theta}_i - \theta)^2/\sigma_i^2\},$$

and the log-likelihood,

$$l = -\tfrac{1}{2}\log(2\pi) - \tfrac{1}{2}\log(\sigma_i^2) - \tfrac{1}{2}\sum_{i=1}^{n}(\hat{\theta}_i - \theta)^2/\sigma_i^2.$$

Differentiating the log-likelihood with respect to $\theta$ and equating to zero gives the likelihood equation for $\hat{\theta}$ as

$$\sum_{i=1}^{n} \hat{\theta}_i/\sigma_i^2 - \theta \sum_{i=1}^{n} 1/\sigma_i^2 = 0,$$

and hence,

$$\hat{\theta} = \frac{\sum_{i=1}^{n} w_i \hat{\theta}_i}{\sum_{i=1}^{n} w_i}, \tag{8.1}$$

where $w_i = 1/\sigma_i^2$.

Thus our estimate, $\hat{\theta}$, of $\theta$, is a weighted sum of the individual study estimates, $\hat{\theta}_i$, with weights proportional to their inverse variances.

The variance of $\hat{\theta}$ is easily seen to be,

$$\text{var}(\hat{\theta}) = \frac{\sum_{i=1}^{n} w_i^2 \sigma_i^2}{\left(\sum_{i=1}^{n} w_i\right)^2} = \frac{1}{\left(\sum_{i=1}^{n} w_i\right)}.$$

We do not know the population values, $\sigma_i^2$, and so they are estimated by the sample variances in the original studies, $\hat{\sigma}_i^2$, but, we leave the "hats" off and write these as $\sigma_i^2$, otherwise we will have too many hats in our meta-analysis equations, causing confusion. So when you see $\sigma_i^2$ in an equation, it is really $\hat{\sigma}_i^2$. Well – most of the time!

We could have used weights other than the inverse variances in the estimate $\hat{\theta}$, for instance we might wish to adjust these weights to allow for age of the study, placing more weight on recent studies and less on older studies – but there has to be an objective reason for doing this.

**Example**

Of the thirteen studies in the oesophageal cancer review, three did not provide enough information on overall survival to be included in a meta-analysis. One did not include any overall survival data, one did not provide enough information on the chemotherapy, surgery and survival, and one did not provide enough detail to calculate the hazard ratio. The ten studies that could be used are labelled as: Roth 1988; Nygaard 1992; Schlag 1992a; Maipang 1994; Law 1997; Kelsen 1998; Baba 2000; Wang 2001; Ancona 2001; MRC Allum 2009; Stilidi 2006; Boonstra 2011; and Ychou 2011.

The authors of the systematic review extracted relevant statistics from the papers and then for some, transformed them to obtain the outcome of interest, log-hazard ratio, LHR. For example, for the Boostra paper the extracted hazard ratio (HR) and 95% confidence interval was 0.86 $[0.75, 0.99]$. These were transformed to LHR $= \log(\text{HR}) = \log(0.86) = -0.151$, with standard error, s.e.(LHR) $= (\log(0.99) - \log(0.75))/(2 \times 1.96) = 0.071$.

The extracted LHR and s.e.(LHR) values for all the studies are:

| Study | Ancona | Boos. | Kels. | Law | Maip. | MRC-A. | Nyga. | Roth | Achl. | Ychou |
|---|---|---|---|---|---|---|---|---|---|---|
| LHR | −0.163 | −0.151 | 0.068 | −0.462 | 0.182 | −0.174 | 0.077 | −0.371 | 0.174 | −0.163 |
| s.e.(LHR) | 0.256 | 0.072 | 0.106 | 0.167 | 0.481 | 0.081 | 0.206 | 0.392 | 0.321 | 0.071 |

The authors of the Cochrane review used Cochrane's REVMAN for their analyses, but we will use R and the libraries metafor[72] and netmeta for our meta-analyses.

Figure 8.1 shows the thirteen studies in a *forest plot*. The plot shows the studies, the numbers of patients in each arm, the hazard ratios $\hat{\theta}_i$, with 95% confidence intervals, the weights given to each study and a plot of the hazard ratios, with horizontal lines indicating 95% confidence intervals. The size of the square indicating the position of the HR of a study on the x-axis is proportional to the weight given to the study. A vertical dotted line is plotted at HR $= 1$ and points to the left of it favour preoperative chemotherapy and points to the right favour immediate surgery.

From Equation 8.1,

$$\hat{\theta} = -0.136 \text{ and s.e.}(\hat{\theta}) = \sqrt{0.001374} = 0.0371.$$

Exponentiating, the combined estimate of the hazard ratio and 95% confidence interval is,

$$\text{HR} = \exp(-0.136) = 0.87, \text{ CI: } \exp(-0.136 \pm 1.96 \times 0.0371), \text{ i.e. } [0.81, 0.94].$$

The z-value for testing $H_0 : \theta = 1$ is $z = -0.136/0.0371 = -3.67$ with $p < 0.001$, and hence the evidence suggests that preoperative chemotherapy is a benefit for survival. The combined estimate is shown on the forest plot as a horizontally stretched diamond, the centre being placed at the value of HR and the left and right hand ends at the lower and upper points of the

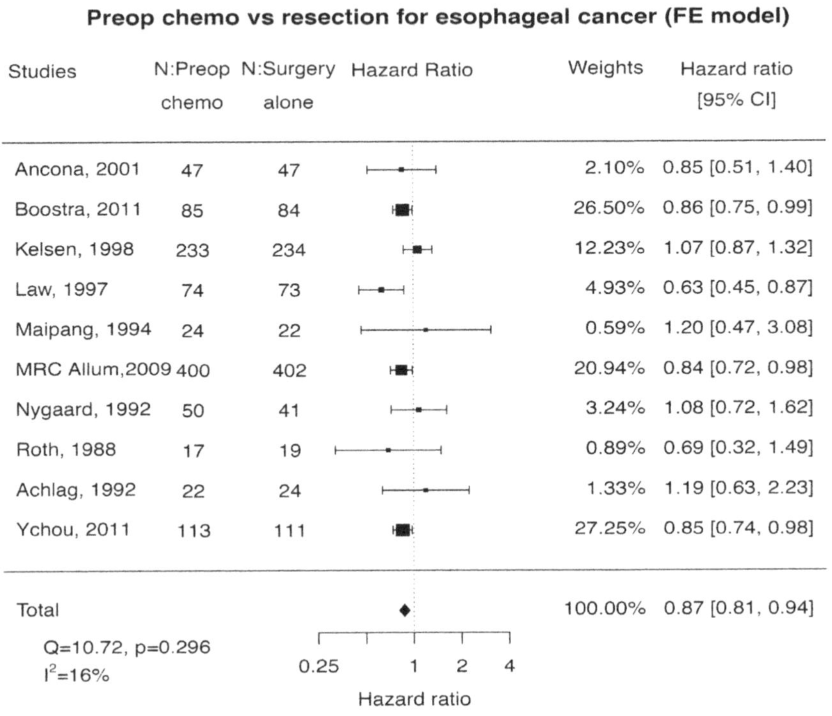

**Preop chemo vs resection for esophageal cancer (FE model)**

| Studies | N:Preop chemo | N:Surgery alone | Hazard Ratio | Weights | Hazard ratio [95% CI] |
|---------|---------------|-----------------|--------------|---------|-----------------------|
| Ancona, 2001 | 47 | 47 | | 2.10% | 0.85 [0.51, 1.40] |
| Boostra, 2011 | 85 | 84 | | 26.50% | 0.86 [0.75, 0.99] |
| Kelsen, 1998 | 233 | 234 | | 12.23% | 1.07 [0.87, 1.32] |
| Law, 1997 | 74 | 73 | | 4.93% | 0.63 [0.45, 0.87] |
| Maipang, 1994 | 24 | 22 | | 0.59% | 1.20 [0.47, 3.08] |
| MRC Allum,2009 | 400 | 402 | | 20.94% | 0.84 [0.72, 0.98] |
| Nygaard, 1992 | 50 | 41 | | 3.24% | 1.08 [0.72, 1.62] |
| Roth, 1988 | 17 | 19 | | 0.89% | 0.69 [0.32, 1.49] |
| Achlag, 1992 | 22 | 24 | | 1.33% | 1.19 [0.63, 2.23] |
| Ychou, 2011 | 113 | 111 | | 27.25% | 0.85 [0.74, 0.98] |
| Total | | | | 100.00% | 0.87 [0.81, 0.94] |

Q=10.72, p=0.296
$I^2$=16%

0.25    1    2    4
Hazard ratio

**FIGURE 8.1**
Forest plot for preoperative chemotherapy and resection versus resection alone for thoracic oesophageal cancer, using a fixed effects model

confidence interval. Here, the confidence interval is rather short and so the diamond here is not particularly stretched.

## Cochrane's measure of heterogeneity

A question we might ask ourselves, is whether the studies in our meta-analysis are homogeneous, each with the same population mean, $\theta$, or alternatively, whether the studies are heterogeneous with differing population means. Cochrane's measure of heterogeneity is calculated as,

$$Q = \sum_{i=1}^{n} w_i(\hat{\theta}_i - \hat{\theta})^2, \tag{8.2}$$

which has an approximate chi-square distribution with $n-1$ degrees of freedom under the hypothesis of no heterogeneity. As the p-values for Q depend on the number of studies, $n$, the $I^2$ statistic can be used instead, which is robust against the size of $n$,

$$I^2 = \frac{Q - (n-1)}{Q}, \text{ if } Q > (n-1) \text{ or } I^2 = 0 \text{ if } Q \le (n-1).$$

Large values of $I^2$ signify heterogeneity, with a suggested scale of: $I^2$ close to 0, no heterogeneity; $I^2 = 25\%$, low heterogeneity; $I^2 = 50\%$, moderate heterogeneity; and $I^2 = 75\%$, high heterogeneity.

For our data, $Q = 10.72$, with $p = 0.296$ and $I^2 = 16\%$, suggesting no, or only a small amount of heterogeneity. These values have been placed at the bottom of the forest plot.

## 8.3   Random effects model

If the estimates, $\hat{\theta}_i$ are more variable than that predicted by the fixed effect model, a meta-analysis random effects model can be used to account for the extra variation. The random effects model is,

$$\hat{\theta}_i = \theta + u_i + \epsilon_i,$$

where $u_i$ is a normal random variable with mean 0 and variance $\tau^2$ and as before, $\epsilon_i$ has a normal distribution, $N(0, \sigma_i^2)$. We also assume that $u_i$ and $\epsilon_i$ are independent. Thus the $u_i$'s are a random sample from a $N(0, \tau^2)$ distribution and this models the extra variation in the mean. The mean and variance of $\theta_i$ are easily seen to be,

$$E(\hat{\theta}_i) = \theta + E(u_i) + E(\epsilon_i) = \theta$$
$$\text{var}(\hat{\theta}_i) = \text{var}(u_i) + \text{var}(\epsilon_i) = \tau^2 + \sigma_i^2,$$

and hence, $\hat{\theta}_i \sim N(\theta, \tau^2 + \sigma_i^2)$. The likelihood, L, and log-likelihood, $l$, are

$$L = \prod_{i=1}^{n} \frac{1}{\sqrt{2\pi(\tau^2 + \sigma_i^2)}} \exp\{-\tfrac{1}{2}(\hat{\theta}_i - \theta)^2/(\tau^2 + \sigma_i^2)\}$$

$$l = -\tfrac{n}{2}\log(2\pi) - \tfrac{1}{2}\sum_{i=1}^{n}\log(\tau^2 + \sigma_i^2) - \tfrac{1}{2}\sum_{i=1}^{n}\{(\hat{\theta}_i - \theta)^2/(\tau^2 + \sigma_i^2)\}.$$

The maximum likelihood estimate of $\theta$ is the same as in Equation 8.1, except that the weights now include $\tau^2$, and so

$$\hat{\theta} = \frac{\sum_{i=1}^{n} w_i^* \hat{\theta}_i}{\sum_{i=1}^{n} w_i^*}, \text{ with } w_i^* = 1/(\hat{\tau}^2 + \sigma_i^2). \tag{8.3}$$

As before, $\sigma_i^2$ is estimated by $\hat{\sigma}_i^2$ from the $i$th original study, leaving $\tau^2$ to be estimated. The likelihood equation for $\tau^2$ (when differentiating the log-likelihood, it is easier to think of the parameter to be estimated as $\tau^2$, rather than $\tau$) is

$$\sum_{i=1}^{n} \frac{1}{\hat{\tau}^2 + \sigma_i^2} - \sum_{i=1}^{n} \frac{(\hat{\theta}_i - \hat{\theta})^2}{(\hat{\tau}^2 + \sigma_i^2)^2} = 0.$$

Writing this equation as

$$\sum_{i=1}^{n} \frac{\hat{\tau}^2 + \sigma_i^2}{(\hat{\tau}^2 + \sigma_i^2)^2} = \sum_{i=1}^{n} \frac{(\hat{\theta}_i - \hat{\theta})^2}{(\hat{\tau}^2 + \sigma_i^2)^2},$$

and then re-arranging,

$$\hat{\tau}^2 = \sum_{i=1}^{n} w_i^{*2} \{ (\hat{\theta}_i - \theta)^2 - \sigma_i^2 \} / \sum_{i=1}^{n} w_i^{*2}. \tag{8.4}$$

Equations 8.3 and 8.4 can be used to find $\hat{\theta}$ and $\hat{\tau}^2$ iteratively, using a starting value for $\hat{\tau}^2$ to find $\hat{\theta}$, then use this to find an updated $\hat{\tau}^2$, an so on. Alternatively, the likelihood or log-likelihood could be maximised directly using numerical methods. Or a REML (REstricted Maximum Likelihood) estimate could be used.

However, another approach is to use a simple estimate of $\tau^2$, the most commonly used being the *DerSimonian and Laird* estimate, based on the expected value of $Q$ for the fixed effects model in Equation 8.2. First, we find the expected value of $Q$,

$$\mathrm{E}[Q] = E[\textstyle\sum_i w_i(\hat{\theta}_i - \hat{\theta})^2] = \textstyle\sum_i w_i \mathrm{E}\left[ \left( \hat{\theta}_i - \frac{\sum_j w_j \hat{\theta}_j}{\sum_j w_j} \right) \left( \hat{\theta}_i - \frac{\sum_k w_k \hat{\theta}_k}{\sum_k w_k} \right) \right].$$

Notice we have split $(\hat{\theta}_i - \hat{\theta})^2$ into two brackets, one using sums over $j$ and one using sums over $k$. Remember that $\sum_i w_i = \sum_j w_j = \sum_k w_k$, the $i$, $j$ and $k$ being arbitrary labels. Continuing,

$$\mathrm{E}[Q] = \textstyle\sum_i w_i E\left[ \left( \frac{\sum_j w_j(\hat{\theta}_i - \hat{\theta}_j)}{\sum_j w_j} \right) \left( \frac{\sum_k w_k(\hat{\theta}_i - \hat{\theta}_k)}{\sum_k w_k} \right) \right]$$

$$= \textstyle\sum_i w_i E\left[ \left( \frac{\sum_j w_j\{(\hat{\theta}_i - \theta) - (\hat{\theta}_j - \theta)\}}{\sum_j w_j} \right) \left( \frac{\sum_k w_k\{(\hat{\theta}_i - \theta) - (\hat{\theta}_k - \theta)\}}{\sum_k w_k} \right) \right]$$

$$= \frac{1}{(\sum_i w_i)^2} \textstyle\sum_i \sum_j \sum_k w_i w_j w_k \mathrm{E}[\{(\hat{\theta}_i - \theta) - (\hat{\theta}_j - \theta)\}\{(\hat{\theta}_i - \theta) - (\hat{\theta}_k - \theta)\}]$$

$$= \frac{1}{(\sum_i w_i)^2} \textstyle\sum_i \sum_j \sum_k w_i w_j w_k \mathrm{E}[\{(\hat{\theta}_i - \theta)^2 - (\hat{\theta}_i - \theta)(\hat{\theta}_j - \theta)$$

$$- (\hat{\theta}_i - \theta)(\hat{\theta}_k - \theta) + (\hat{\theta}_j - \theta)(\hat{\theta}_k - \theta)\}].$$

Now, under the random effects model,

$$E[(\hat{\theta}_i - \theta)^2] = \tau^2 + \sigma_i^2,$$
$$E[(\hat{\theta}_i - \theta)(\hat{\theta}_j - \theta)] = \tau^2 + \sigma_i^2 \text{ if } j = i, \text{ and 0 otherwise,}$$
$$E[(\hat{\theta}_i - \theta)(\hat{\theta}_k - \theta)] = \tau^2 + \sigma_i^2 \text{ if } k = i, \text{ and 0 otherwise,}$$
$$E[(\hat{\theta}_j - \theta)(\hat{\theta}_k - \theta)] = \tau^2 + \sigma_j^2 \text{ if } k = j, \text{ and 0 otherwise.}$$

Substitute these four quantities into the above equation to obtain

$$E[Q] = \frac{1}{(\sum_i w_i)^2} \left[ \sum_i \sum_j \sum_k w_i w_j w_k (\tau^2 + \sigma_i^2) - \sum_i \sum_k w_i^2 w_k (\tau^2 + \sigma_i^2) \right.$$
$$\left. - \sum_i \sum_j w_i^2 w_j (\tau^2 + \sigma_i^2) + \sum_i \sum_j w_i w_j^2 (\tau^2 + \sigma_j^2) \right].$$

Collecting the values of $\tau^2$ together in the above equation, these are

$$\frac{1}{(\sum_i w_i)^2} \left\{ \left( \sum_i w_i \right)^3 - 2 \sum_i w_i^2 \sum_j w_j + \sum_i w_i \sum_j w_j^2 \right\} \tau^2$$
$$= \left( \sum_i w_i - \sum_i w_i^2 / \sum_i w_i \right) \tau^2,$$

and those for $\sigma_i^2$,

$$\frac{1}{(\sum_i w_i)^2} \left\{ \left( \sum_i w_i \right)^2 \sum_i w_i \sigma_i^2 - 2 \sum_i w_i \sum_i w_i^2 \sigma_i^2 + \sum_i w_i \sum_i w_i^2 \sigma_i^2 \right\}$$
$$= \frac{1}{(\sum_i w_i)^2} \left\{ \left( \sum_i w_i \right)^2 \sum_i 1 - 2 \left( \sum_i w_i \right)^2 + \left( \sum_i w_i \right)^2 \right\} = n - 1,$$

where $\sigma_i^2$ has been replaced by $1/w_i$ and indices $j$ and $k$ have been relabelled $i$ as appropriate.

Hence

$$E[Q] = \left( \sum_i w_i - \sum_i w_i^2 / \sum_i w_i \right) \tau^2 + n - 1,$$

and upon rearranging

$$\tau^2 = \frac{E[Q] - (n-1)}{\sum_i w_i - \sum_i w_i^2 / \sum_i w_i}.$$

To obtain $E[Q]$ we had to work with the population variances, $\sigma_i^2$. Replace these by the usual $\hat{\sigma}_i^2$ and replace $E[Q]$ by the observed value of $Q$ to obtain the estimate $\hat{\tau}_{DL}^2$ of $\tau^2$ as

$$\hat{\tau}_{DL}^2 = \max \left( 0, \frac{Q - (n-1)}{\sum_i w_i - \sum_i w_i^2 / \sum_i w_i} \right),$$

noting we cannot let the estimate be negative which will happen if $Q < n - 1$. If $Q < n - 1$, just put $\hat{\tau}_{DL}^2 = 0$, and so the model reverts to the fixed effect model.

Going back to our estimate, $\hat{\theta}$, in equation 8.3, the DerSimonian and Laird estimate is

$$\hat{\theta}_{DL} = \frac{\sum_{i=1}^{n} w_i^* \hat{\theta}_i}{\sum_{i=1}^{n} w_i^*}, \text{ with } w_i^* = 1/(\hat{\tau}_{DL}^2 + s_i^2),$$

The variance of $\hat{\theta}_{DL}$ is $1/\sum_i w_i^*$.

That is the end of the heavy algebra, the like of which we will not be seeing again in this text, but it is good practice to occasionally bury oneself in such algebraic exercises – it can be enlightening.

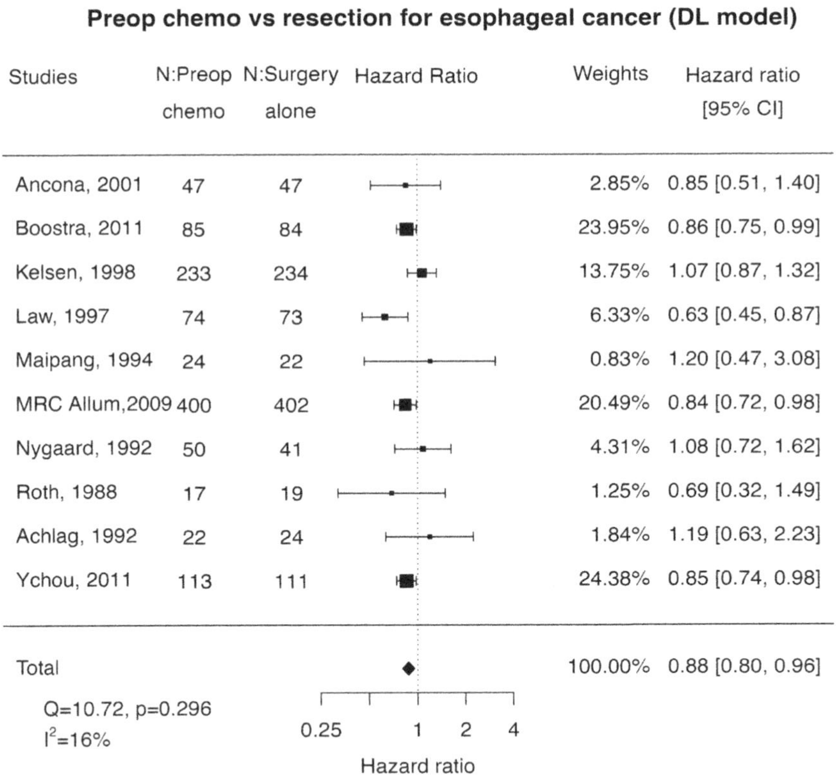

## Preop chemo vs resection for esophageal cancer (DL model)

| Studies | N:Preop chemo | N:Surgery alone | Hazard Ratio | Weights | Hazard ratio [95% CI] |
|---------|---------------|-----------------|--------------|---------|----------------------|
| Ancona, 2001 | 47 | 47 | | 2.85% | 0.85 [0.51, 1.40] |
| Boostra, 2011 | 85 | 84 | | 23.95% | 0.86 [0.75, 0.99] |
| Kelsen, 1998 | 233 | 234 | | 13.75% | 1.07 [0.87, 1.32] |
| Law, 1997 | 74 | 73 | | 6.33% | 0.63 [0.45, 0.87] |
| Maipang, 1994 | 24 | 22 | | 0.83% | 1.20 [0.47, 3.08] |
| MRC Allum,2009 | 400 | 402 | | 20.49% | 0.84 [0.72, 0.98] |
| Nygaard, 1992 | 50 | 41 | | 4.31% | 1.08 [0.72, 1.62] |
| Roth, 1988 | 17 | 19 | | 1.25% | 0.69 [0.32, 1.49] |
| Achlag, 1992 | 22 | 24 | | 1.84% | 1.19 [0.63, 2.23] |
| Ychou, 2011 | 113 | 111 | | 24.38% | 0.85 [0.74, 0.98] |
| Total | | | | 100.00% | 0.88 [0.80, 0.96] |

Q=10.72, p=0.296
$I^2$=16%

0.25    1    2    4
Hazard ratio

**FIGURE 8.2**
Forest plot for preoperative chemotherapy and resection versus resection alone for thoracic oesophageal cancer, using the DerSimonian and Laird random effects model

## Example

Let us return to our oesophageal cancer example, $\hat{\tau}^2 = 2.979 \times 10^{-3}$, $\hat{\theta} = -0.133$, s.e.$(\hat{\theta}) = 0.0442$, $z = -3.000$ and $p = 0.003$. The estimated mean hazard ratio and 95% confidence interval are thus,

$$\text{HR} = \exp(-0.133) = 0.87, \ \text{CI: } \exp(-0.133 \pm 1.96 \times 0.0442), \text{ i.e. } [0.80, 0.95].$$

Figure 8.2 shows the forest plot for the DerSimonian and Laird model. We see that HR is very close to that for the fixed effects model and the 95% confidence interval is slightly wider. Had the heterogeneity been large, this confidence interval would have been much wider than that for the fixed effects model. Remember, the combined hazard ratio for the random effects model is an estimate on the average effect of preoperative chemotherapy versus surgery alone for studies and not the common effect for each study.

There are various other random effect models and estimates of $\tau^2$, which we cannot pursue here.

## 8.3.1 Assessing model fit

Once a model is fitted, an assessment of the fit should be made, but often this step is missed. The R function influence.rma.uni() in the metafor package takes away all the hard work of producing fit statistics and plots. Table 8.1 shows eight different fit statistics which are plotted against study in Figure 8.3.

**TABLE 8.1**
Fit statistics for the DerSimonian and Laird model

| | rstudent | dffits | cook.d | cov.r | tau2.del | QE.del | hat | weight | dfbs | inf |
|---|---|---|---|---|---|---|---|---|---|---|
| Ancona | −0.120 | −0.054 | 0.003 | 1.211 | 0.0048 | 10.7 | 0.029 | 2.9 | −0.050 | |
| Boostra | −0.214 | −0.153 | 0.034 | 1.708 | 0.0067 | 10.7 | 0.240 | 24.0 | −0.160 | |
| Kelsen | 2.052 | 0.809 | 0.517 | 0.800 | 0.0000 | 6.5 | 0.138 | 13.8 | 0.858 | * |
| Law | −2.002 | −0.320 | 0.093 | 0.739 | 0.0000 | 6.7 | 0.063 | 6.3 | −0.363 | |
| Maipang | 0.652 | 0.048 | 0.002 | 1.104 | 0.0040 | 10.3 | 0.008 | 0.8 | 0.046 | |
| MRC Allum | −0.439 | −0.253 | 0.083 | 1.562 | 0.0057 | 10.4 | 0.205 | 20.5 | −0.257 | |
| Nygaard | 1.006 | 0.214 | 0.046 | 1.041 | 0.0029 | 9.6 | 0.043 | 4.3 | 0.214 | |
| Roth | −0.607 | −0.095 | 0.009 | 1.123 | 0.0041 | 10.4 | 0.013 | 1.2 | −0.091 | |
| Achlag | 0.950 | 0.129 | 0.017 | 1.034 | 0.0031 | 9.8 | 0.018 | 1.8 | 0.128 | |
| Ychou | −0.341 | −0.218 | 0.067 | 1.684 | 0.0064 | 10.5 | 0.244 | 24.4 | −0.227 | |

The fitted values for all the studies are $\hat{\theta}_{DL} = -0.133$, as are the predicted values. The residuals are $\hat{\theta}_i - \hat{\theta}_{DL}$, which are usually studentised, internally or externally. Externally, means the $i$th observation is deleted when calculating the standard error for the $i$th residual. Internally, means the $i$th observation is not deleted. Several fit statistics use the deletion method.

The statistics are:

- *rstudent* – (Externally) Studentised residuals.

- *dffits* – This measures the standardised change in predicted value of a data point when it is deleted from the sample.

- *cook.d* – Cook's distance is a measure of the influence of the $i$th data point on the estimate $\hat{\theta}_{DL}$. It is measured by the difference in $\hat{\theta}_{DL}$, estimated with and without the $i$th data point included and then standardised.

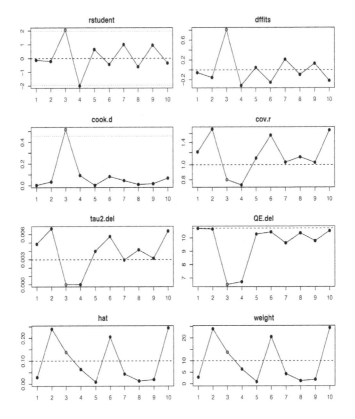

**FIGURE 8.3**
Fit statistics for the DerSimonian and Laird random effects model: 1-Ancona,
2-Boos., 3-Kels., 4-Law, 5-Maip., 6-MRC A., 7-Nyga., 8-Roth, 9-Achl., 10-
Ychou

- *cov.r* – The covariance ratio is the ratio of the variance of $\hat{\theta}_{DL}$ with and
  without the $i$th data point included.

- *tau2.del* – The estimate of $\tau^2$ when the $i$th data point is removed.

- *QE.del* – The estimate of $Q$ when the $i$th data point is removed.

- *hat* – The hat matrix, $\mathbf{H}$. Points of high leverage have undue weight in the
  estimate $\hat{\theta}_{DL}$, with the $i$th data point having high leverage if $h_{ii}$ is large,
  where $h_{ii}$ is the $i$th element of the diagonal of $\mathbf{H}$.

- *weight* – The weights used in $\hat{\theta}_{DL}$.

- *dfbs* – DFBETAS, these calculate the difference in $\hat{\theta}_{DL}$ with and without
  the $i$th data point included and then standardised by the standard error of
  $\hat{\theta}_{DL}$ calculated without the $i$th data point.

- inf – Influential observations. For the R function influence.rma.uni(), an observation is deemed to be influential if one or more of the following is true (note: cut-off values given are calculated for our model)

  - The absolute DFFITS value is larger than $3/\sqrt{k-1} = 0.333$

  - The Cook's distance is greater than the median of the chi-squared distribution with one degree of freedom ($= 0.455$).

  - The hat value is larger than $3/k = 0.3$.

  - Any absolute value of the DFBETAS larger than 1.

From the rstudent values, there are no outliers for us to be concerned about, as we are looking for residuals greater than 3 or 4 to be potential outliers. There is a large value of dffits, 0.809, for the Kelsen study. The Cook's distance for the Kelsen study is also large, 0.517, as is the value of dfbs. Hence the Kelsen study has been flagged as an influential data point. Looking more closely at this study, we see that its hazard ratio of 1.07 (log-hazard ratio, 0.068) is not extreme, being the 3rd largest out of the ten studies, but is extreme when just looking at the four studies with dominant weights, namely the Boostra, Kelsen, MRC Allum and Ychou studies. The hazard ratio for the Boostra, MRC Allum and Ychou studies are 0.86, 0.84 and 0.85, all favouring chemotherapy, whilst for the Kelsen study the hazard ratio is 1.07, favouring surgery alone. The weights, $w_i^*$ for the studies in estimating $\hat{\theta}_{DL}$, with and without the Kelsen study are

| Study | Ancona | Boos. | Kels. | Law | Maip. | MRC A. | Nyga. | Roth | Achl. | Ychou |
|---|---|---|---|---|---|---|---|---|---|---|
| $w^*$ all studies | 14.6 | 122.2 | 70.2 | 32.4 | 4.3 | 104.6 | 22.0 | 6.4 | 9.4 | 124.4 |
| $w^*$ No Kelsen | 15.3 | 192.9 | | 35.9 | 4.3 | 152.4 | 23.6 | 6.5 | 9.7 | 198.4 |

where we see that by excluding the Kelsen study, the weights for the Boostra, MRC Allum and Ychou studies are all increased by about 50%, causing the estimate of $\hat{\theta}_{DL}$ to decrease from $-0.133$ to $-0.165$, and equivalently, the hazard ratio to decrease from 0.88 to 0.85.

The covariance ratios range from 0.80 for the Kelsen study to 1.71 for the Boostra study. Remembering that the weights for the estimate $\hat{\theta}_{DL}$ equal $w_i^* = 1/(\hat{\tau}^2 + s_i^2)$ and $\text{var}(\hat{\theta}_{DL}) = 1/\sum w_i^*$, we see that the Kelsen study (and the Law study) cause the estimate $\hat{\tau}^2$ to become zero when they are removed, i.e. $Q$ has become less than $n-1$. This, in turn, will inflate the values of $w^*$ for the remaining studies, especially those with small values of $\hat{\sigma}_i^2$.

So, have we found that the Kelsen study gives an influential data point? Do we want to exclude it from the meta-analysis? We would have to find a very good reason for doing so, e.g. the trial was so badly run that trial data are unreliable. But this is hardly going to be the case and we cannot use these fit statistics to justify the removal just because it would improve the results! None of the other studies give cause for concern. Results could be reported with and without the Kelsen study, to help readers make informed decisions.

## 8.3.2   Publication bias

Suppose you are the Chief Investigator of a study investigating a drug called Cureit against the accepted standard of care and the primary outcome is the hazard ratio. Suppose the hazard ratio was found to be 1.12 with a 95% confidence interval of [0.98, 1.28] and favoured the accepted standard of care. You called it a negative trial, because the confidence interval straddles the value one, but more accurately, it is described as an inconclusive trial. You tried to get the results of the trial published, but editors of various journals kept rejecting your manuscript.

Other trials had been run with exactly the same treatment and control groups using similar study populations, and three were published with hazard ratios and 95% confidence intervals of

$$0.82 \ [0.70, 0.96], \quad 0.93 \ [0.84, 1.03], \quad 0.85 \ [0.74, 0.97],$$

two positive trials in favour of Cureit and one inconclusive. Later, somebody decided to carry out a fixed effects meta-analysis on these three trials and found the combined hazard ratio and 95% confidence interval to be

$$0.88 \ [0.83, 0.95],$$

and so the evidence is in favour of Cureit as a better treatment.

You now dig deeper into various study reports and carry out a thorough systematic review and find two other studies that had tested Cureit against the same control treatment, but were not published in the popular literature. The hazard ratios and 95% confidence intervals for these studies are 1.15 [0.96, 1.37] and 1.05 [0.93, 1.18]. You do your own meta-analysis using the fixed effects model and obtain the forest plot shown in Figure 8.4.

The conclusion about the effectiveness of Cureit has changed. Including the extra studies has changed the combined hazard ratio and 95% confidence interval to 0.98 [0.89, 1.08], not a significant result any more. Without the extra studies, we have *publication bias*, where trials that do get published can skew the results of the meta-analysis. This is a problem. If your study shows a clear result with one treatment superior than the other, then you will find it easier to get your result published in a popular journal. If your study is inconclusive, then publishers will be reticent in accepting your paper for publication. Your headline statement is rather bland, describing that you found nothing of particular interest in your study, hence the rejections. So effectively, the original meta-analysis is pulling out the studies that "worked" and ignoring the ones that did not.

One method of assessing whether publication bias is potentially present in a meta-analysis is to draw a *funnel plot*, which is a plot of the effect estimate against its standard error for each study, with the standard errors on the $y$-axis, running top to bottom and the effect estimates on the $x$-axis. Figure 8.5 shows the funnel plot for the oesophageal cancer meta-analysis. Also plotted on the funnel plot is a vertical line at the combined effect estimate, together

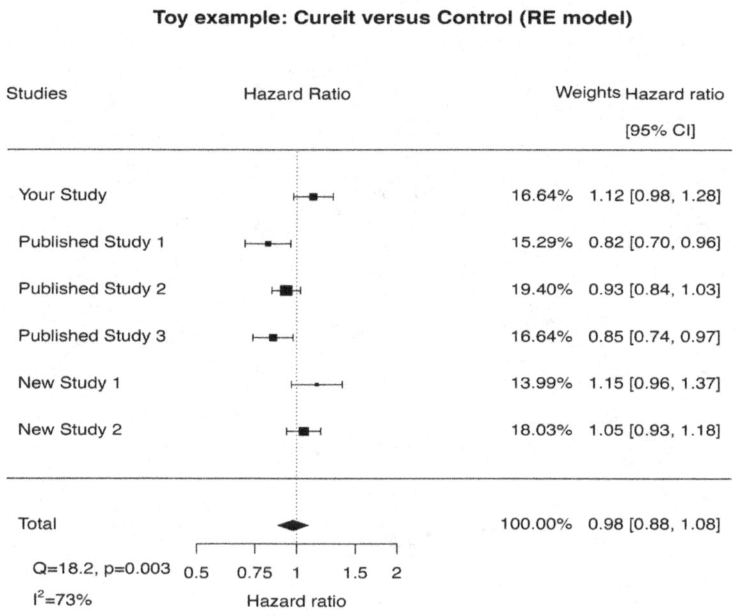

**Toy example: Cureit versus Control (RE model)**

| Studies | Hazard Ratio | Weights | Hazard ratio [95% CI] |
|---|---|---|---|
| Your Study | | 16.64% | 1.12 [0.98, 1.28] |
| Published Study 1 | | 15.29% | 0.82 [0.70, 0.96] |
| Published Study 2 | | 19.40% | 0.93 [0.84, 1.03] |
| Published Study 3 | | 16.64% | 0.85 [0.74, 0.97] |
| New Study 1 | | 13.99% | 1.15 [0.96, 1.37] |
| New Study 2 | | 18.03% | 1.05 [0.93, 1.18] |
| Total | | 100.00% | 0.98 [0.88, 1.08] |

Q=18.2, p=0.003
$I^2$=73%

0.5  0.75  1  1.5  2
Hazard ratio

**FIGURE 8.4**
Forest plot for Cureit vs Control using a fixed effects model

with 95% confidence bands (or other level of confidence can be chosen). The bands are simply the confidence intervals for every value of the standard error which will of course become wider as we travel down the $y$-axis. Also plotted is a vertical dotted line at the $x$-axis value of no effect. The name "funnel plot" stems from the fact that the plot, with some imagination, looks like an inverted funnel.

If there is no publication bias, the points on the funnel plot should be symmetric about the combined effect estimate, and we expect to see about 95% of points within the funnel and that appears to be the case here. Publication bias would lead to asymmetry of the points. However, it is not so simple as that, as other factors can affect the symmetry of the points, notably the presence of heterogeneity between the studies and poor design and analysis of some studies. Note, sometimes sample size or other index of precision is used for the $y$-axis and there are statistical tests for assessing the asymmetry of the funnel plot, such as the "Egger" test. See Sterne[62] for further details.

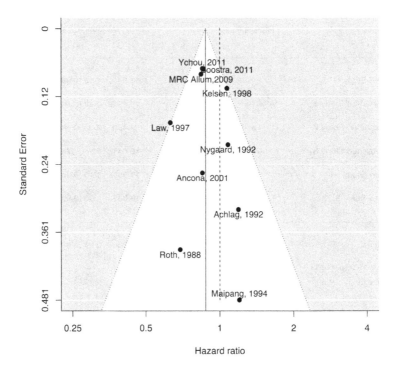

**FIGURE 8.5**
Funnel plot for the studies in the oesophageal cancer meta-analysis

---

## 8.4 Bayesian meta-analysis

In the Appendix, there is a section on Bayesian methods and we will see how these can be used in meta-analyses. Recall, Bayesian inference places *priors* on distribution parameters and then "turns priors into posteriors" using the data. Sometimes, a big problem is the integrations needed to be carried out. Markov Chain Monte Carlo (MCMC) methods have overcome this. First we consider the fixed effects model we have been dealing with.

**Bayesian fixed effects model**
The Bayesian fixed effects model is,

$$\hat{\theta}_i = \theta + \epsilon_i,$$

where $\epsilon_i$ has a normal distribution, $N(0, \sigma_i^2)$. Again we assume the $\sigma_i^2$ are essentially known and replaced by $\hat{\sigma}_i^2$. As Bayesians, we place a prior on $\theta$. We might choose an improper flat prior or perhaps a normal prior. Remembering the posterior is proportional to the likelihood times the prior, then for an improper flat prior,

$$p(\theta|\hat{\boldsymbol{\theta}}) \propto \prod_{i=1}^{n} \frac{1}{\sqrt{2\pi}\sigma_i} \exp\{-\tfrac{1}{2}(\hat{\theta}_i - \theta)^2/\sigma_i^2\} \times 1,$$

$$= C \times \left(\prod_{i=1}^{n} \sqrt{w_i}\right) \exp\left\{-\tfrac{1}{2}\sum_{i=1}^{n} w_i(\hat{\theta}_i - \theta)^2\right\},$$

where $\hat{\boldsymbol{\theta}}$ are the data values $\hat{\theta}_i$, $w_i = 1/\hat{\sigma}_i^2$ as before and $C$ is a constant with respect to $\theta$.

Rearranging and leaving off the indices in the product and summation signs for convenience,

$$p(\theta|\hat{\boldsymbol{\theta}}) = C \times \left(\prod \sqrt{w_i}\right) \exp\left\{-\tfrac{1}{2}\left[\sum w_i\hat{\theta}_i^2 - 2\theta\sum w_i\hat{\theta}_i + \theta^2\sum w_i\right]\right\},$$

$$= C \times \left(\prod \sqrt{w_i}\right) \exp\left\{-\frac{1}{2}\left[\sum w_i\hat{\theta}_i^2 - \frac{\left(\sum w_i\hat{\theta}_i\right)^2}{\sum w_i}\right]\right\} \times$$

$$\exp\left\{-\frac{1}{2}\left(\theta - \frac{\sum w_i\hat{\theta}_i}{\sum w_i}\right)^2 \Big/ \left(1/\sum w_i\right)\right\}.$$

The last term in $p(\theta|\hat{\boldsymbol{\theta}})$ is the only one that involves $\theta$ and we recognise this as the form of the pdf of a normal distribution with mean $\sum w_i\hat{\theta}_i/\sum w_i = \hat{\theta}$ and variance $1/\sum w_i$. So the posterior distribution for $\theta$ is $N(\hat{\theta}, 1/\sum w_i)$, the mean and variance being the fixed effect estimate of $\theta$ and its variance, i.e. $N(-0.136, 0.001374)$.

From the posterior distribution, the mean is $-0.136$ and the 95% credible region is $(-0.209, -0.063)$, leading to the hazard ratio and credible region of $HR = 0.87$ $(0.81, 0.94)$. On reflection, we are not surprised that this is exactly the same as the HR for the fixed effect model, since assuming a flat prior means we are not bringing any prior information into the likelihood.

Let us use an informative prior now. Let the prior be $N(0, 0.0004)$. Then the posterior is

$$p(\theta|\hat{\boldsymbol{\theta}}) \propto \prod_{i=1}^{n} \frac{1}{\sqrt{2\pi}s_i} \exp\{-\tfrac{1}{2}(\hat{\theta}_i - \theta)^2/s_i^2\} \times \exp(-\tfrac{1}{2}\theta^2/0.0004).$$

After some algebra, the posterior becomes,

$$p(\theta|\hat{\boldsymbol{\theta}}) = C \times \exp\left\{-\frac{1}{2}\left(\theta - \frac{\sum w_i\hat{\theta}_i}{(\sum w_i + 2500)}\right)^2 \Big/ \left(1/\left(\sum w_i + 2500\right)\right)\right\},$$

keeping the terms in $\theta$ only and loosing the rest in the constant $C$. Hence the posterior distribution is normal with mean, $\sum w_i \hat{\theta}_i / (\sum w_i + 2500)$, and variance, $1/(\sum w_i + 2500)$. Substituting the data, the posterior is $N(-0.0307, 0.0176^2)$ and then the new hazard ratio and 95% credible region are 0.97 (0.94, 1.00). Compared to the flat prior case, the estimate has moved towards the prior mean of zero and the credible region has shrunk, because of the informative prior. Of course, you have to have a good reason for choosing this particular prior.

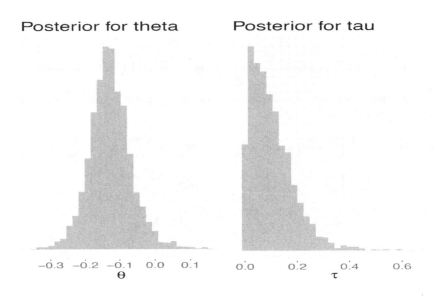

**FIGURE 8.6**
Posterior distributions of $\theta$ and $\tau$ in the random effects model, using flat priors

**Bayesian random effects model**
The Bayesian random effects model is,

$$\hat{\theta}_i = \theta + u_i + \epsilon_i,$$

where $u_i$ is a normal random variable with mean 0 and variance $\tau^2$ and $\epsilon_i$ has a normal distribution, $N(0, \sigma_i^2)$. Now, we not only put a prior on $\theta$, but also on $\tau$ or we can put it on $\tau^2$. Similarly to the fixed effects model, for non informative flat priors on $\theta$ and $\tau$, the posterior distribution for $(\theta, \tau)$ is

$$p(\theta, \tau | \hat{\boldsymbol{\theta}}) \propto \prod_{i=1}^{n} \frac{1}{\sqrt{2\pi(\tau^2 + \sigma_i^2)}} \exp\{-\tfrac{1}{2}(\hat{\theta}_i - \theta)^2/(\tau^2 + \sigma_i^2)\} \times 1$$

$$= C \times \left( \prod_{i=1}^{n} 1/(\tau^2 + w_i^{-1})^{\frac{1}{2}} \right) \exp\left\{ -\tfrac{1}{2} \sum_{i=1}^{n} (\hat{\theta}_i - \theta)^2)/(\tau^2 + w_i^{-1}) \right\}.$$

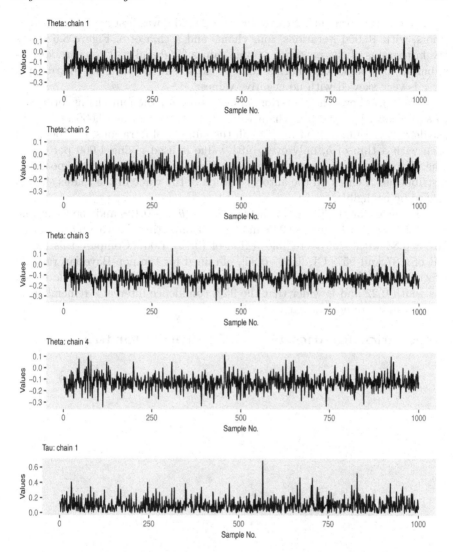

**FIGURE 8.7**
The four MCMC chains for $\theta$ and the first chain for $\tau$ for flat priors

The posterior distribution is a joint distribution of $\theta$ and $\tau$ and to make things easy, we are going to use MCMC to find the marginal posterior distributions of $\theta$ and $\tau$ with the help of the R package, brms (Bayesian regression models using Stan). Stan is a C++ package for Bayesian inference and is called from brms. In order to use it, you must download Rtools from the CRAN website as well as brms.

The function brm() that carries out the MCMC was first run using the flat priors, with 10,000 iterations, four chains and a thin of 5. Figure 8.6 shows the histograms of the posterior sample values for $\theta$ and $\tau$. The posterior distribution for $\theta$ looks very normal and, as expected, the posterior distribution for $\tau$ is very skewed with no negative values.

Figure 8.7 shows the posterior sample values for the four chains for $\theta$ and, to save space, just the first chain for $\tau$. For each chain of 10,000 iterations, the default warmup (or burnin) is half the number of iterations, i.e. 5000, and then with a thin of 5, only every fifth value is used giving 1,000 per chain. The histograms have been drawn using these 4,000 values for the posterior distribution. A check on the autocorrelation of the value in the chains showed this to be negligible.

The mean for the posterior distribution of $\theta$ is $-0.128$ and the lower and upper 2.5% quantiles are $-0.245$ and $0.001$. Converting this to a hazard ratio, HR $= 0.88$ with a 95% credible region of $(0.78, 1.00)$. Compare this to the HR of 0.87 and 95% CI, $[0.80, 0.95]$, obtained for the non-Bayesian random effects model. The hazard ratios are very close, but the 95% credible region has width 0.22 and is wider that the 95% confidence interval – remember, of course, these are not the same entities.

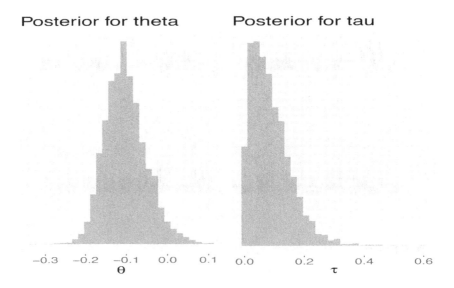

**FIGURE 8.8**
Posterior distributions of $\theta$ and $\tau$ for informative priors

**Informative priors**
Now let us use two informative priors. For $\theta$ we use N(0, 0.1) and for $\tau$, Cauchy(0, 0.2), restricted to the positive axis (pdf: $2/\{\pi\gamma[1 + (x - \mu)^2\gamma^{-2}]\}$, $x \geq 0$). Note, the Cauchy distribution and half-Cauchy distribution do not

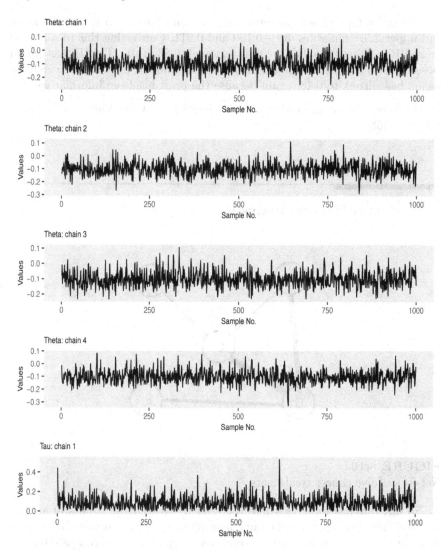

**FIGURE 8.9**
The four MCMC chains for $\theta$ and the first chain for $\tau$ for informative priors

have their mean and variance defined, i.e. you cannot calculate them as a finite number. These priors were chosen as they are very informative about $\theta$ being close to zero with small standard deviation, $\tau$. The function brm() was run again with the same parameter values, but with the new priors. Figure 8.8 shows the histograms of the new posterior values and Figure 8.9 the four chains for $\theta$ and the first chain for $\tau$.

The mean for the new posterior distribution of $\theta$ is $-0.102$ and the lower and upper 2.5% quantiles are $-0.190$ and $0.011$. Converting this to a hazard ratio, HR $= 0.90$ with a 95% credible region of $(0.83, 1.01)$. The HR has moved closer to zero because of the informative prior and the length of the credible region has shrunk to 0.18, again because the prior has been informative.

For more detail on meta-analyses, see Sutton[63], Whitehead[74] and Schwarzer[58].

## 8.5  Network meta-analysis

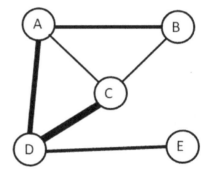

**FIGURE 8.10**
Network of five cancer treatments

So far, our meta-analyses have allowed us to combine results of studies where each study compares the same pair of treatments, A against B. Suppose we have a series of studies where several treatments have been compared, using different pairs of treatments within the various studies. For instance, suppose there are five accepted treatments, A, B, C, D and E for a certain type of cancer and studies 1 and 2 have compared treatment A against treatment B, studies 3 and 4 have compared treatment B against treatment C, study 5 has compared treatment C against treatment D, etc. Figure 8.10 shows a network diagram or graph of the treatments, where a line (edge) joining two particular treatments (nodes) shows that those two treatments have been compared directly in one or more studies. The thickness of the line is proportional to the number of studies directly comparing the two treatments. The treatments not linked have not been compared directly, here A not with E, B not with D, B not with E and C not with E.

The clever thing about network meta-analysis is that *indirect* comparisons can be made between treatments. So, treatment A can be indirectly compared with treatment E, because treatments A and E have both been compared directly with treatment D. So if treatment A was shown to be more efficacious than treatment D in the studies comparing these two treatment directly, and treatment D was shown to be more efficacious than treatment E in the studies comparing these two treatment directly, then we would expect treatment A would be shown to be more efficacious than treatment E if treatment A and treatment E were to be tested, head-to-head in a new study. There are other paths from A to E in this network, e.g. A→B→C→D→E and these all add to the indirect comparison of treatment A with treatment E.

The advantage of this indirect comparison of A with E is that the results obtained will inform us about whether a study comparing A with E directly should be run or not, or the indirect evidence might be sufficient to confidently state that one treatment was more efficacious than the other. Also, in some circumstances where one or more of the treatments is no longer available for testing, for example a placebo used in the past, but now ethically not allowed to be used, then an indirect comparison between a current treatment and the placebo is the best we can ever do.

There are three key underlying assumptions that are made in network meta-analysis models. These are

- *Similarity:* The studies used in the network meta-analysis are similar, in the sense they have all used the same methodology, population, outcome measure, etc. Heterogeneity needs to be assessed.

- *Transitivity:* The effect of the comparison of treatment A with C ($\theta_{AC}$) is the difference between effect of the comparison of treatment A with B ($\theta_{AB}$) and the effect of the comparison of treatment C with B ($\theta_{CB}$), i.e.,

$$\theta_{AC} = \theta_{AB} - \theta_{CB}. \tag{8.5}$$

So, theoretically in an ideal world, you would not need to carry out a study of A versus C if you had carried out studies of A versus B and C versus B, the treatment B linking A and C, as in the network in Figure 8.11. If you did carry out a study of A versus C, you would expect that

$$\hat{\theta}_{AC} \approx \hat{\theta}_{AB} - \hat{\theta}_{CB}.$$

Note, we need to be consistent with the order of the suffices in $\theta_{..}$. For instance, if $\theta_{AC}$ is the effect of A minus the effect of C, then $\theta_{CA} = -\theta_{AC}$, and so Equation 8.5 can be also be written as, $\theta_{AC} = \theta_{BC} - \theta_{BA}$.

- *Consistency:* This is the equivalence of an indirect comparison of two treatments with the direct comparison of the same two treatments, i.e.,

$$\theta_{AB}^{\text{indirect}} = \theta_{AB}^{\text{direct}}.$$

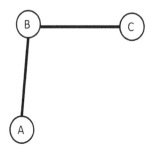

**FIGURE 8.11**
Network of three cancer treatments

For a pair of treatments in a network where there is a direct comparison, their indirect comparison can be compared to their direct comparison.

There are several methods for carrying out a network meta-analysis, e.g. fixed effects methods, random effects methods and Bayesian methods.

### 8.5.1   Fixed effects network meta-analysis

To show how the basic fixed effects method works, we will start with the very simple network shown in Figure 8.11, that has only three treatments. Treatments A and B have been compared directly and so have treatments B and C. We wish to see how treatment A would compare with treatment C in a new study if one were to be carried out. To give some context, we assume the outcome of interest is log relative risk, log(RR).

Regression analysis is used to analyse the data. The two regression equations are:

$$\hat{\theta}_{AB} = \gamma_A - \gamma_B + \epsilon_1,$$
$$\hat{\theta}_{CB} = -\gamma_B + \gamma_C + \epsilon_2,$$

where for the first equation, $\hat{\theta}_{AB}$ is the observed log(RR) comparing treatments A and B, $\gamma_A$ is the effect of treatment A, $\gamma_B$ is the effect of treatment B and $\epsilon_1$ is the random error, and similarly for the second equation.

Putting these two equations together with vectors and matrices,

$$\begin{bmatrix} \hat{\theta}_{AB} \\ \hat{\theta}_{CB} \end{bmatrix} = \begin{bmatrix} 1 & -1 & 0 \\ 0 & -1 & 1 \end{bmatrix}, \begin{bmatrix} \gamma_A \\ \gamma_B \\ \gamma_C \end{bmatrix} + \begin{bmatrix} \epsilon_1 \\ \epsilon_2 \end{bmatrix},$$

or

$$\hat{\boldsymbol{\theta}} = \boldsymbol{\gamma} + \boldsymbol{\epsilon},$$

where

$$\hat{\theta} = \begin{bmatrix} \hat{\theta}_{AB} \\ \hat{\theta}_{CB} \end{bmatrix}, \quad \mathbf{X} = \begin{bmatrix} 1 & -1 & 0 \\ 0 & -1 & 1 \end{bmatrix}, \quad \gamma = \begin{bmatrix} \gamma_A \\ \gamma_B \\ \gamma_C \end{bmatrix} \quad \text{and } \epsilon = \begin{bmatrix} \epsilon_1 \\ \epsilon_2 \end{bmatrix}.$$

Now we cannot use the usual least squares estimate of $\gamma$, $\hat{\gamma} = (\mathbf{X}^{\mathsf{T}}\mathbf{X})^{-1}\mathbf{X}^{\mathsf{T}}\hat{\theta}$, because $\mathbf{X}^{\mathsf{T}}\mathbf{X}$ is singular and hence does not have an ordinary inverse. So we have to use the Moore-Penrose generalised inverse instead. Let

$$\Sigma = \begin{bmatrix} \hat{\sigma}_1^2 & 0 \\ 0 & \hat{\sigma}_2^2 \end{bmatrix}, \mathbf{W} = \begin{bmatrix} w_1 & 0 \\ 0 & w_2 \end{bmatrix} = \begin{bmatrix} 1/\hat{\sigma}_1^2 & 0 \\ 0 & 1/\hat{\sigma}_2^2 \end{bmatrix} = \Sigma^{-1}, \quad \mathbf{L} = \mathbf{X}^{\mathsf{T}}\mathbf{W}\mathbf{X},$$

and let $\mathbf{L}^-$ be the Moore-Penrose generalised inverse of $\mathbf{L}$. Then

$$\hat{\gamma} = \mathbf{L}^-\mathbf{X}^{\mathsf{T}}\mathbf{W}\hat{\theta}, \text{ with } \mathbf{V} = \text{var}(\hat{\gamma}) = \mathbf{L}^-. \tag{8.6}$$

Using a computer symbolic algebra package to help with the algebra,

$$\mathbf{L} = \begin{bmatrix} w_1 & -w_1 & 0 \\ -w_1 & w_1 + w_2 & -w_2 \\ 0 & -w_2 & w_2 \end{bmatrix},$$

$$\mathbf{L}^- = \frac{1}{9w_1w_2} \begin{bmatrix} w_1 + 4w_2 & w_1 - 2w_2 & -2(w_1 + w_2) \\ w_1 - 2w_2 & w_1 + w_2 & -(2w_1 - w_2) \\ -2(w_1 + w_2) & -(2w_1 - w_2) & 4w_1 + w_2 \end{bmatrix},$$

$$\hat{\gamma} = \frac{1}{3} \begin{bmatrix} 2\hat{\theta}_{AB} & -\hat{\theta}_{CB} \\ -\hat{\theta}_{AB} & -\hat{\theta}_{CB} \\ -\hat{\theta}_{AB} & 2\hat{\theta}_{CB} \end{bmatrix}.$$

**Example**

For three treatments, A, B and C, one study compared A with B and another study compared B with C. The log-relative risk for A and B was $-0.741$ with standard error, 0.189 and that for C and B was $-0.793$ with standard error, 0.268. Hence

$$\hat{\theta} = \begin{bmatrix} -0.741 \\ -0.793 \end{bmatrix}, \quad \mathbf{W} = \begin{bmatrix} 27.991 & 0 \\ 0 & 13.92 \end{bmatrix},$$

and then $\hat{\gamma}$ and $\mathbf{V}$ are given by

$$\hat{\gamma} = \begin{bmatrix} -0.230 \\ 0.511 \\ -0.281 \end{bmatrix}, \quad \mathbf{V} = \begin{bmatrix} 0.02386 & 0.00004 & -0.02391 \\ 0.00004 & 0.01195 & -0.01200 \\ -0.02391 & -0.01200 & 0.03590 \end{bmatrix}.$$

So we have our estimates of $\gamma_A$, $\gamma_B$ and $\gamma_C$ as $-0.230$, 0.511 and $-0.281$.

We now need to estimate differences between treatments, A−B, C−B and A−C and these are labelled as $\hat{\theta}_{AB}^{net}$, $\hat{\theta}_{BC}^{net}$ and $\hat{\theta}_{AC}^{net}$. Firstly, A−B

$$\hat{\theta}_{AB}^{net} = \hat{\gamma}_A - \hat{\gamma}_B = -0.741,$$

$$\text{var}(\hat{\theta}_{AB}^{net}) = V_{11} + V_{22} - 2V_{12} = 0.0357, (\text{s.e} = 0.189). \qquad (8.7)$$

On the relative risk scale, $\text{RR}_{AB} = 0.48$ with 95% confidence interval, [0.33, 0.69]. Similarly, $\hat{\theta}_{CB}^{net} = \hat{\gamma}_C - \hat{\gamma}_B = -0.793$, with s.e $= 0.268$ and on the relative risk scale $\text{RR}_{BC} = 0.45$ with 95% confidence interval, [0.27, 0.77]. Note that these estimates of differences are the same as the original data for A compared to B and C compared to B; look at the difference in the first and second elements in the algebraic formula for $\hat{\gamma}$ and in the third and second elements. We are not surprised at this, as the second study has no information about the relative risk of A and B in the first study, and the first study has no information about the relative risk of C and B in the second study. What is of most interest is the comparison of A with C.

A similar calculation gives, $\hat{\theta}_{AC}^{net} = \hat{\gamma}_A - \hat{\gamma}_C = 0.051$, with s.e $= 0.328$, and on the relative risk scale, $\text{RR}_{AC} = 1.05$ with 95% confidence interval, [0.55, 2.00]. Looking back at the algebraic version of $\hat{\gamma}$, we see that $\hat{\theta}_{AC}^{net} = \hat{\theta}_{AB} - \hat{\theta}_{CB}$, which is of no surprise to us, and also $\text{var}(\hat{\theta}_{AC}^{net}) = w_1^{-1} + w_2^{-1} = \hat{\sigma}_1^2 + \hat{\sigma}_2^2$, using the algebraic version of $\mathbf{V}$. No surprises there!

Why have we undertaken all this algebra? If you tried to do the same algebraic exercise for a more complex network like the one in Figure 8.10, you would get into an algebraic muddle, as all we really need is the equation $\hat{\gamma} = \mathbf{L}^-\mathbf{X}^\mathsf{T}\mathbf{W}\hat{\theta}$ and let a computer calculate $\hat{\gamma}$ and $\text{var}(\hat{\gamma})$ for us. But hopefully, this exercise has let you see deeper into the practical workings of the model.

### 8.5.2 Network meta-analysis for breast cancer

Now we look at some real data using the data in another Cochrane review, this one entitled,

"Risk-reducing medications for primary breast cancer: a network meta-analysis".[41]

The review looked at the efficacy of breast cancer prevention agents (CPAs) for the prevention of primary breast cancer in women with above average risk of developing the disease. Toxicity of CPAs was also considered. The literature was searched and after starting with 2,717 records, only six were left after review for a quantitative analysis. The six studies used the treatments: Tamoxifen, Raloxifene, Exemestane, Anastrozole and Placebo in head-to-head comparisons, the main outcomes of interest being: overall breast cancer incidence, invasive breast cancer incidence and severe toxicity. Within a study, comparison between arms is made using relative risk.

Figure 8.12 shows the network of treatments which was drawn using the R package "netmeta". There were three studies comparing Tamoxifen with

Placebo and one study for each comparison of Anastrozole with Placebo, Exemestane with Placebo and Raloxifene with Tamoxifen. Anastrozol and Exemestane are aromatase inhibitors (AI), while Raloxifene and Tamoxifen are selective estrogen receptor modulators (SERM). Note, there is only one non-placebo based comparison, Raloxifene with Tamoxifen. All other non-placebo comparisons have to be through the placebo.

The Cochrane review carried out several comparisons and analyses, for instance a meta-analysis of Tamoxifen versus Placebo, before carrying out a network meta-analysis. Here we will just use a network meta-analysis on overall breast cancer incidence; the data that was extracted from the six studies is shown in Table 8.2.

**TABLE 8.2**
Breast cancer: overall breast cancer incidence

| Study | Treatment 1 | $n_1$ | $N_1$ | Treatment 2 | $n_2$ | $N_2$ | RR | log(RR) | s.e. |
|---|---|---|---|---|---|---|---|---|---|
| Cuzick, 2015 | Tamoxifen | 251 | 3579 | Placebo | 350 | 3575 | 0.72 | −0.33 | 0.079 |
| Fisher, 2005 | Tamoxifen | 212 | 6597 | Placebo | 354 | 6610 | 0.60 | −0.51 | 0.085 |
| Powles, 2007 | Tamoxifen | 96 | 1238 | Placebo | 113 | 1233 | 0.85 | −0.17 | 0.133 |
| Cuzick, 2014 | Anastrozole | 40 | 1920 | Placebo | 85 | 1944 | 0.48 | −0.74 | 0.189 |
| Goss, 2011 | Exemestane | 20 | 2285 | Placebo | 44 | 2275 | 0.45 | −0.79 | 0.268 |
| Vogel, 2006 | Raloxifene | 447 | 9754 | Tamoxifen | 358 | 9736 | 1.25 | 0.22 | 0.069 |

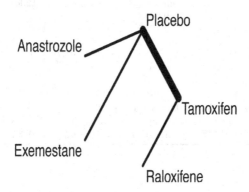

**FIGURE 8.12**
Network of breast cancer treatments

For the first study in Table 8.2, Tamoxifen versus Placebo, the risk, R of developing breast cancer for those women receiving Tamoxifen is estimated as $R = n_1/N_1 = 251/3579 = 0.0701$ and the risk for those women receiving Placebo is estimated as $R = n_2/N_2 = 350/3575 = 0.0979$. The relative risk is $RR = 0.0701/0.0979 = 0.716$. The log relative risk is $\log(RR) = -0.334$, which has standard error $\{n_1/[(N_1 - n_1)N_1] + n_2/[(N_2 - n_2)N_2]\}^{1/2} = 0.079$. The relative risk values and standard errors are calculated and used in the network meta-analysis, using a fixed effects model.

The regression equation for the fixed effects model is

$$
\begin{bmatrix}
\hat{\theta}_{TP_1} \\
\hat{\theta}_{TP_2} \\
\hat{\theta}_{TP_3} \\
\hat{\theta}_{AP} \\
\hat{\theta}_{EP} \\
\hat{\theta}_{RT}
\end{bmatrix}
=
\begin{bmatrix}
0 & 0 & -1 & 0 & 1 \\
0 & 0 & -1 & 0 & 1 \\
0 & 0 & -1 & 0 & 1 \\
1 & 0 & -1 & 0 & 0 \\
0 & 1 & -1 & 0 & 0 \\
0 & 0 & 0 & 1 & -1
\end{bmatrix}
\begin{bmatrix}
\gamma_A \\
\gamma_E \\
\gamma_P \\
\gamma_R \\
\gamma_T
\end{bmatrix}
+
\begin{bmatrix}
\epsilon_1 \\
\epsilon_2 \\
\epsilon_3 \\
\epsilon_4 \\
\epsilon_5 \\
\epsilon_6
\end{bmatrix},
$$

where $\hat{\theta}_{TP_1}$ is the estimated log(RR) of Tamoxifen with Placebo for the first trial, $\hat{\theta}_{TP_2}$ is the estimated log(RR) of Tamoxifen with Placebo for the second trial, etc., $\gamma_A$ is the effect for Anastrozole, $\gamma_E$ is the effect for Exemestane, etc., and the $\epsilon_i$'s are the errors.

The netmeta() function from the netmeta package will fit the model. The output from netmeta() gives the network estimates of the pairwise comparisons of the five treatments, shown in Table 8.3. Exponentiating, Table 8.4 shows the comparisons on the RR scale.

**TABLE 8.3**

Network meta-analysis for the breast cancer treatments: fixed effects model

|  | Anastrozole | Exemestane | Placebo | Raloxifene | Tamoxifen |
|---|---|---|---|---|---|
| Anastrozole | . | 0.05 | −0.74 | −0.59 | −0.37 |
|  |  | [−0.59, 0.69] | [−1.11, −0.37] | [−0.99, −0.18] | [−0.75, 0.02] |
| Exemestane | −0.05 | . | −0.79 | −0.64 | −0.42 |
|  | [−0.69, 0.59] |  | [−1.32, −0.27] | [−1.19, −0.08] | [−0.95, 0.12] |
| Placebo | 0.74 | 0.79 | . | 0.16 | 0.38 |
|  | [0.37, 1.11] | [0.27, 1.32] |  | [−0.02, 0.33] | [0.27, 0.48] |
| Raloxifene | 0.59 | 0.64 | −0.16 | . | 0.22 |
|  | [0.18, 0.99] | [0.08, 1.19] | [−0.33, 0.02] |  | [0.08, 0.36] |
| Tamoxifen | 0.37 | 0.42 | − 0.38 | −0.22 | . |
|  | [−0.02, 0.75] | [−0.12, 0.95] | [−0.48, −0.27] | [−0.36, −0.08] |  |

**TABLE 8.4**

Network meta-analysis on the relative risk scale

|  | Anastrozole | Exemestane | Placebo | Raloxifene | Tamoxifen |
|---|---|---|---|---|---|
| Anastrozole | . | 1.05 | 0.48 | 0.56 | 0.69 |
|  |  | [0.55, 2.00] | [0.33, 0.69] | [0.37, 0.84] | [0.47, 1.02] |
| Exemestane | 0.95 | . | 0.45 | 0.53 | 0.66 |
|  | [0.50, 1.81] |  | [0.27, 0.77] | [0.30, 0.92] | [0.39, 1.13] |
| Placebo | 2.10 | 2.21 | . | 1.17 | 1.46 |
|  | [1.45, 3.04] | [1.31, 3.74] |  | [0.98, 1.39] | [1.31, 1.62] |
| Raloxifene | 1.80 | 1.90 | 3.04 | . | 1.25 |
|  | [1.19, 2.70] | [1.09, 3.29] | [0.72, 1.02] |  | [1.09, 1.43] |
| Tamoxifen | 1.44 | 1.52 | 0.69 | 0.80 | . |
|  | [0.98, 2.12] | [0.89, 2.59] | [0.62, 0.76] | [0.70, 0.92] |  |

From Tables 8.3 and 8.4, we can see that both treatments, Anastrozole and Extremestane, in the aromatase inhibitors (AI) group have a significant RR with respect to Placebo, in favour of the two treatments, but the RR comparing Anastrozole and Extremestane with each other is not significant. Comparing Tamoxifen with Raloxifene, the two treatments in the selective oestrogen receptor modulator (SERM) group, there is a significant RR in favour of Tamoxifen. Tamoxifen also has a significant RR compared to Placebo, but Raloxifene compared with Placebo does not. So Tamoxifen appears to be more efficacious than Raloxifene. Comparing Tamoxifen with Anastrozole and Exemestane, the two RR's are not significant.

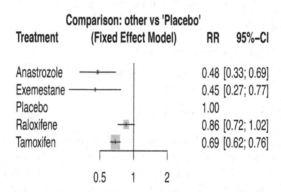

**FIGURE 8.13**
Forest plot for breast cancer network meta-analysis, reference=Placebo

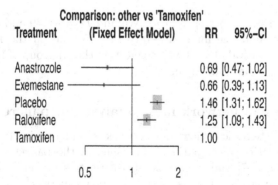

**FIGURE 8.14**
Forest plot for breast cancer network meta-analysis, reference=Tamoxifen

Figure 8.13 shows a forest plot of the treatment comparisons, but we have had to choose a reference treatment, here as Placebo. Figure 8.14 shows the same with Tamoxifen as the reference treatment. The suggestion is that both the AI treatments are more efficacious than Tamoxifen.

Suppose we want to compare the AI group with the SERM group, this is how we can estimate the difference between the two groups. The unknown effects for the five treatments are $\gamma_A$, $\gamma_E$, $\gamma_P$, $\gamma_R$ and $\gamma_T$. The contrast (difference), $D$, between the two groups is

$$D = \tfrac{1}{2}(\gamma_A + \gamma_E) - \tfrac{1}{2}(\gamma_R + \gamma_T) = \tfrac{1}{2}(\gamma_A - \gamma_R) + \tfrac{1}{2}(\gamma_E - \gamma_T) = \tfrac{1}{2}(\theta_{AR} - \theta_{ET}),$$

which is estimated as

$$\hat{D} = \tfrac{1}{2}(\hat{\theta}_{AR} - \hat{\theta}_{ET}).$$

The variance of $\hat{D}$ is

$$
\begin{aligned}
\mathrm{var}(\hat{D}) &= \tfrac{1}{4}\mathrm{var}(\hat{\gamma}_A + \hat{\gamma}_E - \hat{\gamma}_R - \hat{\gamma}_T) \\
&= \tfrac{1}{4}(v_{AA} + v_{EE} + v_{RR} + v_{TT} + 2v_{AE} - 2v_{AR} - 2v_{AT} - 2v_{ER} \\
&\quad - 2v_{ET} + 2v_{RT}) \\
&= \tfrac{1}{4}(1,1,0,-1,-1)\mathbf{V}(1,1,0,-1,-1)^T,
\end{aligned}
$$

where, remember, the covariance matrix, $\mathbf{V}$, is equal to the Moore-Penrose inverse, $\mathbf{L}^-$ which is stored in the netmeta results as "Lplus.matrix". From the R output, $\hat{D} = 0.50$ and $\mathrm{se}(\hat{D}) = 0.18$, giving a 95% confidence interval $[-0.85, -0.16]$ and on the RR scale 0.61 and confidence interval $[0.43, 0.86]$, and so the AI group of treatments is significantly more efficacious than the SERM group.

As Tamoxifen appears more efficacious than Raloxifene, we ought to compare the AI group with Tamoxifen. The contrast is now,

$$D = \tfrac{1}{2}(\gamma_A + \gamma_E) - \gamma_T = \tfrac{1}{2}(\gamma_A - \gamma_T) + \tfrac{1}{2}(\gamma_E - \gamma_T) = \tfrac{1}{2}(\theta_{AT} - \theta_{ET}),$$

and estimated as $\hat{D} = \tfrac{1}{2}(\hat{\theta}_{AT} - \hat{\theta}_{ET})$. We find $\hat{D} = -0.39$ and $\mathrm{se}(\hat{D}) = 0.17$, giving a 95% confidence interval $[-0.73, -0.05]$ and on the RR scale 0.68 and confidence interval $[0.48, 0.95]$, suggesting the AI group of treatments are more efficacious than Tamoxifen.

### 8.5.3   Bayesian network meta-analysis for pancreatic cancer

There was a systematic review and network meta-analysis of treatments for resected pancreatic cancer patients published in the Lancet Oncology journal in 2013, Liao[37] and we are going to use this as our last example for network meta-analysis. There is a later review and network meta-analysis for the same cancer using more treatments and published in Critical Reviews in Oncology/Haematology, Parmar[48] and although this would be more up-to-date, we use the first analysis here, as the network is more useful to us for illustrative purposes.

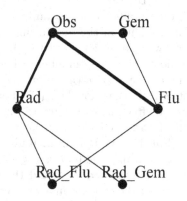 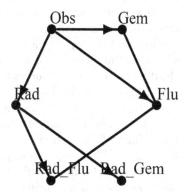

**FIGURE 8.15**

Network of pancreatic cancer treatments. Left: network showing trials as edges; Right: network with arrows showing the parameters of the network meta-analysis model

The Liao review started with 2251 potential articles which were reduced down to eight for a network meta-analysis. The six cancer treatments were: Observation (Obs), Fluorouacil chemotherapy (Flu), Gemcitabine chemotherapy (Gem), Fluorouracil-based chemoradiation (Rad), Chemoradiation plus fluoroucil chemotherapy (Rad_Flu) and Chemoradiation plus gemcitabine chemotherapy (Rad_Gem). There were thirteen pairwise comparisons of treatments. The authors carried out pairwise meta-analyses using Stata and a Bayesian meta-analysis using WinBUGS. We use the hazard ratios and 95% confidence intervals displayed in a forest plot in the published paper for our own Bayesian meta-analysis using the R package gemtc.

The left hand plot of Figure 8.15 shows the treatment network with the thickness of the edges proportional to the number of comparisons between pairs of treatments.

The random effects model we use for the Bayesian network analysis is as follows:

$$\hat{\theta}_{ABi} \sim N(\delta_{ABi}, s_i^2)$$

$$\delta_{ABi} \sim N(d_{AB}, \sigma^2)$$

$$d_{AB} \text{ and } \sigma \sim \text{vague uniform priors,}$$

although informative priors for $d_{AB}$ and for one of $\sigma^2$, $1/\sigma^2$ or $\sigma$ can be set. For the first line of the model, $\hat{\theta}_{ABi}$ is the observed log-hazard ratio comparing treatment A with treatment B in clinical trial $i$, which has a normal distribution with mean $\delta_{ABi}$ and variance $s_i^2$, where $\delta_{ABi}$ is the population log-hazard ratio comparing treatment A with treatment B in clinical trial $i$. For example, for the first trial, $i = 1$, A=Obs and B=Flu.

Now as we have a random effects model, the parameter, $\delta_{ABi}$, varies between studies and so this is given a distribution. Thus in the second line of the model, i.e. $\delta_{ABi}$ has a normal distribution with mean $d_{AB}$ and variance $\sigma^2$. Note, the mean for of $\delta_{ABi}$ does not depend on the study number $i$. The third line of the model shows the prior distributions put on $d_{AB}$ and $\sigma^2$.

The posterior distributions we are interested in are those for the $d_{AB}$'s. For our data, we have six treatments and so there are fifteen pairs of comparisons, with the series of trials covering just seven of these: Obs vs Flu, Obs vs Geb, Flu vs Gem, Obs vs Rad, Rad+Flu vs Flu, Rad vs Rad+Flu, Rad vs Rad+Gem. These are the $d_{AB}$'s for which we have data and so you would think we could put prior distributions on each of these and then find their posterior distributions using MCMC. But this is not the case because we need *transitivity*, implying $d_{AB} = d_{AC} - d_{BC}$ for all treatments $A$, $B$ and $C$. For example, looking at the network graph we see that "Obs", "Gem" and "Flu" form a closed loop, and so we must have

$$d_{Flu,Gem} = d_{Obs,Gem} - d_{Obs,Flu},$$

and thus we we can use only two of $d_{Flu,Gem}$, $d_{Obs,Gem}$ and $d_{Obs,Flu}$. The R function mtc.model() chooses the particular $d_{AB}$'s to use in the model. The edges with arrows in the right hand network in Figure 8.15 indicate the $d_{AB}$'s used in the model, $d_{Obs,Gem}$, $d_{Obs,Flu}$, $d_{Obs,Rad}$, $d_{Rad,Rad\_Gem}$ and $d_{Rad,Rad\_Flu}$. We can see the network is covered by these five parameters with no closed loops of the parameter arrows.

**TABLE 8.5**
Estimated parameters of the pancreatic cancer network

| Parameter | mean | sd | 2.5% | median | 97.5% |
|---|---|---|---|---|---|
| $d_{Obs-Flu}$ | $-0.411$ | $0.143$ | $-0.714$ | $-0.409$ | $-0.134$ |
| $d_{Obs-Gem}$ | $-0.359$ | $0.176$ | $-0.704$ | $-0.365$ | $0.005$ |
| $d_{Obs-Rad}$ | $-0.007$ | $0.194$ | $-0.404$ | $-0.002$ | $0.360$ |
| $d_{Rad-Rad\_Flu}$ | $-0.253$ | $0.312$ | $-0.870$ | $-0.250$ | $0.372$ |
| $d_{Rad-Rad\_Gem}$ | $-0.202$ | $0.360$ | $-0.900$ | $-0.208$ | $0.495$ |
| $\sigma$ | | $0.231$ | $0.123$ | $0.031$ | $0.216$ | $0.529$ |

The posterior distributions of our the parameters is given by

$$P(\boldsymbol{\delta}, \boldsymbol{d}, \sigma^2|\hat{\boldsymbol{\theta}}) \propto P(\hat{\boldsymbol{\theta}}|\boldsymbol{d})P(\boldsymbol{\delta}|\boldsymbol{d}, \sigma^2)P(\boldsymbol{d})P(\sigma^2),$$

where $\boldsymbol{\delta}$ is a vector of the $\delta_{AB}$'s, $\boldsymbol{d}$ is a vector of the $d_{AB}$'s and $\hat{\boldsymbol{\theta}}$ is a vector of the $\hat{\theta}_{ABi}$'s, remembering that as we only have five $d_{AB}$'s in the model, with some written in terms of the others, e.g. $d_{Flu,Gem} = d_{Obs,Gem} - d_{Obs,Flu}$, noting the direction of the arrows in Figure 8.15.

The model was run generating 50,000 sample values, for each of four chains, from the joint posterior distribution, and using a thin of 50 and a burn-in of

## TABLE 8.6
Comparison of treatments in the pancreatic cancer network

| | Flu | Gem | Obs | Rad | Rad_Flu | Rad_Gem |
|---|---|---|---|---|---|---|
| Flu | . | 0.96 (0.63,1.37) | 0.66 (0.49,0.87) | 0.66 (0.42,1.04) | 0.86 (0.48,1.55) | 0.81 (0.35,1.88) |
| Gem | 1.04 (0.73,1.58) | . | 0.69 (0.49,1.01) | 0.69 (0.43,1.19) | 0.9 (0.46,1.79) | 0.85 (0.36,2.06) |
| Obs | 1.5 (1.14,2.03) | 1.44 (0.99,2.05) | . | 1.00 (0.69,1.48) | 1.29 (0.71,2.40) | 1.22 (0.55,2.75) |
| Rad | 1.51 (0.96,2.35) | 1.44 (0.84,2.34) | 1.00 (0.68,1.44) | . | 1.29 (0.7,2.39) | 1.22 (0.59,2.46) |
| Rad_Flu | 1.17 (0.65,2.10) | 1.12 (0.56,2.16) | 0.77 (0.42,1.41) | 0.78 (0.42,1.43) | . | 0.95 (0.37,2.41) |
| Rad_Gem | 1.23 (0.53,2.85) | 1.18 (0.49,2.76) | 0.82 (0.36,1.81) | 0.82 (0.41,1.69) | 1.05 (0.41,2.72) | . |

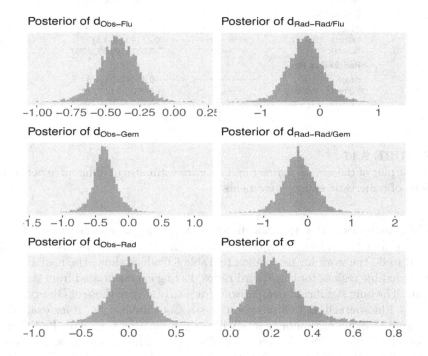

## FIGURE 8.16
Histograms of posterior sample values for the distributions of the Bayesian network meta-analysis model for pancreatic cancer

5000. Figure 8.16 shows the histograms of the posterior samples for the six parameters. All look typically normal, apart from that for the parameter $\sigma$, as would be expected. Table 8.5 shows the mean, sd, median and the 2.5% and

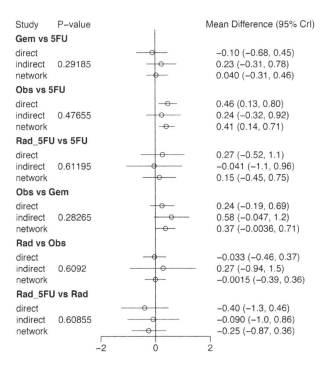

**FIGURE 8.17**

Forest plot of the *direct, indirect* and *network* estimates of differences between pairs of pancreatic cancer treatments

97.5% quantiles of the posterior distributions of the parameters. From these, we can find the indirect comparisons between pairs of treatments – we will let R to do the work for us, leading to Table 8.6 which shows the median and 95% credible regions for the hazard ratios, having exponentiated from the log-scale. The only significant comparisons (or nearly so) are those of Observation versus Fluorouracil and Observation versus Gemcitabine. So, if for example, we were thinking of running a study directly comparing Gem with Rad_Gem, then we would consider this very carefully, as the indirect comparison of these two treatments has hazard ratio, 0.85, with credible region, (0.36, 3.06). We would probably be wasting time and money.

To check consistency, we can use a *node-splitting* model, details of which we will not go into. Basically, node-splitting gives us three estimates of the difference in a pair of treatments: $d_{AB}^{direct}$, the estimate from the direct comparison between the two treatments; $d_{AB}^{indirect}$, from the indirect comparison; and $d_{AB}^{network}$, the overall network estimate. Of course, this can only happen for the pairs of treatments that have been directly compared. Figure 8.17

shows a forest plot for the pairs of treatment that have the three estimates. Consistency appears to be upheld.

---

## 8.6  Individual patient data

We finish this chapter with a short discussion on individual patient data (IPD), another name being individual participant data. If the original individual patient data are available for the studies that are going to be included in a meta-analysis, you can make a more accurate and deeper data synthesis and meta-analysis. This is obviously true, since you would have in your possession, so much more information than the summaries of aggregate data (AD) published in journals or elsewhere. You could easily re-create the aggregate data results for each study from the individual patient data for that study. You can go further, you can: (i) check the published results of each study; (ii) obtain the summary statistic of interest for a study if the study has published a different statistic; (iii) use baseline data as a covariate; (iv) carry out sub-group analyses and other pertinent analyses; and more besides.

However, there are problems in attempting individual patient data analyses, including: (i) it might be very difficult to obtain the individual patient data; you will need the cooperation of the authors of the individual studies – many are very protective of their data; (ii) a lot of patience and time resource is needed in obtaining the study data sets, making sense of them, checking variable names are consistent (with luck, the studies may have used CDISC standards) and then amalgamating the data sets.

If you decide to carry out an individual patient data meta-analysis, you have a choice of a 1-step or a 2-step approach. The 2-step approach analyses the patient data for each study separately, estimating effects of interest for each; this is the first step. Then in the second step these effects are combined using an aggregate data meta-analysis. If the individual study analyses are the same as those reported in the literature, then the results of the individual patient data meta-analysis will be the same as those that would be obtained, simply using a meta-analysis of the reported results, without having the extra work of working with the individual patient data. However, in the separate study analyses, you may have used a covariate, or you may have made allowances for missing data or used a slightly different endpoint, in which case your individual patient data meta-analysis will be different from the simple meta-analysis of the reported results. Modelling in the first step might be simple linear regression, general and generalised linear regression, Cox regression, etc. depending on the endpoint being studied.

The 1-step approach models take study directly into account within the model. For instance the following model might be used when looking at hazard functions and hazard ratios across studies,

$$h_{ij}(t) = h_0(t)\exp(\beta_{0i} + \beta_1\text{treat}_{ij}),$$

where $h_{ij}(t)$ is the hazard function for the $j$th patient in the $i$th study, $h_0(t)$ is the common baseline hazard function, $\beta_{0i}$ is the fixed trial effect, $\beta_1$ is the fixed treatment effect and $\text{treat}_{ij}$ is an indicator variable showing which treatment group the $j$th patient on the $i$th study belongs to.

Another example is the following mixed regression model,

$$Y_{ij} = \beta_{0i} + \beta_1\text{treat}_{ij} + \beta_{2i}x_{ij} + u_i\text{treat}_{ij} + \epsilon_{ij},$$

where $Y_{ij}$ is the outcome for the $j$th patient in the $i$th study, $\beta_{0i}$ is the fixed intercept for the $i$th study, $\beta_1$ is the fixed effect for treatment, $\text{treat}_{ij}$ is an indicator variable showing which treatment group the $j$th patient on the $i$th study belongs to, $\beta_{2i}$ is the fixed covariate effect for the $i$th study, $x_{ij}$ is the covariate value for the $j$th patient in the $i$th study, $u_i$ is a random-effect for treatment and $\epsilon_{ij}$ is the error for the $j$th patient on the $i$ study. Assume, $\epsilon_{ij} \sim N(0, \sigma_i^2)$ and $u_i \sim N(0, \tau^2)$. For example, $Y_{ij}$ could be overall quality of life at the end of clinical intervention, treatment could be radiotherapy versus chemotherapy and $x_{ij}$ is the overall quality of life before treatment began.

Should you carry out an aggregate data (AD) meta-analysis or an individual patient data (IPD) meta-analysis? The answer depends on whether you can obtain all the data and whether you have the resource and time to do an IPD meta analysis. Also, the importance of the research question you are trying to answer needs to be taken into account. Is there more pressing research that should be given higher priority? Often the results of an IPD meta-analysis will not give significant extra information than a quicker and easier AD meta-analysis. But it is your decision – or, more likely, a decision for the research team. See Riley[51], Kontopantelis[33] and Salanti[54] for more detail.

# 9

## Cancer Biomarkers

Biomarkers are associated with *translational medicine*. From *bench to bedside* research, involving clinicians, medical laboratory scientists, bioinformaticians and pharmaceutical companies, that brings new treatments to patients.

The FDA and NIH have published a series of articles relating to biomarkers, BEST (Biomarkers, EndpointS, and other Tools)[12] In the glossary, their definition of a biomarker is,

*"A defined characteristic that is measured as an indicator of normal biological processes, pathogenic processes, or biological responses to an exposure or intervention, including therapeutic interventions. Biomarkers may include molecular, histologic, radiographic, or physiologic characteristics. A biomarker is not a measure of how an individual feels, functions, or survives".*

In simple terms, a biomarker is any measurement made on the body, whether it involves blood, serum, tissue, DNA, X-rays, or other, that helps with the management of diseases.

BEST classifies biomarkers into seven types:

- *Diagnostic biomarkers*: detect or confirm the presence of a disease, e.g. does a patient have liver cancer?

- *Prognostic biomarkers*: inform the patient's outcome from the disease diagnosed, regardless of treatment, e.g. does the patient have a reasonable chance of long-term suvival?

- *Predictive biomarkers*: inform the patient's outcome from the disease diagnosed according to treatment; e.g. will the biomarker predict a better outcome for a liver cancer patient if treated with FOLFOX (Folinic acid, Fluorouracil and Oxaliplatin), rather than GEMOX (gemcitabine and oxaliplatin).

- *Susceptibility/risk biomarkers*: inform the risk of a non-diseased person, developing the disease

- *Pharmacodynamic/response biomarkers*: indicate whether a response to treatment has occurred

- *Monitoring biomarkers*: used repeatedly to assess the status of the disease

DOI: 10.1201/9781003041931-9

- *Safety biomarkers*: check on the safety, before and/or after treatment, e.g is the chemotherapy severely affecting the patient's kidneys?

We will concentrate on diagnostic, prognostic and predictive biomarkers.

Statistical analysis of biomarker data encompasses a variety of techniques, regression analysis, survival analysis, multivariate analysis – in fact any appropriate statistical method. We will use data sets concerning prostate cancer, hepatocellular carcinoma (HCC, liver cancer) and pancreatic cancer.

## 9.1 Diagnostic biomarkers

We consider binary biomarkers, where a medical test using a biomarker provides a yes/no diagnosis. This could be cancer/no-cancer, or presence/absence of a feature of the cancer, for example, lymph node involvement. It could also be a continuous measurement, dichotomised into yes/no or diseased/healthy, for example, it might be that a PSA level over 4 ng/ml indicates prostate cancer and levels below, no cancer. Misdiagnosis can, and does happen. A man with a PSA level of 10 ng/ml might not have prostate cancer and man with a PSA level of 2 ng/ml might have prostate cancer.

In order to assess the performance of a biomarker in diagnosing cancer, the true state of a patient has to be known, the patient has the cancer or does not have the cancer. This will have to be assessed somehow before the biomarker measurement is made, for example by using CT/MRI scans or by biopsies. This test is called the *gold standard test* and will definitively categorise a patient as having cancer or not having cancer. Well, that's the theory, but in practice 100% accuracy for a gold standard is not usually achievable. For example, in prostate cancer MRI scans do not detect all cancers.

We need to think about the population we are dealing with and there is a difference between a biomarker for screening a population where individuals do not have symptoms of cancer and a population of individuals who do have symptoms of possible cancer. For example, the UK bowel cancer screening programme screens all (consenting) adults over the age of 50 for bowel cancer, with each person tested every two years until the age of 75. There is no screening programme for prostate cancer. The population for testing a biomarker for prostate cancer might be all men who have symptoms that make them contact their GP for discussion about prostate cancer. A sample of men from the population might be selected to test a new potential biomarker, where each man has the biomarker measured as well a gold standard test to establish the cancer/ no cancer state. This would be a *cohort study* (see Chapter 7). Another type of study would be a *case-control study* (see Chapter 7), where a sample of men with prostate cancer is selected (cases) and a sample of men without prostate cancer is selected (controls). One problem with a case-control study is that if you only select men with well established prostate cancer, the gold-standard test being nearly 100% reliable in the detection of the cancer, then if you show that your new biomarker is good at detecting the cancer. That is excellent news for men in the more advanced stages of the cancer, but will it work for those men in the early stages of prostate cancer?

Given we are happy with the gold standard test, we can commence a study of our biomarker, once the sample size has been established, the protocol written and all the preliminary tasks completed. For each participant in the study, there are four possibilities relating to the correctness of their diagnosis by the biomarker. The diagnosis of a man with cancer, as established by the gold standard test, who tests positive in the biomarker test (i.e. cancer is indicated) is a *true positive* (TP). The diagnosis of a man without cancer, who tests negative in the biomarker test, is a *true negative* (TN). Now for the mistakes. The diagnosis of a man with cancer, who tests negative with the biomarker test, is a *false negative* (FN). The diagnosis of a man without cancer, who tests positive with the biomarker test, is a *false positive* (FP). These are summarised:

|  | **Man with cancer** | **Man without cancer** |
|---|---|---|
| **Test positive:** | True Positive (TP) | False Positive (FP) |
| **Test negative:** | False Negative (FN) | True Negative (TN) |

The *True Positive Rate* (TPR) is $TP/(TP + FN)$ and the *False Positive Rate* (FPR) is $FP/(FP + TN)$.

### Example: two biomarkers for prostate cancer

Our example uses data, taken from Vickers[71], who compared two diagnostic biomarkers for prostate cancer using some of the original data coming from the European randomisation study of prostate cancer screening[57]. The two potential biomarkers that were assessed are: (i) FT (free-to-total PSA ratio), with a positive test if FT < 20%, and (ii) HK (human kallikrein), with a positive test if HK > 0.075 ng/mL. The gold-standard test was by biopsy. There were 192 cases of cancer and 548 non-cancer. Table 9.1 shows a $2 \times 2$ table of test results against the gold standard for FT, and similarly, Table 9.2 shows the results for HK. Tables 9.3 and 9.4 shows the notation we will be using.

Before we go further, we need to introduce some important measurements of biomarker performance, where C in the following probabilities represents cancer and $\widetilde{C}$ denotes non-cancer,

- *Sensitivity* (Se): $\Pr(Y = 1|C)$, the probability the biomarker indicates cancer when the patient has cancer. $Se = TP/(TP + FN) = n_{11}/n_1 = TPR$.

- *Specificity* (Sp): $\Pr(Y = 0|\widetilde{C})$, the probability the biomarker indicates no cancer when the patient does not cancer. $Sp = TN/(TN + FP) = n_{00}/n_0 = 1 - FPR$.

- *Positive Predictive Value* (PPV): $\Pr(C|Y = 1)$, the probability the patient has cancer, given the biomarker indicates cancer. $\text{PPV} = \text{TP}/(\text{TP} + \text{FP}) = n_{11}/m_1$.

- *Negative Predictive Value* (NPV): $\Pr(\widetilde{C}|Y = 0)$, the probability the patient does not have has cancer, given the biomarker indicates no cancer. $\text{NPV} = \text{TN}/(\text{TN} + \text{FN}) = n_{00}/m_0$.

We would like our biomarker to have high sensitivity, so it is excellent at diagnosing cancer, when a person has cancer. We would also like it to have high specificity, so it is also excellent at diagnosing non-cancer for a person without cancer. But, generally, it has to be a trade off between the sensitivity and the specificity. The better one of them, the worse for the other, and a balance has to be made depending on context. For example, if you attempt to achieve 99% sensitivity, the specificity might be as low as 30%, and you are falsely identifying cancer, possibly for many patients. On the other hand, increasing specificity to 70% may mean the sensitivity is only 50%, and you are missing half the cancers.

The positive predicted value is a useful number to know when a person tests positive and similarly for the negative predicted value.

**TABLE 9.1**
Biomarker FT

|       | Cancer | No Cancer | Total |
|-------|--------|-----------|-------|
| FT +  | 174    | 330       | 504   |
| FT −  | 18     | 218       | 236   |
| Total | 192    | 548       | 740   |

**TABLE 9.2**
Biomarker HK

|       | Cancer | No Cancer | Total |
|-------|--------|-----------|-------|
| HK +  | 97     | 118       | 215   |
| HK −  | 95     | 430       | 525   |
| Total | 192    | 548       | 740   |

**TABLE 9.3**
TP, FP, TN, FN

|       | Cancer | No Cancer | Total   |
|-------|--------|-----------|---------|
| Y=1   | TP     | FP        | TP+FP   |
| Y=0   | FN     | TN        | FN+TN   |
| Total | TP+FN  | FP+TN     | N       |

**TABLE 9.4**
Mathematical notation

|       | Cancer   | No Cancer | Total |
|-------|----------|-----------|-------|
| Y=1   | $n_{11}$ | $n_{01}$  | $m_1$ |
| Y=0   | $n_{10}$ | $n_{00}$  | $m_0$ |
| Total | $n_1$    | $n_0$     | $N$   |

Table 9.5 shows the Se, Sp, PPV and NPV for the two prostate cancer biomarkers, FT and HK, together with 95% confidence intervals. Before discussing the values, let us see how we have calculated the confidence intervals for these.

Conditioned on $n_1$, Se has a binomial distribution , $\text{Bin}(n_1, p_1)$, with $p_1$ estimated as $\hat{p}_1 = \widehat{\text{Se}} = n_{11}/n_1$. An approximate confidence interval would be $\widehat{\text{Se}} \pm z_{\alpha/2}\sqrt{\widehat{\text{Se}}(1 - \widehat{\text{Se}})/n_1}$, using the usual normal approximation. However, if we have the sensitivity or specificity close to zero or unity, this

**TABLE 9.5**

Estimates os Se, Sp PPV and NPV for the prostate cancer biomarkers

| Biomarker | FT | | | HK | | |
|---|---|---|---|---|---|---|
| | Est. | S.E. | 95% CI | Est. | S.E. | 95% CI |
| Se | 0.91 | 0.021 | $[0.87, 0.95]$ | 0.51 | 0.036 | $[0.43, 0.58]$ |
| Sp | 0.40 | 0.021 | $[0.36, 0.44]$ | 0.78 | 0.018 | $[0.75, 0.82]$ |
| PPV | 0.35 | 0.021 | $[0.30, 0.39]$ | 0.45 | 0.034 | $[0.38, 0.52]$ |
| NPV | 0.92 | 0.017 | $[0.89, 0.96]$ | 0.82 | 0.017 | $[0.79, 0.85]$ |

approximation is too crude and better methods are needed. We have several choices, for example the Clopper-Pearson and the Wilson approximations. We will use the Clopper-Pearson method, which for a random variable, $X \sim \text{Bin}(n, p)$ and observed value, $x$, finds the endpoints of the CI as the largest $p_L$ and smallest $p_U$, such that $\text{Pr}(X \leq x) = \sum_{i=0}^{x} \binom{n}{i} p_L^i (1 - p_L)^{n-i} > \alpha/2$, and $\text{Pr}(X \geq x) = \sum_{i=x}^{n} \binom{n}{i} p_U^i (1 - p_U)^{n-i} > \alpha/2$, and hence the CI is $[p_l, p_U]$. If conditioned on, $m_1$ and $m_0$, the CIs for PPV and NPV can be found using similar methods as those for Se and Sp.

Returning to our data, the sensitivity for the FT test is 0.91, compared to 0.51 for the HK test, but in contrast, the specificity of the HK test is 0.78 compared to 0.40 for FT, clearly showing that choosing a test with high sensitivity will often require the acceptance of low specificity. It is good to identify patients that potentially have the cancer and then subjecting them to further definitive tests, but not good to subject many patients who do not have the cancer to the same tests, especially if the further tests are invasive. Also, time and money are being wasted.

The positive predictive value for FT is 0.35, so only about a third of the tests give a positive result for patients with cancer, the other two thirds of positive tests indicate cancer for patients who do not have cancer. The positive predictive value for HK is 0.45. The negative predicted values for the two tests are 0.92 and 0.82 respectively, showing negative tests do quite well at indicating that a patient does not have cancer.

## 9.1.1 Measuring prevalence

Can we estimate the prevalence of prostate cancer in the population from our data? Yes, but only if the sample of patients is a true representation of the population from which it was drawn; here it is those men reporting prostate cancer symptoms, not men as a whole. Essentially, we have a *cross-sectional study*, from which, the prevalence can be estimated. Also, for a cohort study, where a sample is chosen, without regard to the state of the disease, prevalence can be estimated.

The *prevalence*, $\tau$, of the disease is estimated by $\hat{\tau} = n_1/N$, and a CI can be found as above. For our prostate cancer example, $\hat{\tau} = 0.26$, with 95% CI, $[0.23, 0.29]$, using the Clopper-Pearson method.

For a case-control study, we cannot obtain an estimate of the prevalence of the cancer, because we have chosen the sample sizes for both the cases and controls, and then $\hat{\tau}$ will not be a sensible estimate of $\tau$. The estimates of Se and Sp are not affected by the prevalence and are calculated as usual. However, PPV and NPV do depend on the prevalence, but as long as the prevalence can be estimated by other means, these can be estimated using Se and Sp. To see this, using Bayes theorem,

$$\text{PPV} = \Pr(C|Y=1) = \frac{\Pr(Y=1|C)\Pr(C)}{\Pr(Y=1)},$$

$$= \frac{\Pr(Y=1|C)\Pr(C)}{\Pr(Y=1|C)\Pr(C) + \Pr(Y=1|\widetilde{C})\Pr(\widetilde{C})},$$

$$= \frac{\text{Se} \times \tau}{\text{Se} \times \tau + (1 - \text{Sp}) \times (1 - \tau)},$$

and hence PPV can be estimated using the estimates of Sp and Se together with the estimate of $\tau$. A similar calculation shows that NPV can be estimated using the equation,

$$\text{NPV} = \frac{\text{Sp} \times (1 - \tau)}{\text{Sp} \times (1 - \tau) + (1 - \text{Se}) \times \tau}.$$

Variances of PPV and NPV can be estimated using the variances of Sp and Se and by making use of the delta method (see Appendix) on the equations above.

### 9.1.2  Comparing two tests

We can summarise the accuracy of a binary diagnostic test using sensitivity, specificity, positive and negative predictive values and other measures, three of which are:

- Probability of a correct decision: $\Pr(\text{correct decision}) = \text{PCD} = \Pr(Y=1|C)\Pr(C) + \Pr(Y=0|\widetilde{C})\Pr(\widetilde{C}) = \text{Se} \times \tau + \text{Sp} \times (1 - \tau)$.

- The odds ratio (OR): the odds of a positive test relative to a negative test for patients with cancer divided by the odds of a positive test relative to a negative test for patients without cancer, $\text{OR} = \frac{\Pr(Y=1|C)/\Pr(Y=0|C)}{\Pr(Y=1|\widetilde{C})/\Pr(Y=0|\widetilde{C})}$.

- Youden Index (YI): $\text{YI} = \text{Se} + \text{Sp} - 1$.

A problem with using PCD is that the true positives are given as much importance as the true negatives, or equivalently, the false positives are given as much importance as the false negatives, which is usually not desirable. Also, if the cancer is rare then you could fix your test so that *every* patient has a negative test result. You will never detect any of the rare cancers, but you will correctly diagnose all the non-cancer cases. Your overall PCD will be very high!

If $OR = 1$, the likelihood of a positive test is the same for patients with or without cancer. For $OR > 1$, the odds of a positive test result is greater for patients with cancer than those without cancer. The YI also suffers from giving equal importance to true positives and true negatives.

When there are two binary diagnostic tests, these can be compared on the above measures. We just look at comparing Se and Sp for two tests, and test the hypotheses,

$$H_0 : Se_A = Se_B \quad vs \quad H_1 : Se_A \neq Se_B,$$

where $Se_A$ and $Se_B$ are the sensitivities for tests A and B. Similar hypotheses can be written for $Sp_A$ and $Sp_B$.

Like the two sample t-test, there are two situations, one where the two tests are carried out on separate samples of patients and the other where the two tests have been performed on each patient, i.e. we have paired data. For separate samples, the test statistic is,

$$Z = \frac{(\widehat{Se}_A - \widehat{Se}_B)}{\{\widehat{Se}_A(1 - \widehat{Se}_A)/n_{A1} + \widehat{Se}_B(1 - \widehat{Se}_B)/n_{B1}\}^{1/2}},$$

where $n_{A1}$ and $n_{B1}$ are numbers of patients with cancer for the two tests. If sample sizes are small or $\widehat{Se}_A$ or $\widehat{Se}_B$ are close to 0 or 1, Fisher's exact test is used.

For paired data, McNemar's test is used. Let $n_{11}$ be the number of patients who have a positive result for both diagnostic tests, $n_{00}$ the number of patients who a negative result for both tests, $n_{10}$ be the number of patients who have a positive result for the first test and a negative result for the second, and $n_{01}$ be the number of patients who have a negative result for the first test and a positive result for the second. Then the test statistic is,

$$X^2 = \frac{(n_{10} - n_{01})^2}{n_{10} + n_{01}},$$

which, asymptotically, has a chi-squared distribution, with 1 df, under $H_0$. There is an exact test if the sample size is small.

The R package DTCompair was used to analyse the data for the prostate cancer tests. For testing the equality of the sensitivities, an exact binomial test was used, rather than a normal approximation. The difference in sensitivities is $-0.401$ with p-value $p < 0.001$, not surprisingly. For the specificities, the difference is 0.387, with p-value $p < 0.001$, again not surprising us. The difference in the Youden indices is $-0.014$, with approximate variance, 0.00267, giving a Z-test statistic of 0.275, with p-value, $p = 0.784$.

### 9.1.3  Receiver Operating Characteristic (ROC) curves

When we have a continuous biomarker, we usually choose a cut-point, where patients with biomarker values greater than a cut-point are placed in the cancer category, and patients with biomarker values less than or equal to the cut point are placed in the non-cancer category (or vice versa, depending on the biomarker). The cut-point is chosen based on the sensitivity and specificity, which leads us to *receiver operating characteristic curves*, or ROC curves, which emanated from electrical/radar engineering during World War II. We will illustrate these curves using a liver cancer example.

**Example: hepatocellular carcinoma (HCC)**
Patients with chronic liver disease (CLD) have an increased risk of hepato-cellular carcinoma (HCC) – liver cancer. Surveillance of CLD patients using ultrasound examination can help to detect HCC early. If the ultrasound examination detects a potential lesion, then further examinations are carried out using CT/MRI scans and possibly biopsies. It would be helpful if biomarkers could be used to detect HCC in the surveillance of CLD patients and more widely.

We will make use of two unique liver cancer datasets, one from the UK and one from Japan. The UK dataset was kindly supplied by Prof. Johnson at Liverpool University. The Japan dataset by Drs. Toyoda and Kumada at Ogaki Municipal Hospital, Ogaki, Japan. Their generous collaboration has formed the basis of GALAD and BALAD scores, that will be described in due course. Dr. Berhane at Liverpool University played a key role in their development.

A case-control study carried out at the Queen Elizabeth Hospital, Birmingham[27], using 670 patients, 331 with confirmed HCC and 339 with CLD. Various serological biomarkers were measured. We will use the data from the study, but only looking at: albumin (ALB); serum $\alpha$-fetoprotein (AFP); an isoform, AFP_L3, of AFP, expressed as a percentage (L3); des-carboxy-prothrombin (DCP), along with gender and age.

We start with ALB as a potential diagnostic biomarker for HCC. Its range is from 20 to 53 with a median value of 42. Let us use a cut-point, $c$, of the biomarker values, so that a patient is classified as HCC if ALB $\leq c$ and as CLD if ALB $> c$. We take a sequence of cut-point ALB values from 20 to 42 and for each value classify the patients into HCC and CLD groups. Then we calculate Se and Sp for each cut-point. As the cut-points increase, Se decreases and Sp increases. We could plot Se against Sp to visualise the relationship between them. We do not do that, but plot Se against 1-Sp, equivalent to the TPR against the FPR, as this is the convention. The resulting curve is the receiver operating characteristic curve (ROC curve). To produce the curve, it is best to use the ordered ALB values in the total dataset as the sequence of cut-points to be used.

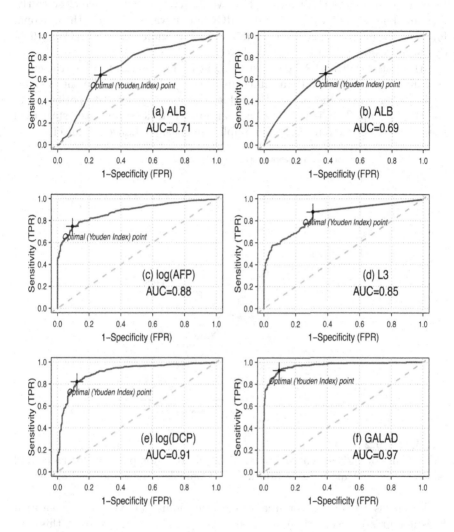

**FIGURE 9.1**
ROC curves for five biomarkers and the GALAD set of biomarkers

Figure 9.1(a) shows the ROC curve for ALB. The ROC curve starts at
(1,1), where all CLD patients are misclassified and all HCC patients correctly
classified, and travels down to the origin, where all CLD patients are correctly
classified and all HCC patients misclassified. For a biomarker that perfectly
diagnoses HCC, and never misclassifies any patient, the ROC curve would
start at (1,1), move across to the point $(0, 1)$ and then down to the origin. In
practice this cannot usually be achieved, but a good biomarker would have
a curve that stayed close to the y-axis and the line from $(0, 1)$ to $(1, 1)$. A

biomarker that can not diagnose HCC would have a ROC curve close to the diagonal line from $(0,0)$ to $(1,1)$. If a ROC curve was well below the diagonal line, then a mistake has been made, because it indicates a situation where diagnosis is worse than by chance alone.

Alternative to the ROC curve starting at $(1,1)$, you can think of the curve starting at the origin and travelling up to $(1,1)$. This would be the case for a biomarker where the higher the value, the more chance the patient is in the HCC group. This is the usual convention when developing the statistical theory: increasing biomarker value implies increasing chance of cancer.

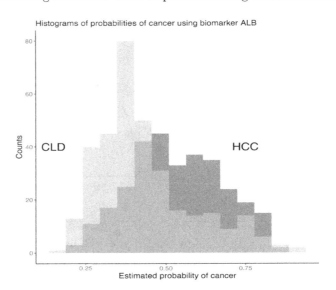

**FIGURE 9.2**
Histograms of the estimated probabilities of HCC for patients for cases and controls. Cases – medium and dark grey, Controls – light and medium grey

To assess ALB as a potential diagnostic biomarker, we could have used logistic regression, sometimes referred to as logistic discrimination in this context, with the model for the probability that a patient has HCC, based on ALB,

$$\Pr(\text{HCC}) = \exp(\beta_0 + \beta_1 \times \text{ALB})/\{1 + \exp(\beta_0 + \beta_1 \times \text{ALB})\} = \frac{\exp(D)}{1 + \exp(D)},$$

where $D = \beta_0 + \beta_1 \times \text{ALB}$. Often, $D$ is referred to as a *discriminant*.

The estimated coefficients are: $\hat{\beta}_0 = 4.465$, $p < 0.001$, and $\hat{\beta}_1 = -0.113$, $p < 0.001$, showing ALB could potentially be used as a biomarker. The estimated probability of HCC for the $i$th patient, is $\hat{p}_i = \exp(\hat{\beta}_0 + \hat{\beta}_1 \times \text{ALB}_i)/\{1 + \exp(\hat{\beta}_0 + \hat{\beta}_1 \times \text{ALB}_i)\}$ where $\text{ALB}_i$ is the level of ALB for the $i$th patient. Figure 9.2 shows histograms of the $\hat{p}_i$'s split by group, CLD vs HCC. The histograms are opaque, so the area of one histogram that would be hidden behind the other, shows through; the light grey area combined with the medium grey area is the histogram for CLD, the medium grey area combined with the dark grey area is the histogram for HCC. Ideally, the histograms would not overlap, but that never happens.

We need to choose a cut-point for use in practice. We could use $\Pr(\text{HCC}) = 0.5$ from the logistic model and say all future patients, of unknown group, with $\hat{p}_i \leq 0.5$ are placed in the HCC group and in the CLD group if $\hat{p}_i > 0.5$. This equates to an ALB cut-point of $\text{ALB}_{\text{cut}} = (\log(0.5) - \hat{\beta}_0)/\hat{\beta}_1 = 39.6$. If we use this cut-point to reclassify our patients, 165 of the 308 HCC patients were reclassified as HCC, giving a sensitivity of 0.54, and 260 of the 339 CLD patients were reclassified as CLD, giving a specificity of 0.77. Another approach is to use the cut-point that gives the maximum value of Youden's index, this point being shown on the ALB ROC curve in Figure 9.1. It is best to use clinical/epidemiological reasoning where desired Se and Sp values are to be set, balancing the trade-off between the two. Do you want a very high Sp at the expense of a low Se, or will you accept a lower Sp, with an acceptable Se?

### 9.1.3.1 ROC curve theory

Let $Y$ be the biomarker value and suppose $Y$ has cumulative distribution function, $F_C(y)$ for HCC patients (Group $C$) and cumulative distribution function, $F_{\widetilde{C}}(y)$ for CLD patients (Group $\widetilde{C}$). Then,

$$\text{Se} = \Pr(Y \geq c|C) = 1 - F_C(c), \quad 1 - \text{Sp} = \Pr(Y \geq c|\widetilde{C}) = 1 - F_{\widetilde{C}}(c),$$

where $c$ is the cut-point (not confusing $c$ with $C$). A ROC curve is a plot of $(1 - F_{\widetilde{C}}(c), 1 - F_C(c))$ for all values of $c$. Let $x = 1 - F_{\widetilde{C}}(c)$, and so $c = F_{\widetilde{C}}^{-1}(1 - x)$, where $F^{-1}$ is the inverse function of $F$. Let $z = 1 - F_C(c)$, and so,

$$z = 1 - F_C(F_{\widetilde{C}}^{-1}(1 - x)), \quad (0 \leq x \leq 1), \tag{9.1}$$

so a plot of $z$ against $x$ gives the ROC curve. Let us look at some properties of ROC curves.

*Transformations of the biomarker*
A strictly increasing transformation, $t()$, of the biomarker does not change the ROC curve. This is easily seen, since $\text{Se} = \Pr(t(Y) \geq t(c)|C) = \Pr(Y \geq c|C) = 1 - F_C(c)$, and similarly for Sp. This shows that basing the cut-point on the logistic regression probabilities will give the same ROC curve for ALB as we had previously.

*Area under the ROC curve*
The area under the ROC curve (AUC) is given by $\int_0^1 z(x)dx$. Thus,

$$\text{AUC} = \int_0^1 \{1 - F_C(F_{\widetilde{C}}^{-1}(1 - x)\}dx.$$

Let $u = F_{\widetilde{C}}^{-1}(1-x)$, and so,

$$\text{AUC} = 1 + \int_{\infty}^{-\infty} F_C(u) f_{\widetilde{C}}(u) du,$$

$$= 1 - \int_{-\infty}^{\infty} F_C(u) f_{\widetilde{C}}(u) du = \int_{-\infty}^{\infty} (1 - F_C(u)) f_{\widetilde{C}}(u) du,$$

$$= \int_{-\infty}^{\infty} \Pr(Y > u | C) \Pr(Y = u | \widetilde{C}) du,$$

$$= \int_{-\infty}^{\infty} \Pr(Y_C > u, Y_{\widetilde{C}} = u) du, \quad \text{as } Y_C, Y_{\widetilde{C}} \text{ are independent,}$$

$$= \Pr(Y_C > Y_{\widetilde{C}}). \tag{9.2}$$

Commonly, normal distributions for $Y_C$ and $Y_{\widetilde{C}}$ are used, the binormal case, with $Y_C \sim N(\mu_C, \sigma_C^2)$ and $Y_{\widetilde{C}} \sim N(\mu_{\widetilde{C}}, \sigma_{\widetilde{C}}^2)$. From $u = F_{\widetilde{C}}^{-1}(1-x)$ above, $\Phi\left(\frac{u - \mu_{\widetilde{C}}}{\sigma_{\widetilde{C}}}\right) = 1 - x$, and hence $\left(\frac{\mu_{\widetilde{C}} - u}{\sigma_{\widetilde{C}}}\right) = z_x$, where $z_x$ is the standard normal value for probability $x$, and hence $u = \mu_{\widetilde{C}} - \sigma_{\widetilde{C}} z_x$. (Do not confuse $z_x$ with the $z$ in Equation 9.1.) Using this in Equation 9.1,

$$z = 1 - \Phi\left(\frac{u - \mu_C}{\sigma_C}\right) = \Phi\left(\frac{\mu_C - u}{\sigma_C}\right) = \Phi\left(\frac{\mu_C - \mu_{\widetilde{C}} + \sigma_{\widetilde{C}} y_x}{\sigma_C}\right).$$

Hence a plot of $z$ against $x$ using these formulas gives the ROC curve for the binormal case. The two normal distributions can be fitted using maximum likelihood and then the ROC curve plotted. Figure 9.1(b) shows this for the ALB biomarker.

For the area inder the ROC curve, AUC, $Y_C - Y_{\widetilde{C}} \sim N(\mu_C - \mu_{\widetilde{C}}, \sigma_C^2 + \sigma_{\widetilde{C}}^2)$, and hence

$$\text{AUC} = \Pr(Y_C - Y_{\widetilde{C}} > 0) = \Phi\left(\frac{\mu_C - \mu_{\widetilde{C}}}{(\sigma_C^2 + \sigma_{\widetilde{C}}^2)^{1/2}}\right).$$

The AUC for the ROC curve in Figure 9.1(b) is AUC = 0.69. The AUC for the corresponding empirical ROC curve in Figure 9.1(a) is calculated by summing the areas of the trapeziums below the ROC curve, giving, AUC = 0.71. The two AUCs are very close, but there is a difference in their shape.

### 9.1.4  HCC revisited

Figures 9.1, (a)-(e) shows ROC curves for the four biomarkers. The best, in terms of AUC, is log(DCP) with AUC = 0.91. Close behind are log(AFP) and L3. But there is no reason why we cannot combine biomarkers. We do this again using logistic regression. Using a stepwise selection approach, the variables, **G**ender, **A**ge, **L**3, log(**A**FP) and log(**D**CP) (GALAD) were selected for an optimal model. The estimated discriminant, $D$, also known as the GALAD score, is

$$D = -10.26 + 0.10 \, \text{Age} + 1.39 \, \text{Gender} + 2.42 \log(\text{AFP}) + 0.04 \, \text{L3} + 1.44 \log(\text{DCP}).$$

Each patient has combined biomarker score $D_i$ and this is used in the same way as a single biomarker. Figure 9.1(f) shows the ROC curve for $D$. It has AUC = 0.97, an exceptionally large value, not often seen with diagnostic biomarkers. Note, there have been later adaptions of the GALAD model.

### 9.1.4.1 Validation

We should validate our diagnostic model. Of the various options, we will carry out a k-fold validation, with k chosen as 10, fitting the logistic model to 90% of the data at a time and then evaluating its accuracy in predicting cancer/no cancer on the remaining 10% of the data. The accuracy is measured as the proportion of correct classifications (and termed "accuracy") and also by Cohen's kappa, $\kappa$. Accuracy is $(TP + TN)/N$. However, this accuracy can be misleading if the number of cases and controls are not balanced. Cohen's $\kappa$ attempts to adjust for this by allowing for the number of correct classifications that would arise by chance, $\kappa = (p_0 - p_e)/(1 - p_e)$, where $p_0$ is the estimated probability of a correct classifications (equals accuracy) and $p_e$ the probability of a correct classification arising by chance, and using the notation of Table 9.4, $p_e = m_1 n_1/N^2 + m_0 n_0/N^2$.

The R package caret can be used for validation methods on a variety of models and was used here. The accuracy values for the ten models fitted were:

0.95, 0.94, 0.89, 0.83, 0.95, 0.89, 0.89, 0.94, 0.89, 0.89.

The kappa values were:

0.91, 0.88, 0.79, 0.66, 0.91, 0.79, 0.78, 0.88, 0.78, 0.78.

The mean accuracy was 0.91 and the mean kappa value was 0.81. I think we are very happy with those numbers.

Using the model on an external data set gives a more objective validation and so this was done classifying a similar set of patients from a hospital in Newcastle. Of 158 patients, TP = 56, TN = 84, FP = 11, FN = 7, Se = 0.89 and Sp = 0.88. As these were excellent results, the final model in the GALAD paper was based on the combined Birmingham and Newcastle data.

Zhou[76] is recommended reading for diagnostic biomarkers.

## 9.2 Prognostic biomarkers

Now it is the turn of prognostic biomarkers. These attempt to give an indication of the future progress of the cancer for a patient. So, for example, can the ESPAC4 multivariate Cox PH fitted model of Table 5.2, be used to assess the survival prospects of a patient? As a reminder, for the Cox PH model, the linear predictor for the $i$th patient is, $\boldsymbol{\beta}^{\mathsf{T}}\boldsymbol{x_i}$, giving their estimated hazard function, $h_i(t) = h_0(t)\exp(\boldsymbol{\beta}^{\mathsf{T}}\boldsymbol{x_i})$. The baseline hazard function, $h_0(t)$ can be estimated as $\hat{h}_0(t)$, thus leading to the patient's estimated survival function, $S_i(t) = \exp(-\hat{H}_0(t)\exp(\boldsymbol{\beta}^{\mathsf{T}}\boldsymbol{x_i}))$. (See Section 5.2.3.)

We could predict how long a patient has left to live by the median of $S_i(t)$, but most predictions would be inaccurate. The best we can do it to calculate $S_i(t)$ and calculate a few particular survival probabilities, for example, $S_i(12)$, $S_i(24)$, $S_i(36)$, $S_i(60)$, corresponding to one year, two years, etc. It is fairer to discuss these with the patient, than just give them the median survival time.

However, how do we know that the Cox PH linear predictor is good at predicting survival time? A statistic for checking this is *Harrell's concordance index*, or *Harrell's c-index*. See the article by Therneau and Atkinson, https://cran.r-project.org/web/packages/survival/vignettes/concordance.pdf.

Harrell's $c$ is defined by, $c = \Pr(\boldsymbol{\beta}^{\mathsf{T}}\boldsymbol{x_i} < \boldsymbol{\beta}^{\mathsf{T}}\boldsymbol{x_j}|t_i > t_j)$, for two randomly chosen patients, $i$ and $j$, i.e the probability the linear predictor will predict a better survival curve for the $i$th patient than for the $j$th patient, given that the actual survival time for the $i$th patient is longer than that for the $j$th patient. The estimate of $c$ looks at every possible pair of patients and checks whether the two survival times and linear predictors agree with the orderings of this probability, i.e. they are *concordant*. If they do not agree, they are *discordant*. Let $C$ be the number of discordant pairs of patients and $D$ the number of discordant pairs. Then $c = C/(C+D)$. But note, censored survival times must be allowed for, not every pair can be counted. If both patients in a particular pair have censored survival times, the pair cannot be counted. If one patient in a pair has a censored survival time, say, $t_j^{cen}$ and $t_i$ is not censored, then the pair can be counted if $t_i < t_j^{cen}$, but not if, $t_i > t_j^{cen}$. This also assumes there are no ties in the survival times, nor in the linear predictors. If there are, $c$ has to be adjusted. The standard error of $c$ can be estimated but we will not go into this.

The R function concordance() will calculate Harrell's $c$. For the ESPAC4 multivariate predictor (see Section 5.2), $c = 0.70$ with s.e. $= 0.014$. The values of $c$ must lie between 0 and 1, but values much less than 0.5 should not happen since $c = 0.5$ indicates no predictive value. It is hard to give a guide to the interpretation of the $c$ value, some say $c = 0.7$ implies a good model and $c = 0.8$ a very good model.

Note, the model should be validated. Can the model be successfully used on a new dataset? We will not take this further for the ESPAC4 data, but return to hepatocellular carcinoma data.

### 9.2.1 Prognostic biomakers for HCC

Using similar data to that for the diagnostic GALAD model, the BALAD model for progression of HCC for patients diagnosed with HCC was developed.

Progression of HCC is mainly based on CT and MRI imaging, using the number and size of tumours and if there is portal vein invasion, together with a staging measure such as the Child-Pugh score. The Child-Pugh score is a measure of liver function, using the amounts of bilirubin and albumin in the blood, prothrombin time (how quickly the blood clots), ascites (is there is fluid in the abdomen) and encephalopathy (is the brain being affected).

The BALAD model uses Bilirubin, Albumin, L3, AFP and DCP as biomarkers for HCC, the last three being those used in the GALAD model. The original BALAD model[67] used cut-points for the five biomarkers, whereas the refined BALAD model[14] uses the continuous values of the biomarkers.

HCC data from the UK and Japan were used to develop the refined BALAD model. The Japan data were split into training and test sets and a multivariate model using $\sqrt{\text{bil}}$, alb, AFP ($\times 10^{-3}$), L3 and log(DCP), was taken forward as the potential scoring system for progression of HCC. A flexible parametric model (see Chapter 5) was fitted to the Japan training dataset, which was then assessed on the Japan test dataset and also the UK dataset. For the UK dataset, an adjusted baseline hazard function had to be used since survival times in the Japanese chronic liver disease population are generally longer than those in the UK population. We mimic the analysis here using a slightly different dataset to that in the BALAD publication. Again, for brevity, only the key analyses are included here with the standard data and model checking left out.

The BALAD model uses a Royston-Parmar (RP) restricted cubic splines model for the log of the baseline cumulative hazard function,

$$\log\{H_0(t)\} = a_0 + a_1 x + \sum_{j=2}^{k+1} a_j \left\{ (x-k_i)_+^3 - \xi_i (x-k_{min})_+^3 - (1-\xi_i)(x-k_{max})_+^3 \right\},$$

where, $x = \log(t)$, $k$ is the number of knots and $\xi_i = (k_{max} - k_i)/(k_{max} - k_{min})$.

Let this model formula be shortened to, $\log\{H_0(t)\} = g(t, \boldsymbol{a})$, where $g(t, \boldsymbol{a})$ is the right hand side of the previous equation. For a proportional hazards model, $h(t) = \gamma h_0(t)$, or $H(t) = \gamma H_0(t)$, and so $\log\{H(t)\} = \{H_0(t)\} + \log \gamma$. The BALAD model is a proportional hazards model with $\log\{H_0(t)\} = g(t, \boldsymbol{a}) + B(\boldsymbol{\beta}^\mathsf{T} \boldsymbol{x})$. Hence, the survival and hazard functions are,

$$S(t) = \exp(-H(t)) = \exp[-\exp\{g(t, \boldsymbol{a}) + B(\boldsymbol{\beta}^\mathsf{T} \boldsymbol{x})\}], \qquad (9.3a)$$

$$h(t) = \frac{dH(t)}{dt} = \exp\{g(t, \boldsymbol{a}) + B(\boldsymbol{\beta}^\mathsf{T} \boldsymbol{x})\} \frac{dg(t, \boldsymbol{a})}{dt}, \qquad (9.3b)$$

where we will not write out the differential of $g(t, \boldsymbol{a})$ here.

The model was fitted using the R flexsurvspline() function in the flexsurv package. The function allows for two other models, the cumulative odds model and an inverse normal model, as well as the transformation to log(survival time).

The Japan data were randomly split into training and test sets. The split could have taken account some stratification factors.

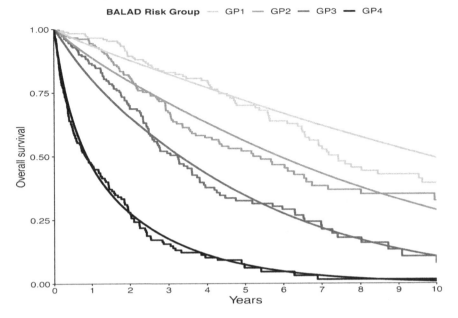

**FIGURE 9.3**

Japanese data: survival curves predicted from the BALAD model, patients split by the quartiles of the BALAD scores

Two knots were chosen for the spline model. We will not give details of the fitted spline part of the model, but just the BALAD part. The fitted model is

$$B = 0.19\,(\sqrt{\text{bil}} - 4.18) - (0.08\,(\text{alb} - 35.3) + 0.02\,(\text{AFP} - 3.17)$$
$$+ 0.01\,(\text{L3} - 15.4) + 0.18\,(\log(\text{DCP} - 0.68),$$

noting the model has been fitted to mean-corrected data. We see that $B$ increases, and hence survival worsens, as any of $\sqrt{\text{bil}}$, AFP, L3 and $\log(\text{DCP})$ increase, and as alb decreases. Harrell's $c$ is equal to 0.76.

The BALAD $B$-score was calculated for each patient in the Japan training set and then were placed into four groups according to the quartiles of the set of $B$-scores. Group 1 patients had $B$-score less than $-0.83$, Group 2 patients had $B$-score between $-0.83$ and $-0.27$, Group 2 patients had $B$-score between $-0.27$ and $0.38$, and Group 4 patients had $B$-score greater than $0.38$. The estimated survival function for each patient was calculated using Equation 9.3a and then the mean estimated survival function was found for each group. Figure 9.3 shows these overlaid on the corresponding Kaplan-Meier curves. The agreement between the two sets of curves is excellent.

Can we successfully use the BALAD $B$-score as a prognostic indicator of survival for patients from the UK? Does the $B$-score translate across to another country? The answer is yes but the baseline hazard function does not. The $B$-score was calculated for each patient in the UK dataset. These were used to calculate Harrell's $c$, as $c = 0.73$, an excellent value considering this is the test set and is almost equal to the value for the training set.

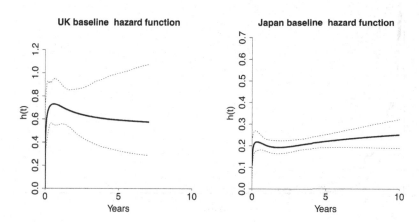

**FIGURE 9.4**

Baseline hazard functions from the UK and Japan BALAD models

The patients were then split into four groups using the quartiles of the set of UK $B$-scores, but we cannot use Equation 9.3a directly to estimate survival curves. This is because survival for UK patients is not so good as that for the Japan patients – their baseline hazard functions are different. Note, we are accepting that the proportional hazards model is appropriate across the two countries. When the RP model is fitted to the UK dataset, the estimated coefficients obtained are similar to those of the Japan model. Figure 9.4 shows the estimated baseline hazard functions using Equation 9.3b for the two countries, based on BALAD models. They are similar, but the scales are different. The easiest way to scale the Japan baseline hazard function to be appropriate for the UK data, is to use a Cox PH model to estimate the hazard rate, $\gamma$, for the variable country, and use that as the scaling factor. Then, $S_{UK}(t) = \{S_{JP}(t)\}^{\gamma}$, which can be used to calculate the estimated survival function for each of the UK patients. Figure 9.5 shows the mean survival curve for each of the four groups of UK patients based on the quartiles of $B$-scores, overlaid on the Kaplan-Meier curves. The fit is good apart for Group 4.

Why did we not have to worry about different baseline hazard functions for the two countries when we calculated Harrell's $c$? Because Harrell's $c$ only depends on ranks of the BALAD score and the ranks of the survival times.

In conclusion, we would say that the BALAD score is a good biomarker for prognosis of HCC within the population of patients with liver disease.

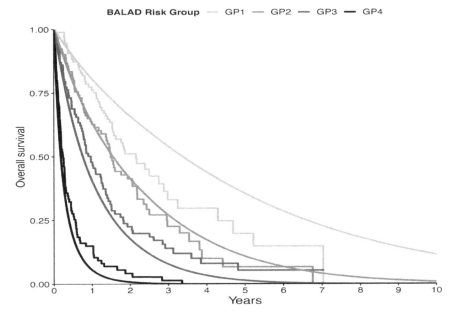

**FIGURE 9.5**
UK data: survival curves predicted from the BALAD model, patients split by
the quartiles of the BALAD scores

## 9.3   Predictive biomarkers for pancreatic cancer

Predictive biomarkers are similar to prognostic biomarkers, but concentrate
on predicting response to treatments. As part of *personalised medicine*, can a
biomarker help in the choice of treatments for a patient, can it indicate that
treatment A suits this particular patient better than treatment B?

We look at a study of two predictive biomarkers for pancreatic cancer,
dihydropyrimidine dehydrogenase (DPD) and human equilibrative nucleoside
transporter-1 (hENT1)[19]. Tumour tissue samples had been taken, prior to
treatment, from patients who received chemotherapy treatment by GEM or
5FU in the ESPAC3 trial and stored in freezers at very low temperature. Cores
taken from these tissue samples were placed in tissue microarrays, from which
$3\mu$m sections were cut and placed on slides, there being between four and
eight sections per patient. The sections were stained so that DPD could be
measured and then each was scored on a 0–3 point system by two pathologists:
0 – no staining, 1 – weak, 2 – moderate, 3 – strong staining. The more staining
there is, the higher the DPD expression in the tumour. A patient's final score

was taken as the mean, to the nearest integer, over all their section scores. This was converted to low and high DPD expression status; low-scores were 0,1; high-scores were 2,3. The hENT1 scores had been found for the patients in a previous translational study, using staining and determining a cutpoint to assess hENT1 expression as low and high. We will carry out some statistical analyses similar to those that appear in the DPD/hENT1 publication.

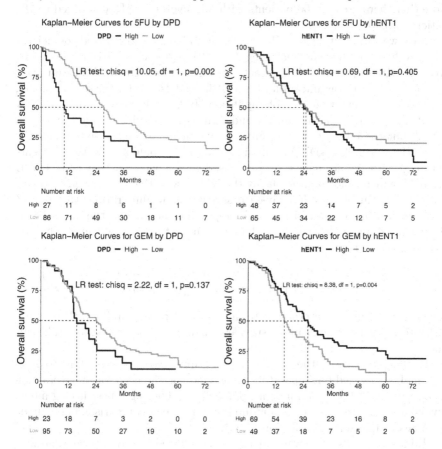

**FIGURE 9.6**

Kaplan-Meier curves for 5FU and GEM patients, split first by DPD = high/low and then hENT1= high/low

**TABLE 9.6**

Cox PH models for DPD and hENT1

| Model: | DPD | | Arm*DPD | | Model: | hENT1 | | Arm*hENT1 | |
|---|---|---|---|---|---|---|---|---|---|
| | coef | p-val | coef | p-val | | coef | p-val | coef | p-val |
| DPD (L) | −0.56 | 0.001 | −0.75 | 0.002 | hENT1 (L) | 0.18 | 0.230 | −0.18 | 0.408 |
| Arm (Gem) | | | −0.19 | 0.523 | Arm (Gem) | | | −0.22 | 0.238 |
| Arm*DPD | | | 0.39 | 0.267 | Arm*hENT1 | | | 0.74 | 0.012 |

Figure 9.6 shows the Kaplan-Meier curves for 5FU and GEM patients, split first by DPD = high/low and then hENT1= high/low. It appears that patients receiving 5FU with low levels of DPD have a survival advantage over patients with high levels, but hENT1 levels do not affect survival. For patients receiving GEM, DPD does not particularly affect survival, but patients with high levels of hENT1 have a survival advantage over patients with low levels. Note that there are only 50 patients with low levels of DPD, compared to 181 patients with high levels.

Now, we carry out a Cox PH, regression. Table 9.6 shows fitted coefficients and p-values for four models using Cox PH regression: DPD alone; DPD, Arm (treatment) and their interaction; GEM alone; GEM, Arm and their interaction. DPD is significant for both DPD models, but for the second model, Arm and the interaction are not significant. In the other two models, hENT1 is not significant for either model and Arm is not significant for the second model, but the interaction is significant. We might be disappointed that the interaction term, Arm*DPD is not significant as it would appear that patients with high DPD values have a survival disadvantage whether they receive 5FU or GEM, which is not bourne out by the KM curves. Had there been a greater sample size, things may have been different. As an illustration, the dataset was increased in size to 924 simply by repeating each row of the dataset four times and adding a very small random amount to the survival times. The Cox PH model then had

$$\text{DPD: coefft.} = -0.76 \ (p < 0.001); \ \text{Arm: coefft.} = -0.19 \ (p = 0.203);$$
$$\text{Arm*DPD: coefft.} = 0.39 \ (p = 0.025),$$

showing a significant interaction term. We need to be careful with p-values and interpretation when sample sizes are small in exploratory studies like this one!

In the publication, the data were split into two sets according to Arm and analyses carried out separately, thus removing the need for an interaction term. But, we will take a different approach, by choosing a best fitting model involving Arm, the biomarkers and interactions, the fit comparison being made on Akaike's information criterion (AIC).

Table 9.7 shows only some of the models fitted to save space. The first five models have terms in Arm, DPD and hENT1; the last three models incorporate the significant demographic and clinical variables selected in the multivariate Cox PH model in the publication. Note, country and resection margin are included as they clinical trial data were stratified on these two variables. For the models based on Arm and the biomarkers themselves, model 4 has the lowest AIC. It used both biomarkers and has an interaction term of Arm with hENT1, but not Arm with DPD. Harrell's $c$ is rather low for all models. Model 6 shows the results when the only terms in the model are the demographic and clinical variables, and immediately, AIC decreases significantly and $c$ increases, and so you would not choose to use the biomarkers for survival prediction above the demographic and clinical variables.

**TABLE 9.7**

Cox PH models for DPD and hENT1

| Model | −loglik. | AIC | Harrell's C |
|---|---|---|---|
| 1. A+D+A∗D | 891.7 | 1789.4 | 0.58 |
| 2. A+H+A∗H | 892.9 | 1791.8 | 0.55 |
| 3. A+D+H+A∗D | 890.7 | 1789.5 | 0.58 |
| 4. A+D+H+A∗H | 888.2 | 1784.3 | 0.58 |
| 5. A+D+H+A∗D+A∗H | 887.5 | 1785.0 | 0.59 |
| 6. C+RM+W+LN+A | 878.4 | 1776.9 | 0.64 |
| 7. C+RM+W+LN+A+D+H+A∗H | 867.7 | 1761.4 | 0.66 |
| 8. C+RM+W+LN+A+D+H+A∗D+A∗H | 866.9 | 1761.8 | 0.66 |

A=Arm, D=DPD, H=hENT1,
C=Country, RM=Resection Margin, W=WHO, LN=Lymph Node

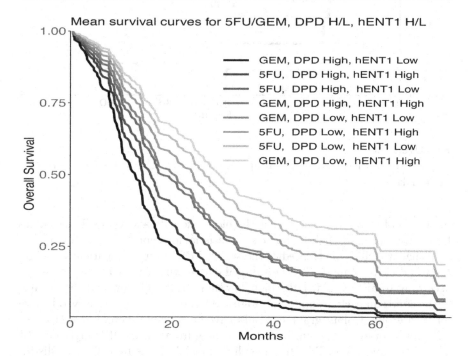

**FIGURE 9.7**

Mean predicted survival curves for eight groups of patients: GEM/5FU ×
DPD H/L × hENT1 H/L

If the biomarkers are to be considered as predictive biomarkers for use in
practice, then they must add value to the demographic and clinical variables.
Models 7 and 8 combine both types of variable and the AIC is reduced even

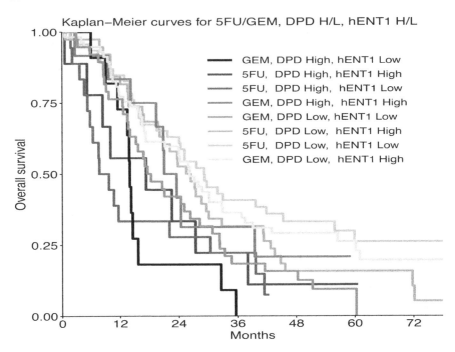

**FIGURE 9.8**

Kaplan-Meier curves for eight groups of patients: GEM/5FU × DPD H/L × hENT1 H/L

further, and so we can accept that the biomarkers do add value. The $c$ values have only increased slightly, but the $c$ statistic does not tell the full story.

What we will do is to predict the survival curve for each patient based on the fitted Cox PH model 7. The survival curves will be split into eight groups according to 5FU/GEM, DPD high/low and GEM high/low and the average of the curves in each group found. Figure 9.7 shows the average survival curve for each group. Colour graphics would have been helpful here, but we can see a hierarchy of predicted survival curves, going from GEM: DPD high; hENT1 low as the worst, to GEM: DPD low; hENT1 high as the best. Clearly, distinguishing the order of two curves that are close is not particularly meaningful, but we can see a pattern. For 5FU patients, high DPD gives a poor outcome, regardless of hENT1 values. For GEM patients, high DPD and low hENT1 gives poor survival, but for high DPD and high hENT1, survival is improved and is about the same as for low DPD and low hENT1. For high DPD and high hENT1 survival is good (in relative terms).

Figure 9.8 shows the corresponding Kaplan-Meier curves for the eight groups of patients. In contrast to the neatly ordered curves from the Cox PH model, these cross each other many times, not surprisingly. One curve had only nine patients with eight deaths. Median survival times from the KM curves are:

**TABLE 9.8**
Median survival times (months) for the eight groups of patients

| | | | |
|---|---|---|---|
| 5FU DH hL: **9** | GEM DH hL: **14** | 5FU DH hH: **17** | GEM DL hL: **18** |
| GEM DH hH: **22** | 5FU DL hH: **26** | GEM DL hH: **26** | 5FU DL hL: **29** |

Overall conclusions are: high hENT1 patients would benefit from GEM treatment; low hENT1 and low DPD patients would benefit from 5FU treatment. But, this is an exploratory study and so more research would be needed to make a definitive statement on which treatment to be given to patients based on these biomarkers. One problem in biomarker research is that the world moves on and treatments improve, making previous treatments obsolete, for instance you might now be given a combination of drugs such as Folfirinox. Then biomarkers studies have to be updated or new ones instigated.

## 9.4 Biomarker trial design

With biomarker research expanding and the drive for personalised medicine, the demand for more and more Phase II and III clinical trials would easily grow exponentially. Lack of money, resources and the number of patients from which to recruit, means more efficient designs have to be considered. There are various biomarker trial designs that have been used to date, for example:

- *Biomarker stratified design*: The patients are stratified into two groups based on the biomarker value, let us say, Biomarker+ and Biomarker−; then the patients in each group are randomised to the treatments under investigation.

- *Enrichment design*: Only patients that are Biomarker+ are selected for the trial.

- *Basket design*: A single treatment is investigated across a range of cancers.

- *Umbrella design*: Several treatments are studied within a single cancer population

- *Adaptive design*: A design that allows the trial to be modified during the course of the trial, for example one biomarker or treatment stops being used and another takes its place.

- *A combination of some of the above*

We give two examples.

## Basket trial

The Jupiter trial was a basket trial of combination therapy with trastuzumab and pertuzumab for patients with solid cancers harbouring human epidermal growth factor receptor 2 amplification (HER2). (HER2 is a gene associated with the repair of breast cells and growing more of them.)[65] The trial was a multicenter, single-arm, basket phase II study based in Japan. Patients who had bile duct, urothelial, uterine, ovarian, and other solid cancers were recruited.

## Adaptive, umbrella, biomarker stratified trial

The FOCUS4 trial is an adaptive, biomarker stratified trial for colorectal cancer, that opened in 2014 and is still recruiting patients. To date is has opened 94 centres and recruited 1349 patients. Patients are recruited, are given 16 weeks of standard chemotherapy and then are allocated to one of four sub-trials based on a biomarker panel of genes. The sub-trials can go seamlessly from phase II to phase III. The trial can also be described as *multi-arm, multi-stage* (MAMS). Further details can be found at http://www.focus4trial.org/.

## Reporting biomarker studies (REMARK)

Clinical trials should be reported using the CONSORT guidelines. Likewise, biomarker trials should also follow the *Reporting Recommendations for Tumour Marker Prognostic Studies* (REMARK) which has a 20 item checklist, covering: an introduction, materials and methods, results and a discussion. See Sauerbrei[55].

# 10

## Cancer Informatics

The term *Informatics* has different connotations and definitions in different countries. For some it would simply be "computer science". I like the definition of, "*The science of information and the practice of information processing*". When applied to various disciplines, we have subjects like *Engineering Informatics, Chemoinformatics, Financial Informatics, Bioinformatics* and for us, *Cancer Informatics* which is a subset of *Bioinformatics*. *Big Data* and *Big Data Analytics* are about storing, analysing and extracting information from very large datasets. How large does a large dataset have to be, to be called "Big Data"? The cancer informatics data we will use relate to several thousand genes on, say, fifty patients and would be a small fraction of the size a large bank's total financial transactions over a period of a year. The term, informatics emerged in the 1950's and big data in the 1990's.

In this chapter we give a very basic introduction to cancer informatics, concentrating on genetic data, but the subject area can cover any data relating to cancer, for instance cancer meta-analyses could be classed as cancer informatics. The delightful thing about cancer informatics and bioinformatics is that is usually requires close team work, with members such as clinicians, geneticists, medical scientists, computer scientists, informaticians and statisticians being involved.

Cancer informatics and bioinformatics in general, covers various genetic areas and gives us some *"omics"* technologies. Those relevant to the study of cancers are:

- *Genomics*: the study of the genome

- *Transcriptomics*: the study of RNA molecules

- *Proteomics*: the study of the proteome (all the proteins produced by the genes)

- *Metabolomics*: the study of chemical processes involving metabolites (substances involved in metabolism)

We will mainly be looking at transcriptomics using two data sets, one based on microarrays and one on next generation sequencing. There is a lot of jargon in cancer informatics. You may wish to return to Chapter 2 to remind yourself of some of it. Also, the R programs that can be downloaded from the publisher's website contains more detail for this chapter than for the others, since it can be difficult to get started with cancer informatics.

First, we briefly describe the technologies that generate the genetic data.

DOI: 10.1201/9781003041931-10

## 10.1   Producing genetic data

Generating genetic data needs a sample of DNA, but sometimes the sample is too small to be used. For example, a forensic scientist may only have a tiny spec of blood from which to extract the genetic data. However, this problem can be overcome using the clever method of *polymerase chain reaction* (PCR) that amplifies DNA.

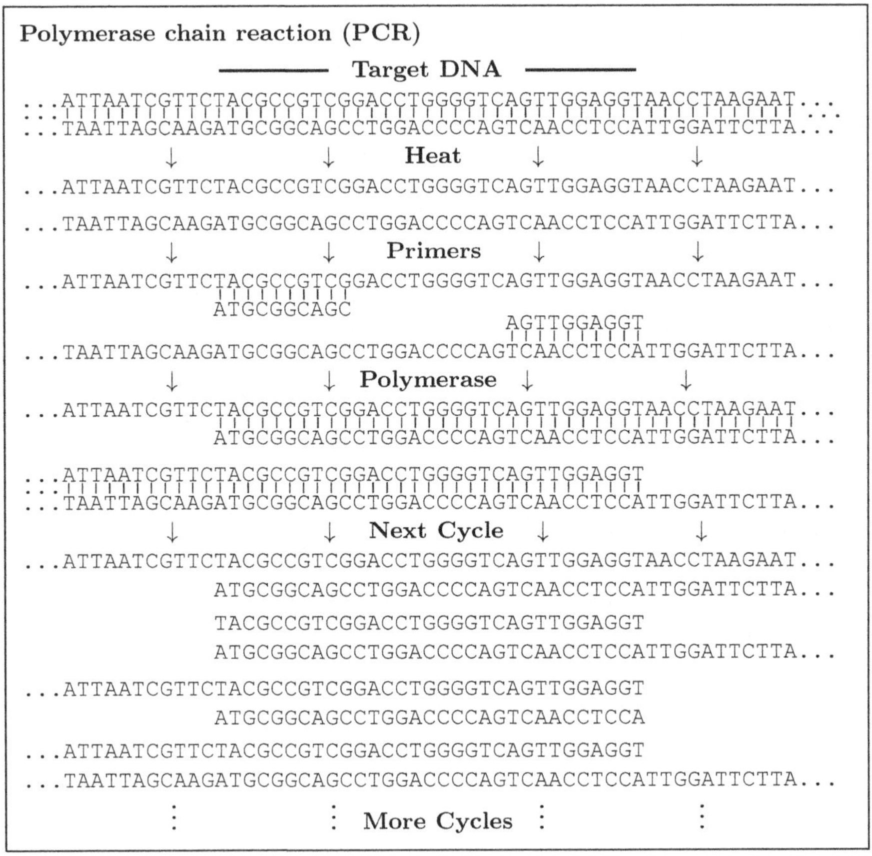

```
Polymerase chain reaction (PCR)
              ———— Target DNA ————
...ATTAATCGTTCTACGCCGTCGGACCTGGGGTCAGTTGGAGGTAACCTAAGAAT...
   |||||||||||||||||||||||||||||||||||||||||||||||||||||
...TAATTAGCAAGATGCGGCAGCCTGGACCCCAGTCAACCTCCATTGGATTCTTA...
        ↓            ↓     Heat     ↓            ↓
...ATTAATCGTTCTACGCCGTCGGACCTGGGGTCAGTTGGAGGTAACCTAAGAAT...

...TAATTAGCAAGATGCGGCAGCCTGGACCCCAGTCAACCTCCATTGGATTCTTA...
        ↓            ↓    Primers   ↓            ↓
...ATTAATCGTTCTACGCCGTCGGACCTGGGGTCAGTTGGAGGTAACCTAAGAAT...
            ATGCGGCAGC
                                   AGTTGGAGGT
                                   ||||||||||
...TAATTAGCAAGATGCGGCAGCCTGGACCCCAGTCAACCTCCATTGGATTCTTA...
        ↓            ↓  Polymerase  ↓            ↓
...ATTAATCGTTCTACGCCGTCGGACCTGGGGTCAGTTGGAGGTAACCTAAGAAT...
            ||||||||||||||||||||||||||||||||||||||||||||
            ATGCGGCAGCCTGGACCCCAGTCAACCTCCATTGGATTCTTA...
...ATTAATCGTTCTACGCCGTCGGACCTGGGGTCAGTTGGAGGT
   ||||||||||||||||||||||||||||||||||||||||||
...TAATTAGCAAGATGCGGCAGCCTGGACCCCAGTCAACCTCCATTGGATTCTTA...
        ↓            ↓  Next Cycle  ↓            ↓
...ATTAATCGTTCTACGCCGTCGGACCTGGGGTCAGTTGGAGGTAACCTAAGAAT...
            ATGCGGCAGCCTGGACCCCAGTCAACCTCCATTGGATTCTTA...
            TACGCCGTCGGACCTGGGGTCAGTTGGAGGT
            ATGCGGCAGCCTGGACCCCAGTCAACCTCCATTGGATTCTTA...
...ATTAATCGTTCTACGCCGTCGGACCTGGGGTCAGTTGGAGGT
            ATGCGGCAGCCTGGACCCCAGTCAACCTCCA
...ATTAATCGTTCTACGCCGTCGGACCTGGGGTCAGTTGGAGGT
...TAATTAGCAAGATGCGGCAGCCTGGACCCCAGTCAACCTCCATTGGATTCTTA...
        ⋮            ⋮  More Cycles ⋮            ⋮
```

**FIGURE 10.1**
Illustration of polymerase chain reaction (PCR)

## 10.1.1 Polymerase chain reaction (PCR)

To amplify DNA using PCR, a segment of DNA is placed in solution containing primers (short sequence of single stranded DNA), free DNA nucleotides, DNA polymerase (an enzyme that catalyses the synthesis of DNA molecules) and other ingredients. Figure 10.1 illustrates the process where a target section of DNA is to be amplified. The solution is heated to over 90°C to separate the DNA strands. The temperature is then lowered and primers attach themselves to the ends of the target section of DNA. The polymerase builds two new strands starting at the primers, but note only one end is extended. Heating again separates the strands into four separate ones. On lowering the temperature, primers again attach themselves and the polymerase extends the strands. There are now four strands of DNA. The next step will produce eight strands of the target DNA. The process continuously repeats, producing $16, 32, \ldots, 2^n$ strands and so easily making many copies of the target DNA, which can then be used in experimentation.

RNA can also be amplified using PCR, by first transcribing it to complementary DNA (cDNA) by an enzyme, reverse transcriptase.

## 10.1.2 DNA microarrays

The development of DNA microarrays started in the 1970s and by the 1990s and 2000s they had become to main tool for concurrently measuring the expression of thousands of genes. A typical gene expression microarray (also known as a gene array or DNA chip) is made with a flat piece of material, such as glass or silicone, with spots of known DNA sequence or genes attached in a grid pattern. These are known as probes.

An experiment to look for difference in the gene expression of tumour cells and normal cells within an organ, starts with a sample of mRNA from the tumour cells and a sample from the normal cells. The two mRNA samples are converted into complementary DNA (cDNA) and labelled with fluorescent dyes, usually green for one sample and red for the other. Let us assume red is used for the tumour sample and green for the normal tissue sample. The samples are mixed together and then applied to the microarray, where the cDNA molecules bind to the probes (hybridisation). An image of the microarray is taken using fluorescent imaging. For a particular probe (gene), if the colour is more red than green, then the gene is expressed more in the tumour cells, than in the normal cells (the gene is *overexpressed* in the tumour cells). If the colour is more green than red, then the gene is expressed more in the normal cells, than in the tumour cells (the gene is *underexpressed* in the tumour cells). If the colour is yellow, the gene for the tumour cells is neither overexpressed nor underexpressed. The image gives the relative gene expression difference for each gene.

This sounds straightforward, but as with all experiments, experimental design is essential. How many tumour and corresponding normal cell samples should we use?

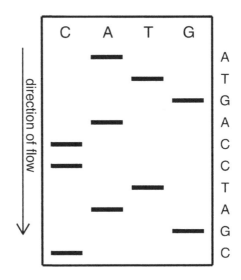

**FIGURE 10.2**
Gel electrophoresis of four PCR samples for sequencing target DNA

### 10.1.3   Next generation sequencing (NGS)

Nowadays, microarray technology have been overtaken by next *generation sequencing* (NGS) technology, where whole genomes can be sequenced very quickly. Microarrays are still used for some experiments and also there is a huge amount of historical microarray data that can be used for current research and so microarrays and their data are not obsolete yet. To understand next generation sequencing, it is worth describing *first generation sequencing*, when sequencing of genomes started in the 1970s. There were two technologies at the time, the Maxam-Gilbert method and the Sanger method, the latter becoming the most popular. We describe Sanger sequencing.

In Sanger sequencing, four inhibitors of chain elongation in the process of DNA polymerisation are used. They are dideoxynucleotides (modified nucleotides), denoted by ddATP, ddCTP, ddGTP, ddTTP, and they stop a chain from further elongation at an A nucleotide (denoted dATP), a G nucleotide (dGTP), a C nucleotide (dCTP) and a T nucleotide (dTTP) respectively. Polymerase chain reaction is carried out with four separate samples of the DNA as described above, but with one of dideoxynucleotides added to each sample. So the first sample might have ddCTP added but in a smaller concentration than the "C" nucleotide, and similarly for the other three samples.

During PCR for the strand (or fragrant) building caused by the polymerase in the first sample, each fragment will build as normal with the dATPs, dCTPs, dGTPs, dTTPs being added until, at random, a ddCTP is added and so the fragment will terminate at that point, and similarly for the other three samples. After the PCR is finished, each of the four samples will contain fragments of various lengths, all terminating in their added dideoxynucleotide.

Next, the samples are subjected to gel electrophoresis, which separates the fragments by size. Figure 10.2 illustrates this. The gel has four wells for the four samples and an electric current makes the fragments move across the gel. When the current is switched off, the shortest fragment will have travelled the furthest distance and the longest fragment the shortest. The position of the various length fragments are shown as bands on the gel. From these, the target sequence can be determined. For illustration, suppose the target fragment is actually CGATCCAGTA, but we do not know this at the start of the experiment. During the PCR, the dideoxynucleotide, ddCTP, will have produced the fragments: C, CGATC and CGATCC; ddATP will have produced the fragments CGA, CGATCCA, CGATCCAGTA; ddTTP will have produced the fragments CGAT, CGATCCAGT and ddGTP will have produced the fragments CG, CGATCCAG. These all appear on the gel of Figure 10.2 and the target fragment can be read on the left hand side, from the bottom, as C-G-A-T-C-C-A-G-T-A.

Sequencing technology quickly improved from first generation sequencing, the dideoxynucleotides were labelled using fluorescent dyes and a single sample of fragments after PCR are run through a capillary gel electrophoresis tube. As each fragment reaches a particular point, a laser reads the dideoxynucleotide of the fragment, based on its dye colour and hence the sequence of colours detected will give the DNA sequence. Technologies were then invented to speed up the DNA sequencing, giving us *next generation sequencing*. These technologies have names like Illumina Sequencing, Ion Torrent Sequencing and Oxford Nanopore Sequencing.[30]

The raw data from digital imaging of microarrays and from gene sequencing has to be preprocessed before final analysis and we will only outline the steps needed to do this. We will use two examples to illustrate various analyses of genetic data, the first uses data from a microarray experiment on metastatic melanoma and the second data from NGS of clear cell renal carcinoma tumours.

## 10.2   Analysis of microarray data

We start with a brief description of the dataset we will use for illustration of how microarray data are analysed.

**Metastatic melanoma**
The data we use for illustration comes from a microarray experiment on metastatic melanoma[49]. Microarray data from eight patients who had metastatic recurrences over a period of years were used to study the evolution of the metastases. There were a total of 26 recurrences. Data for another 22 patients with rapid disease course and only one metastases, were used for comparison, giving a total of 48 microarrays. DNA copy number, alterations and DNA expression levels were measured, but we will only use the expression levels. The patients with recurrent metastases were labelled, "A", "B", ..., "H", with individual ordered metastases labelled as "A1", "A2", "A3", "A4", "B1", "B2", etc. and "A2a", "A2b", etc. if there were more than one metastasis at a particular time point.

## 10.2.1   Preprocessing microarray data, low-level analysis

The main steps for preprocessing microarray data, a *low-level analysis*, are: *background correction, normalisation* and *summarisation*. There are various methods available to carry out these steps. As an example, we describe the steps for an Affymetrix GeneChip. The term low-level analysis does not imply the methods used are simple!

**Background correction**
The probes on the chip are in pairs and each is 25 bases long. Within a particular pair, one of the probes is a perfect match (PM) to a portion of a particular gene's mRNA and the other is a mismatch (MM) where the centre base is changed. A gene is typically represented by 11–20 pairs of probes. The idea of having mismatches is that the portion of mRNA that will hybridise to a probe is going to "choose" the PM probe and not the MM probe. However, the hybridisation of the MM probe is not necessarily zero, and is taken to be a measure of the *background noise* of non-specific hybridisation and general noise.

Estimation of the background noise can be based on PM−MM values, but more sophisticated methods are commonly used. For instance, Wu[75] discusses a *robust multi-array analysis* model (RMA), where for each probe pair,

$$PM = O_{PM} + N_{PM} + S,$$
$$MM = O_{MM} + N_{MM} + \phi S,$$

where $O$ represents optical noise, $N$ represents nonspecific hybridization and $S$ (the "signal") is a quantity proportional to RNA expression. As the MM can detect some signal, it includes a proportion, $\phi, (0 < \phi < 1)$ of $S$. It is assumed that $O$ follows a lognormal distribution and $N_{PM}$ and $N_{MM}$ follow a bivariate lognormal distribution. The data for each probe pair on each array are used to estimate parameters for the distributions and $\phi$. The signal $S$ can then be estimated. Other assumptions can be made to give different models, e.g. $\phi = 0$,

$O$ is a constant dependent on array, and $S$ is taken as a random variable. We go no further than to say that whatever model is used, background correction is carried out.

## Normalisation

Normalisation of the arrays is needed to remove bias and systematic measurement errors such as differences in dye incorporation and RNA quality across the arrays. There are various methods to carry out the normalisation. We assume most genes are equally expressed across all samples and only a relatively small proportion are differentially expressed. For the *scaling* method, we assume that total hybridisation is the same across all samples, and so we can adjust the expression values on each array simply by scaling them so that all means (or medians) are the same across the arrays.

The *quantile* method attempts to normalise the whole distributions of expression across the arrays. Let there be $N$ expression values on each array. These are ordered for each array. The standard distribution is found by averaging the ordered values across the arrays. So the minimum value of the standard distribution is the mean of the minimum values across the arrays. The second smallest value is the mean of the second smallest values of all the arrays, etc. Then for a particular array, the gene with the minimum expression value has that value replaced by the minimum value of the standard distribution, and so forth. Also house-keeping genes, where the samples should have equal expression of these can be used for normalisation, this could involve non-human genes.

## Summarisation

After normalisation, we need to summarise the expression values for probes within each array for each gene. The simplest way would be to use the mean of the PM−MM values, possibly using a robust mean that discards outliers. However, more sophisticated approaches have been suggested. Following Irizarry[23], let the $j$th probe for the $n$th gene on the $i$th array have log-transformed expression value, $Y_{ijn}$, having completed the background correction and the normalisation. A linear model for $Y$ is,

$$Y_{ijn} = \mu_{in} + \alpha_{jn} + \epsilon_{ijn},$$

where $\mu_{in}$ is the population mean expression for gene $n$ for array $i$, $\alpha_{jn}$ is a probe effect for the $n$th gene and $\epsilon_{ijn}$ is the mean zero error term. The model is fitted using a robust regression method, *median polish*, which is an iterative method involving the medians of the rows and columns. The method is robust to outliers.

Once fitted, the estimates, $\hat{\mu}_{in}$ give the $I$ expression values to be taken forward for the high level statistical analysis. However, at this stage the genes are labelled by probeset names, not their gene name.

## Gene annotation

To give the genes their names, we have to map the probeset names to genes, using a mapping library. Some arrays have one probe matching to one gene, but others can have several probes that match to a particular gene. In this case, what can we do? For a particular gene, we could take the mean or median expression of all the probes that match to that gene, or we could take the maximum value.

**Quality control and filtering**

Quality control checks can be carried out before or after preprocessing, for instance to look for arrays that might be outliers and could be discarded from the analysis. Gene filtering removes genes from the analysis based on particular criteria, e.g. the variance of the expression level of a gene across arrays is too large, or it does not help to differentiate cases from controls.

## 10.2.2    Preprocessing the melanoma microarray data

We use *Bioconductor* software, which is based around R, for the analysis of our genetic datasets. Bioconductor can be installed from the Bioconductor website, *www.bioconductor.org* and Bioconductor packages downloaded. There are many packages, some of the ones we use are Biobase, limma, oligo and TGAbiolinks. Two documents worth looking at are, "An Introduction to Bioconductor's ExpressionSet Class"[11] and "An end to end workflow for differential gene expression using Affymetrix microarrays, Version 2"[32]. The first describes ExpressionSets, which are objects that hold the different types of data for a microarray experiment, such as assay data, phenotypic data, feature annotations and meta-data. The second shows how to carry out a complete analysis of the data. The flow of our work follows much of this.

**Downloading the data**

The data we will use are from the European Bioinformatics Institute, which is part of the European Molecular Biology Institute (EMBL-EBI). Their *ArrayExpress* database holds a wealth of genetic data (https://www.ebi.ac.uk/arrayexpress/). Browsing ArrayExpress, our dataset is in the project: E-GEOD-38119–Gene expression analysis of recurrent metastatic melanoma lesions. You can download the raw data or the processed data, here we use the raw data. The microarrays used were the type: A-AFFY-141–Affymetrix GeneChip Human Gene 1.0 ST Array [HuGene-1_0-st-v1].

The Bioconductor package ArrayExpress can be used to download the data, but unfortunately, that did not work for us and so the raw data was simply downloaded directly from the website. The gene expression data are contained in CEL files, e.g. GSM934837_1_HuGene-1_0-st-v1_.CEL. The SDRF (Sample and Data Relationship Format) file contains information about the project, such as the data files, the characteristics of the samples and the relationship between samples, arrays, etc. Note, when using this in your R program, make the folder containing the data your working directory.

The function oligoread.celfiles() is used to read in the CEL files and analyse the arrays at the probe level. The function read.delim("xxxx.sdrf.txt") reads the SDRF file (Sample and Data Relationship Format), which contains details of the samples and the relationship between samples, arrays, etc.

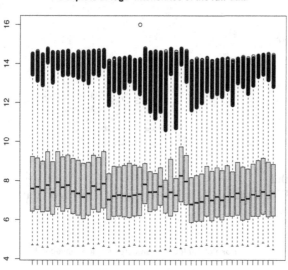

**FIGURE 10.3**

Boxplots of $\log_2$ transformed raw gene expression data (intensities)

**Some quality control**

First we look at boxplots of the raw probe data, but with a $\log_2$ transformation. Without transforming, the function struggled with these particular data because of many outliers. Figure 10.3 shows the boxplots. The medians are varied and there are still many outliers at the top end for all the microarrays. The function oligo::boxplots() has decided it cannot squeeze all the array labels onto the x-axis, but the missing labels are easily figured out as they are in a particular order.

Next, using the function isoMDS(), a non-metric multidimensional scaling (MDS, see the Appendix) for the microarrays, using the raw data, is shown in Figure 10.4. The *STRESS* was 15.4%. The arrays form three groups, one containing "F", "G" and "H" arrays, one containing mostly non-recurrent arrays and one containing mostly "A", "B", "C", "D" and "E" arrays.

**Background correction, normalisation and summarisation**

Next we carry out the pre-processing, RMA, with the function, oligo::rma(), followed by boxplots of the normalised data, Figure 10.5 and another MDS,

Nonmetric MDS plot of arrays (raw data)

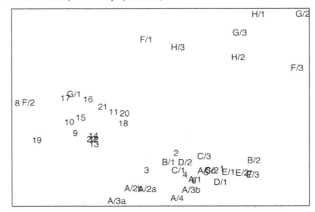

**FIGURE 10.4**
Nonmetric MDS of arrays based on raw gene expression data

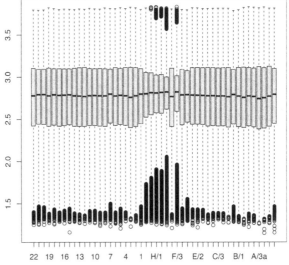

**FIGURE 10.5**
Boxplots of normalised gene expression data (intensities)

Figure 10.6. The boxplots show that the medians are more consistent, but there are a group of recurrent metastases arrays in the centre that give cause for concern. These are again seen in the MDS plot, which had *STRESS* of 18.5%.

Nonmetric MDS plot of arrays (processed data)

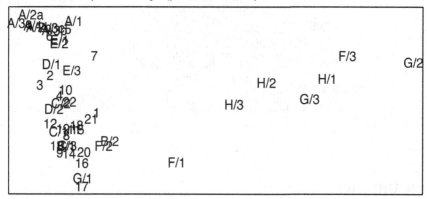

**FIGURE 10.6**

Nonmetric MDS of arrays using normalised gene expression data

### Gene annotation

Now we can match the probes to genes. The gene library for our microarray data is "pd.hugene.1.0.st.v1", which can be downloaded from the Bioconductor website. The function oligo::getNetAffx() will sort out the gene annotation. Once annotated, for each gene, we take the median of the expression values for all the probes matching to the particular gene. For example, the gene, "PTEN" has two probesets, 7928959 and 8160718 associated with it, and so the median or mean of the two expression values is used for the gene. We are left with 23,307 genes that have expression values.

### Filtering

We now filter out genes with low expression values measured as the median across all arrays. This is because the level of expression of the low-expression genes is of the same size as the background noise. Figure 10.7 shows a histogram of the medians for all the genes. We choose a cut-off of 4.3 and remove all genes with expression values below this. We are left with 21,501 genes.

Finally we make a $\log_2$ transform of the expression values, ready for the high-level analysis.

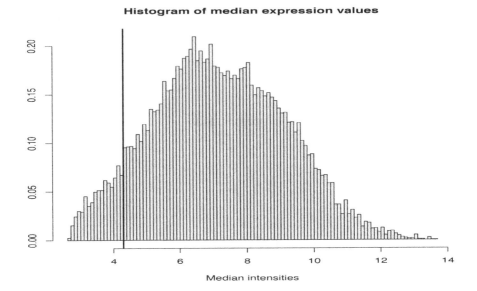

**FIGURE 10.7**
Histogram of median expression values for all genes

## 10.2.3   High-level analysis of the melanoma data

Now we are ready for a high-level analysis of the gene expression data. The following gives a flavour of what can be done. We launch into the execution of *modified* t-tests to see which genes are differentially expressed between the recurrent and non-recurrent metastases groups. This is achieved using the package limma and the function lmFit() to fit a linear model to each gene,

$$y = \beta_0 + \beta_1 x,$$

where $y$ is the $\log_2$ expression value and $x$ is an indicator variable, with values 0 for the non-recurrent group and 1 for the recurrent group. Note, we have ignored the fact that for the recurrent group, each patient has more than one expression value. We could take that into account, but for simplicity and the fact that resulting p-values are only a guide, we do not do so. Note, in other situations, our linear model could be more complicated with various factors added.

We now use the function limma::eBayes() to adjust the p-values of the t-tests. Without going into details, eBayes adjusts standard errors towards a common value across all genes, increasing low standard errors, hence increasing the corresponding p-value, and decreasing high standard errors, hence decreasing the p-value. Next, the function limma::decideTests() orders the adjusted p-values and uses a method to allow for the *false discovery rate* (FDR).

## False discovery rate (FDR)

We digress to explain false discovery rate. The problem with carrying out so many t-tests or other hypothesis tests is that there will be many Type I errors made. So, we will have a high false discovery rate. We have so many very small p-values and we need to adjust these to allow for the FDR.

Suppose we have $m$ p-values from $m$ tests, and that an unknown number of these, $m_0$, have null hypotheses that are true, i.e. no difference. Let the p-values for the tests be denoted by $p_1, p_2, \ldots, p_m$, and their ordered values by $p_{(1)}, p_{(2)}, \ldots, p_{(m)}$. Let there be $R$ tests where the p-values indicate a significant result. Let there be $V$ of these where the actual null hypothesis is true (no difference) but of course we do not know this number. Then the false discovery rate, $Q$, is $Q = V/R$. We need a way of selecting tests that we will deem to be significant.

A popular method to control the FDR is use the *Benjamini-Hochberg* procedure, where the largest $k$ is found such that $p_{(k)} \le k\alpha/m$, where $\alpha$ is a chosen FDR. The $k$ tests with the p-values, $p_i \le p_{(k)}$ are declared significant. It can be shown the $E(Q) \le m_0\alpha/m \le \alpha$. An adjusted p-value can be quoted as $p_{(i)}m/k$.

We might be able to assess the number of null-hypotheses that are false, i.e there is a difference. If all $m$ null hypotheses were true, then the p-values would follow a uniform distribution on the interval, $[0, 1]$. So we would expect about $m\alpha$ of the $m$ p-values to be less than or equal to $\alpha$, a chosen significance level. Hence, an estimate of the number of true significant differences is $R - m\alpha$. Again, this does assume the tests are independent.

## TABLE 10.1

Genes ordered by differential expression p-value

| Order | Gene | logFC | AveExpr | t | p-value | adj. p-value |
|---|---|---|---|---|---|---|
| 1 | MT-TA | 0.14 | 3.59 | 7.366 | $1.748e^{-09}$ | < 0.001 |
| 2 | RNU6-762P | −0.21 | 2.19 | −6.832 | $1.18e^{-08}$ | < 0.001 |
| 3 | MT-TQ | 0.19 | 3.61 | 6.800 | $1.32e^{-08}$ | < 0.001 |
| 4 | MT-TM | 0.13 | 3.40 | 6.630 | $2.42e^{-08}$ | < 0.001 |
| 5 | NOTCH2NL | −0.08 | 3.41 | −5.943 | $2.82e^{-07}$ | 0.001 |
| $\vdots$ | $\vdots$ | $\vdots$ | $\vdots$ | $\vdots$ | $\vdots$ | $\vdots$ |
| 819 | OR5B12 | 0.09 | 2.33 | 3.311 | 0.002 | 0.04962 |
| 820 | RNU6-773P | 0.09 | 2.81 | 3.311 | 0.002 | 0.04962 |
| 821 | FOXP2 | 0.25 | 2.38 | 3.308 | 0.002 | 0.05004 |
| $\vdots$ | $\vdots$ | $\vdots$ | $\vdots$ | $\vdots$ | $\vdots$ | $\vdots$ |

We return to the data. The function limma::decideTests() uses the p-values from the lmFit() function and carries out an FDR adjustment, the default method being the Benjamini-Hochberg procedure. The result shows how many

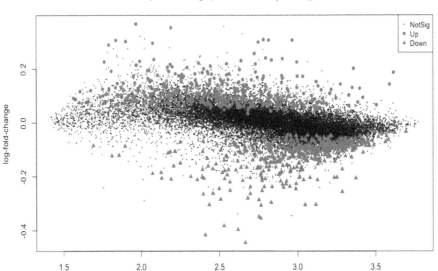

**FIGURE 10.8**
Mean-difference plot of log expression values for genes

up- and down-regulated genes there are. For the melanoma data, there are 385 up-regulated, 435 down-regulated and 22,487 non-significant genes.

The function limma::topTable() orders the p-values from the lmFit() function, carries out an FDR adjustment, and displays a specified number of them. Table 10.1 shows a small portion of the table produced by topTable(), giving the estimate of log fold change, mean log expression across all arrays, the modified t-values and its p-value, and the adjusted p-value using the Benjamini-Hochberg procedure with $\alpha = 0.05$. Note, the fold change for two expression values is the ratio of one expression value to the other. Usually $\log_2$ fold change is used, e.g. for values of 2 and 16, the fold change is $16/2 = 8$, and $\log_2(8) = 3$, i.e. "2" has to be doubled three times to reach "16".

We see there are 820 significant genes. Figure 10.8 shows a plot of log fold-change against mean log expression (i.e. difference versus mean of log expression values), with the up- and down-regulated genes emphasised in grey, which is inferior to a plot with the red/blue highlight colours that we would usually choose.

The expression differences appear to be more or less constant over the mean log expression values. The figure was produced using the function limma::plotMD. The function limma::volcanoplot plots a "volcano plot", as

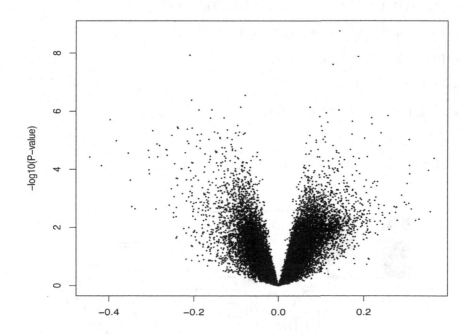

**FIGURE 10.9**
Volcano plot of differential expression p-values

shown in Figure 10.9. Scaled p-values are plotted against $\log_2$ fold change, giving a visual summary, where we would be particularly looking at the p-values for large fold changes.

Next we are going to consider a group of 30 known melanoma related genes, mentioned in the original paper: *ALCAM, ANGPT1, B2M, BRAF, BRCA1, CASP8, CDH1, CDK2, CDKN1A, CDKN2A, CDKN2B, CIITA, EGF, FAS, ITGA4, KIT, MAGEA1, MAGEA4, MAGEA9B, MAGEB2, MAGEC2, MGMT, MMP15, MMP19, MYC, NRAS, PTEN, RBL2, TEP1* and *TP53BP2*. Figure 10.10 shows a heatmap for the 30 genes and 48 arrays. We have had to use a grey-scale, whereas normally a colour scale, e.g. red-blue, would be used. We cannot distinguish a large negative x-score from a large positive one. We could use the heatmap to select clusters of genes and arrays. Looking at the dendrograms, suggested clusters for genes and arrays might be,

- {MAGEA4, MAGEC2, B2M, MAGEA1, MAGEA9B, CIITA, MMP19, MMP15, ANGPT1, ITGA4, CASP8, EGF}

- {CDKN2A, CDKN2B, MYC, MGMT, BRAF, TEP1, PTEN, BRCA1, CDK2, NRAS, RBL2, TP53BP2 }

- {ALCAM, CDKN1A, FAS, MAGEB2, CDH1, KIT }

- {A/2a, A/3a, A/4, A/3b, A/2b, A/3c, A/1}

- {20, 9, D/2, D/1, G/1, 14, C/3 }

- {2, E/2, E/1, 10, E/3, 6, 5, 22, 3 }

- {7, 4, F/1, G/3, G/2, H/2, H/1 }

- {17, 13, 11, C/1, C/2, 19, 1 }

- {F/2, H/3, F/3, 21, 16, 12, 18, 8, 15, B/2, B/1 },

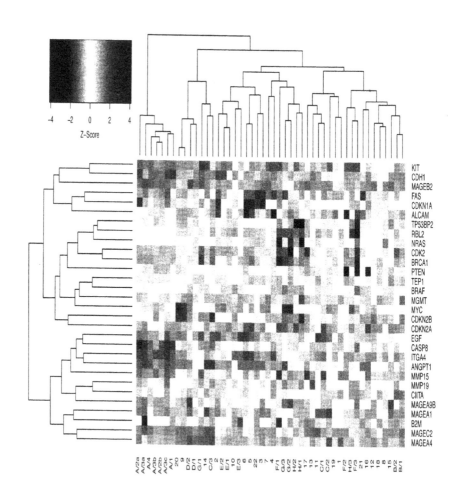

**FIGURE 10.10**
Heatmap for 30 melanoma-related genes

or perhaps finer divisions may be appropriate. Note, the second cluster of arrays contains the "rogue" arrays that have stood out from the others – this still worries us and now is the time we should talk to the geneticists.

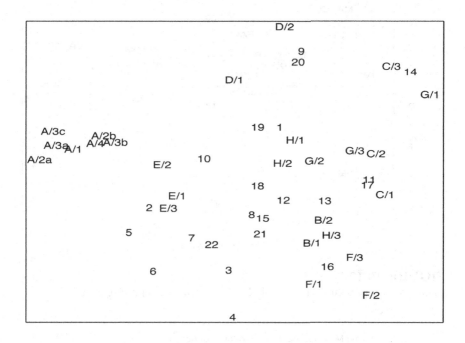

**FIGURE 10.11**
Multidimensional scaling of the arrays based on 25 melanoma-related genes

Figure 10.11 shows a non-metric MDS plot for arrays based on Euclidean distance using the 30 melanoma-related genes. The STRESS is 17.1%. We see that the "A" arrays cluster together and are separated from the others, you could argue the "C"s and "G"s are close to each other, as are the "B"s and "F"s. The "E"s are close to the "A"s and the "D"s are on their own with "9" and "20". This, of course, is somewhat subjective.

Figure10.12 shows a principal components analysis (PCA) plot of the genes, using the first and second principal components. Figure 10.13 shows non-metric MDS plot for the genes. The plots are similar when the y-axis of one of them is reversed, i.e. y-values multiplied by $-1$. Looking at the MDS plot, we note that the genes on the left hand side of the plot, MAGE genes, ANGPT1, EGF, MMP19 and ITGA4, are associated with cell matrix remodelling and surface adhesion (cell scaffolding and the binding of cells) and angiogenesis (growing a blood supply for the cell). At the top, KIT is a proto-oncogene involved with cell signalling. Close by, CDH1 is a tumour suppressor gene involved with cell adhesion, away from EGF which is also

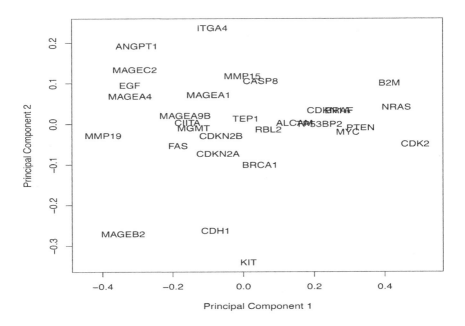

**FIGURE 10.12**
Principal components analysis of the 30 melanoma-related genes

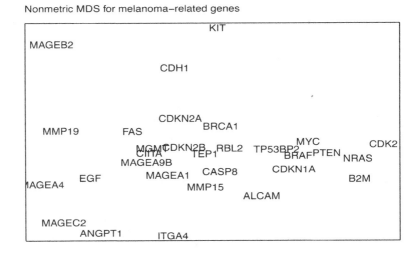

**FIGURE 10.13**
Multidimensional scaling of the 30 melanoma-related genes

involved with cell adhesion, but is a proto-oncogene. On the right hand side of the plot, NRAS, CDK and PTEN are involved with the cell cycle, and B2M makes a protein found on cell surfaces mainly on tumour cells.

**Longitudinal analysis**

Lastly, we consider how gene expression might change over time for the recurrent melanoma group. We only show details for the MGMT gene, where the longitudinal model is,

$$expr = time + patientID + error,$$

where time is the recurrence time, 1,2,3 or 4, patientID is A,B,..., H, a random effect and error is a normal error term. The model was fitted using the lmer() function.

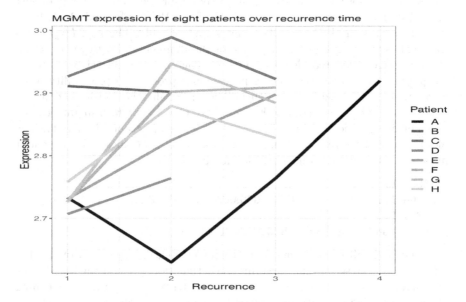

**FIGURE 10.14**
MGMT expression plotted against recurrence time for the melanoma patients

Figure 10.14 shows a plot of the expression values of MGMT for the eight patients against their recurrences. It looks as if the expression values increase from time 1 for six of the patients. Note, time = 4 was excluded from the model fitting, because only one patient had a value at this time point. The estimated fixed effects were:

| | Estimate | Std. Error | t-value | 95% CI |
|---|---|---|---|---|
| Intercept | 2.78 | 0.033 | 84.48 | [2.71, 2.84] |
| Time2 | 0.077 | 0.032 | 2.45 | [0.016, 0.139] |
| Time3 | 0.091 | 0.035 | 2.62 | [0.024, 0.159] |

The standard deviation for the patient random effect was 0.068 and the error standard deviation was 0.063.

We can say there is evidence that the expression for MGMT does increase from the first recurrence, but of course, we have limited data to make a strong case.

## 10.3    Pre-processing NGS data

DNA and RNA sequencing can be carried out by various methods, for example there is SOLID sequencing, Ion Torrent sequencing, Illumina (Solexa) sequencing and others. The sequencing may be for the whole genome or just part of it. Every short-strand (fragment) of the PCR enhanced DNA/RNA is sequenced, with all the results being stored in a file. The *read length* of a fragment is the length of the sequence of the fragment. The *fragment length* is the average of the read lengths. The sequencing is not perfect, with errors creeping in, and so the sequencing process also produces a probability of an error, $p$, for each base in a sequence. The probability is transformed into a *quality score*, $Q = -10 \log_{10}(p)$. For $p = 0.05$, $Q = 13.0$, which might be used as a cut-off for poor quality, higher values being of acceptable quality. The files produced have various formats, for example, *FASTQ* files which have four lines for each sequence fragment, for example,

```
@SEQ_ID
TTGGGACTCATATCTGGATTCCCCTGGGAAACCAGGAATGGCCTTT
+SEQ_ID
EEEEEGGGCDE9895++++++BCCCCFFGHIIIII????D////--
```

The first and third lines contain the fragment identifier, plus optional other information. The second line contains the actual sequence and the fourth line, one character for each base representing its quality score, with the encoding of the scores differing with the different sequencing platforms.

### Alignment of sequence fragments
Once the RNA fragments have been sequenced, we then have to align them to a reference sequence, see the sketch below. This is not a trivial task because of the amount of data involved.

**Short fragment sequences**

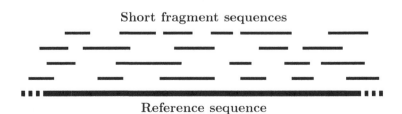

Reference sequence

Depending on the task at hand, the alignment might be to the whole genome, to the exome (all the exons in the genome) or to a particular target region of the genome. Various algorithms have been developed for doing this, some using dynamic programming methods, some using database search methods, plus other methods, resulting in many packages for carrying out the task, packages such as BLAST (Basic Local Alignment Search Tool) and Bowtie.

The aligned fragment sequences (short reads) are stored in *SAM* and *BAM* files. SAM files are in text format and BAM files are compressed versions of SAM files. Different applications of NGS technologies require different number of reads from individual samples. In the case of genomic DNA sequencing, the depth of coverage tells how many times a target nucleotide is sequenced in a sample.

We are not going to give an example of reading and processing FASTAQ and other sequence files, but use sequence data that has already been processed. The example is from the National Cancer Institute's Cancer Genome Atlas Program (TCGA), which collects genomic, epigenomic, transcriptomic and proteomic data that can be used by the public for research purposes (https://portal.gdc.cancer.gov). The data are placed under "projects", which cover 33 different cancers. Some of the data need permission to be downloaded and analysed, especially sequence files. We are going to use the project labelled "TCGA-KIRC" which contains $16,225$ files relating to kidney renal clear cell carcinoma data on 537 cases. The data cover: sequencing reads, transcriptome profiling, simple nucleotide variation, copy number variation, DNA methylation, clinical and biospecimen. We are interested in the transcriptome profiling and simple nucleotide variation and its relationship to the clinical data.

---

## 10.4  TCGA-KIRC: Renal clear cell carcinoma

The Bioconductor package TCGAbiolinks can be used to download TCGA data and carry out analyses. The package TCGAWorkflow is a good starting point for help with the TCGAbiolinks package. Other packages such as limma can of course be used instead to analyse the data. We rely heavily on the two TCGA packages in this section. The following functions download and prepare the data,

- Use TCGAbiolinks:::getProjectSummary("TCGA-KIRC") to obtain a summary of the TCGA-KIRC project. .

- Use GDCquery() to "enquire" about the gene expression data, the simple nucleotide variation data, the clinical data and other data types.

- Use GDCdownload() to download and save data files.

- Use GDCprepare() to prepare the data.

Now, what links the different datasets are the patient barcodes, e.g. there is a patient with the barcode: "TCGA-A3-3358" which is found in the data frame of the clinical data. The expression data, after processing, is stored in a matrix, with row names giving the particular genes and column names giving the patient barcodes, plus other information. For instance, for the TCGA-A3-3358 barcode, there are two columns of expression data, "TCGA-A3-3358-01A-01R-1541-07" and "TCGA-A3-3358-11A-01R-1541-07", the first giving the expression data for a sample of tissue from the patient's tumour and the second giving the expression data for a sample of normal tissue from the same patient (note the 01A and 11A in the two column names to distinguish the two). Of the 533 patients with tumour expression data, 72 of these had expression data for normal tissue.

### 10.4.1   Differential expression

First the gene expression data has to be preprocessed.

- Use the function TCGAanalyze_Preprocessing() to look for patients' gene expression sets that might be outliers, using Pearson correlation between the pairs of patient expression sets.

- Use the function TCGAanalyze_Normalization() to normalise the patients' expression sets. As with microarrays, bias occurs via various sources, such as gene length, lane effect, where the sequencing machine has several processing lanes and sequencing depth.

- Use the function TCGAanalyze_Filtering() to filter out some of the genes, based on mean quantile values, to obtain a gene expression set.

- Make a $\log_2$ transformation on the resulting expression set.

The function TCGAanalyze_Preprocessing() produces a plot of pairwise correlations between the patients' gene expression sets and for each patient, a boxplot of the correlations between that patient's expression data and each of the other patients' expression data (not shown here). Figure 10.15 shows a histogram of all the $\log_2$ transformed, normalised and filtered gene expression values.

Next we use the patients with tumour and normal tissue samples to carry out a differential expression analysis. Note, the function TCGAanalyze_DEA() within TCGAbiolinks has an error with paired data, and so we will use limma, as we did for the microarray data. Out of 14,889 genes, 11,864 were either under- or overexpressed in the tumour tissue compared to the normal tissue. Of the 11,864 genes, we focus on those with fold-change greater than 2 (or less than 0.5), leaving us with 3,832 genes in our expression set. To save space, we do not show MD or volcano plots here, nor MDS and cluster plots as we did for the microarray data.

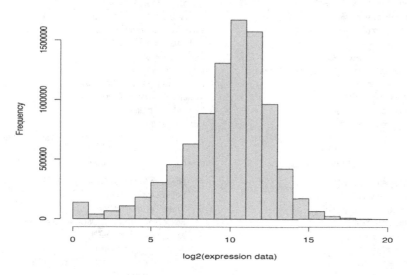

**FIGURE 10.15**
Histogram of the normalised and filtered $\log_2$ transformed gene expression data

The top ten genes, in terms of p-values are *PIK3C2G*, *C16orf11*, *PVT1*, *GABRD*, *CA10*, *BSND*, *PAK6*, *SLC9A4*, *ACPP* and *NDUFA4L2*. It seems almost pointless listing the first few genes like this, as the order is somewhat arbitrary since we are ordering many p-values of order $10^{-10}$ to $10^{-35}$. Instead we will carry out an Enrichment Analysis (EA) of the 3, 382 differentially expressed genes, using the function TCGAanalyze_EAcomplete().

### Enrichment Analysis (EA)
So far, we have looked at one gene at a time in our analyses. Gene Set Enrichment Analysis (GSEA) looks at sets of genes at a time. The idea is to see if a set of genes associated with a particular gene pathway, or biological function, or process are well represented in the over or under expressed genes that have been identified in the samples. We do not go into the details of how this in done.

Figure 10.16 shows the gene enrichment analysis of the 3,382 GENE expression set. The top ten of each of the biological processes, molecular functions, cellular components and pathways that contain genes from the expression set, are shown as barcharts, the length of a bar being equal to $-\log_{10}(\text{FDR})$, a log-transformation of the false discovery rate. Also plotted is the ratio of the number of genes in the list that are in a particular biological process, molecular function, cellular component, or pathway compared to the total number

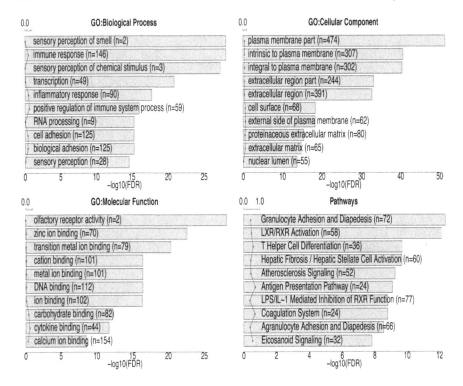

**FIGURE 10.16**

Enrichment analysis for the genes differentially expressed

of genes in the particular biological process, molecular function, cellular component, or pathway.

Note, several of the processes, cellular components, functions and pathways do not have many genes from the gene set involved. For those that do: the main biological processes are adhesion and immune response; the main cellular component is the plasma membrane; the main molecular functions are to do with binding; the main pathways include adhesion and signalling, plus others.

## 10.4.2 Genes as biomarkers for survival

A quick way of seeing whether a gene might be a potential biomarker for survival, is to split the patients into two groups, the first having a low expression value for the gene, and those having a high expression value for the gene. We will use the cutoffs as the bottom and top tertiles of the expression values to define the two groups. The middle third of patients are ignored. Then a

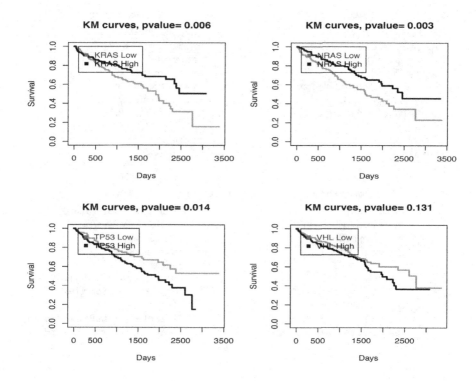

**FIGURE 10.17**

Kaplan-Meier curves, split by low and high expression for four genes

log-rank test is applied to the survival times for the two groups. This can be carried out using the function TCGAanalyze_SurvivalKM().

The log-rank p-values for all genes was found, 7,690 were less than 0.05. The same function can plot the Kaplan-Meier curves for each gene (not recommended for thousands of genes). Figure 10.17 shows the KM curves for the three genes, KRAS, NRAS, TP53, which had p-value less than 0.05 and also for VHL, a gene highly associated with renal clear cell carcinoma, where there is no significant survival difference. KRAS and NRAS are proto-oncogenes and high expression of them is associated with better survival, while TP53 is a tumour suppressor gene and high expression is associated with worse survival.

The 5,652 genes that had log-rank test p-value less than 0.01 were subjected to enrichment analysis; the results are shown in Figure 10.18. the main biological processes relate to the cell cycle and signalling; the main cellular components are organelle lumina (the inside space of organelles); the main molecular functions are to do with binding; the main pathways are about cell signalling.

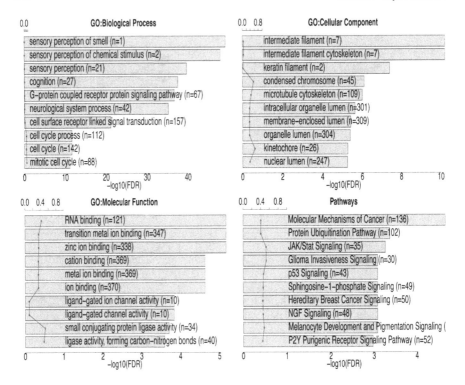

**FIGURE 10.18**
Enrichment analysis for the genes associated with survival

## 10.4.3 Single nucleotide variation

In our last analysis, we look at mutations in the genes. The mutation data is stored in *MAF* files (Mutation Annotation Format) which are downloaded in like manner to the expression data. The Bioconductor package maftools is used to analyse the data. The functions getGeneSummary() and getSampleSummary() summarise the mutations. There are five types of mutation recorded:

- *Missense mutation*: a change in a base pair that leads to a different amino acid being used in the resulting protein.

- *Nonsense mutation*: a change in a base pair that leads to a premature stop codon.

- *Nonstop mutation*: a change in a base pair in a stop codon leading to further inappropriate translation.

- *Splice site*: an alteration at the boundary af an exon and an intron.

- *Translation start site*: an alteration of the start site of translation of a gene.

Figure 10.19 shows the summaries obtained, note the colours that the plots would normally use, have had to be put into greyscale. At the top left there is a barchart of the numbers of mutations, the missense mutations dominating. The top middle plot is redundant here. The top right plot is a barchart of the particular changes in base pairs, with C>T being the most common. The bottom left plot shows the number of mutations in each sample, the median being 36. The bottom middle plot shows box and whisker plots of the data in the top left plot. Lastly, the bottom right plot shows the 23 genes that had the most mutations, the numbers being split by type of mutation. The scale of the plot has only allowed half of the gene names to be written, the full list is: VHL, PBRM1, TTN, SETD2, MTOR, BAP1, MUC16, DNAH9, LRP2, SPEN, HMCN1, CSMD3, KMT2C, DST, FBN2, RYR3, ANK3, DNAH2, SMARCA4, AKAP9, ATM, ERBB4 and BRCA2. The VHL gene has the most mutations, PBRM1 has the most nonsense mutations.

Figure 10.20 shows Kaplan-Meier plots for four of the genes in the top 23 genes, using the function mafSurvival() and with samples split into mutated gene versus wild type gene. The VHL gene has been chosen for the display as it is the gene most associated with this cancer – it shows no disadvantage in survival for mutated genes. The other three genes were chosen because they had small p-values, but note the small number of mutated genes. These three mutated genes give a survival disadvantage.

Finally, we carried out an enrichment analysis using a gene list of all the genes that were mutated in at least three samples. Figure 10.21 shows the results: cell adhesion and organisation features highly in the affected biological precesses; nucleotide binding features in the affected molecular functions; the cilia/axoneme are the cell component affected by the mutations; the pathways affected by the mutations are those of signalling.

The reader may find the following texts useful: [4], [21], [36], [39], [7] and [47].

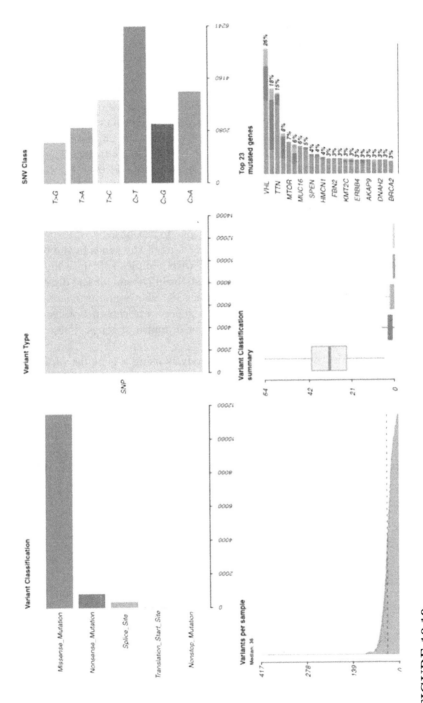

**FIGURE 10.19**
Summary of mutations

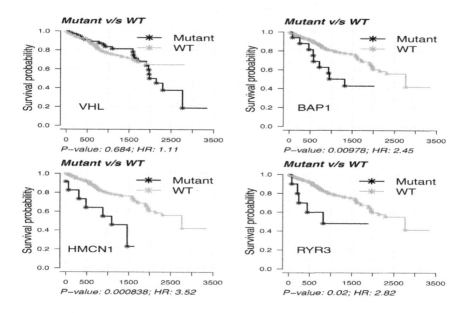

**FIGURE 10.20**

Survival curves for mutated and non-mutated genes for four genes

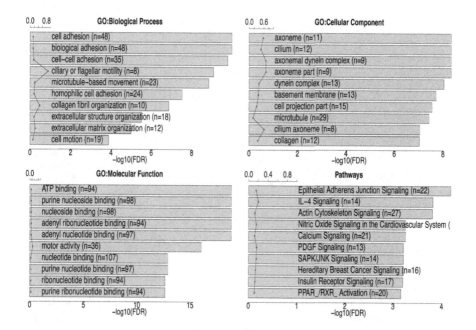

**FIGURE 10.21**

Enrichment analysis for the genes that were mutated in at least three samples

# Appendices

# A

## Statistics Revision

Here, we briefly review some statistical methods. This might be useful for some readers. Only the section on Markov Chain Monte Carlo (MCMC) Bayesian analysis is covered in more detail, as this in not widely familiar.

## A.1 Some distribution theory

The two commonest distributions are probably the binomial distribution and the normal distribution. The binomial distribution, $B(N, p)$, has probability mass function, $p(i)$,

$$p(i) = \binom{N}{i} p^i (1-p)^{N-i}, \quad (0 \le i \le N),$$

the mean, $\mu$, is given by, $\mu = Np$ and the variance, $\sigma^2$, by $\sigma^2 = Np(1-p)$. The normal distribution, $N(\mu, \sigma^2)$, has probability density function (pdf), $f(x)$,

$$f(x) = \frac{1}{\sqrt{2\pi}\sigma} \exp(-\tfrac{1}{2}(x-\mu)^2/\sigma^2), (-\infty < x < \infty),$$

the mean is $\mu$ and variance, $\sigma^2$. The standard normal distribution has $\mu = 0$ and $\sigma^2 = 1$. It is usually labelled, $Z$, so $Z \sim N(0,1)$. The upper $100\alpha$ percentage point of $Z$ is $z_\alpha$, where $\Pr(Z \le z_\alpha) = \alpha$.

Table 3.1 gives some distributions used in survival analysis.

The multivariate normal distribution has probability density function, $f(\mathbf{x})$,

$$f(\mathbf{x}) = \frac{1}{(2\pi)^{p/2}|\boldsymbol{\Sigma}|} \exp[-\tfrac{1}{2}(\mathbf{x}-\boldsymbol{\mu})^{\mathsf{T}}\boldsymbol{\Sigma}^{-1}(\mathbf{x}-\boldsymbol{\mu})],$$

where, $\mathbf{x} = (x_1, x_2, \ldots, x_p)$. The vector mean is $\boldsymbol{\mu}$, and the covariance matrix is $\boldsymbol{\Sigma}$.

### A.1.1 The central limit theorem

The central limit theorem (CLT) is a key theorem in statistics. Basically, it says that if $X_1, X_2, \ldots, X_n$ is a sample of independent random variables, from any type of distribution (more or less), that has mean, $\mu$, and variance $\sigma^2$, then the mean, $\bar{X}_n$, of the sample, for large $n$, has approximately a normal distribution, $N(\mu, \sigma^2/n)$. You can see how useful this will be, as you do not have to worry about whether the sample has come from an exponential distribution, a chi-squared distribution, a binomial distribution, or any other.

As a proper definition of the CLT, this is not acceptable. To be statistically correct, we have to write something like,

$$\sqrt{n}(\bar{X}_n - \mu) \xrightarrow{d} N(0, \sigma^2),$$

where $\xrightarrow{d}$ means, "converges in distribution". Under certain conditions, "independence" can be relaxed.

As an example, let $X_i$ have a $Bin(1, p)$ distribution (the Bernoulli distribution, $Bern(p)$), which has mean $p$ and variance $p(1 - p)$. By the CLT, $\bar{X} = n^{-1}\sum_{i=1}^{n} X_i$ has approximately a $N(p, p(1 - p)/n)$ distribution. But, $\sum_{i=1}^{n} X_i$ has a $Bin(n, p)$ distribution, thus showing for large $n$, a binomial distribution can be approximated with a normal distribution.

## A.1.2  Transformations of random variables

Suppose we wish to make a transformation, $g(X)$, of a random variable, for example, $g(X) = \log(X)$, Sometimes we might find the exact distribution of $g(X)$ using the pdf, $f(x)$, of $X$. We can try the distribution function approach. For example, let $X$ have an exponential distribution with pdf, $f(x) = \lambda \exp(-\lambda x)$, and suppose we are interested in the transformation, $g(X) = \sqrt{X}$, writing it as $Y = \sqrt{X}$, $Y$ having pdf, $h(y)$, and distribution function, $H(y)$. Then succinctly, starting from the distribution function of $X$,

$$F(x) = 1 - \exp(-\lambda x); \quad x = y^2; \quad H(y) = 1 - \exp(-\lambda y^2); \quad h(y) = 2\lambda y \exp(-\lambda y^2).$$

Alternatively, we can try the pdf approach, where $h(y) = f(g^{-1}(y)) \times f\left(\frac{dg^{-1}(y)}{dy}\right)$, where $g^{-1}(.)$ is the inverse function of $g(.)$, and where $g$ is a monotonic function. For example, let $X$ have a standard normal distribution, with pdf $\frac{1}{\sqrt{2\pi}} \exp(-\frac{1}{2}x^2)$. Let $g(X) = X^2$ and so $g^{-1}(y) = \sqrt{y}$. But, we have to be careful as the domain of $X$ is $(-\infty, \infty)$ and $g(.)$ is not monotonic over this domain. So we can just consider $X$ on $(0, \infty)$, where $g(.)$ is monotonic, and then multiply by 2 since $f(.)$ is symmetric about zero; this is a *half-normal* distribution. Then succinctly,

$$\frac{dg^{-1}(y)}{dy} = \tfrac{1}{2}y^{-\frac{1}{2}}; \quad h(y) = 2 \times \frac{1}{\sqrt{2\pi}} \exp(-\tfrac{1}{2}y) \times \tfrac{1}{2}y^{-\frac{1}{2}} = \frac{1}{2^{\frac{1}{2}}\Gamma(\frac{1}{2})}y^{-\frac{1}{2}} \exp(-\tfrac{1}{2}y),$$

where $\sqrt{\pi}$ is written as $\Gamma(\frac{1}{2})$, and noting $h(.)$ is written in the form usually used for the chi-squared distribution with one degree of freedom. Thus, $X^2 \sim \chi_1^2$.

Next, we see how Taylor's series and the *delta method* can help us with some approximations for the distribution of $g(X)$. We outline these methods leaving out convergence issues.

## Taylor's series

As long as the appropriate derivatives exist, the Taylor's series for $g(X)$ about the mean $\mu$ of $X$, is

$$g(X) = g(\mu) + \tfrac{1}{1!}g'(\mu)(X - \mu) + \tfrac{1}{2!}g''(\mu)(X - \mu)^2 + \tfrac{1}{3!}g'''(\mu)(X - \mu)^3 + \ldots.$$

So an approximation for $g(X)$ is,

$$g(X) \approx g(\mu) + g'(\mu)(X - \mu),$$

and hence $\mathrm{E}[g(X)] \approx g(\mu) + g'(\mu)\mathrm{E}[X - \mu] = g(\mu)$. Then

$$\mathrm{var}[g(X)] \approx \mathrm{E}[\{g(X) - g(\mu)\}^2] = \mathrm{E}[\{g'(\mu)(X - \mu)\}^2] = \{g'(\mu)\}^2\sigma_X^2.$$

So, the approximate mean and variance of $g(X)$ are $g(\mu)$ and $\{g'(\mu)\}^2\sigma_X^2$.

Let us try this out for the transformation, $g(X) = \sqrt{X}$, on the exponential distribution, which has mean, $\mu = 1/\lambda$, and variance, $\sigma_X^2 = 1/\lambda^2$. We have $g'(X) = \tfrac{1}{2}/\sqrt{X}$ and so the mean and variance of $Y = \sqrt{X}$ are approximated by $g(\mu) = 1/\sqrt{\lambda}$ and $g'(\mu)^2\sigma_X^2 = \tfrac{1}{4}/\lambda$. Simple integration shows the true mean and variance of $Y$ to be $\tfrac{\sqrt{\pi}}{2}/\sqrt{\lambda} = 0.886/\sqrt{\lambda}$ and $(1 - \pi/4)/\lambda = 0.215/\lambda$.

## Delta method

The delta method is a generalisation of the CLT using the Taylor's series method above. It states that if $X_1, X_2, \ldots$ is a sequence of random variables that has a limiting normal distribution, $\sqrt{n}(X_n - \mu) \to \mathrm{N}(0, \sigma^2)$, then the transformation, $g(X_n)$, also has a limiting normal distribution,

$$\sqrt{n}(g(X_n) - g(\mu)) \to \mathrm{N}(0, g'(\mu)^2\sigma^2).$$

So, $g(X_n)$ has an approximate normal distribution, $\mathrm{N}(g(\mu), g'(\mu)^2\sigma^2/n)$, for large $n$.

As an example, let us look at the $X \sim \mathrm{Bin}(n, p)$. We are interested in the transformation $g(X) = X/(n - X)$. Now $\sqrt{n}(X - np) \to \mathrm{N}(0, np(1 - p))$ and $g'(X) = n/(n - X)^2$. Hence $Y = X/(n - X)$ has an approximate normal distribution $\mathrm{N}(p/(1 - p), p/\{n(1 - p)^3\})$.

The ratio $R = p/(1 - p)$ is the *odds-ratio*, $(p : 1 - p)$, or sometimes written $p : q$, where $q = 1 - p$. For a sample of $n$ cancer patients where interest is whether there is a complete response (CR) using the RECIST criteria, $X$, the number of patients with CR follows a binomial distribution, with $p$ the probability of a CR for a patient. For large $n$, $X/n$ estimates $p$ and has approximate distribution, $\mathrm{N}(p, p(1 - p)/n)$. The ratio of the odds is estimated by $X/(n - X)$ and has approximate distribution, $\mathrm{N}(p/(1 - p), p/\{n(1 - p)^3\})$. If we use the logarithm of $R$, $\log(R) = \log(p/(1 - p)) = \log(p) - \log(1 - p)$, which is estimated by $\log(X) - \log(n - X)$. For the delta method with transformation, $g(X) = \log(X) - \log(n - X)$, we have $g'(X) = 1/X + 1/(n - X) = n/\{X(n - X)\}$, and hence $\log(\mathrm{R})$ has an approximate $\mathrm{N}(\log(p/(1 - p)), 1/\{np(1 - p)\})$ distribution. The normal approximation to $\log(R)$ is better than that for $R$, since the true distribution of $R$ is skewed.

## A.2    Statistical inference

### A.2.1    Parameter estimation

For a random sample, $x_1, \ldots, x_n$, from a particular distribution, the likelihood, $L$, is the joint probability distribution function (or probability mass function) of $x_1, \ldots, x_n$. Assuming, the $x_i$'s are independent, $L$ is the product of the individual pdfs. The likelihood is seen as a function of the parameters, not as a function of the $x_i$'s. The log-likelihood, $l$, is the log of the likelihood. For example, the Bernoulli distribution has $p(x) = p^x(1-p)^{1-x}$ and,

$$L = \prod_{i=1}^{n} p^{x_i}(1-p)^{(1-x_i)}, \quad l = \log(p)\sum_{i=1}^{n} x_i + \log(1-p)\sum_{i=1}^{n}(1-x_i).$$

The maximum likelihood estimator $\hat{p}$, of $p$ is that value of $p$ which maximises the likelihood or, equivalently, maximises the log-likelihood. Differentiating the log-likelihood with respect to $p$ and equating to zero,

$$\frac{dl}{dp} = \frac{1}{p}\sum_{i=1}^{n} x_i - \frac{1}{1-p}\sum_{i=1}^{n}(1-x_i) = 0.$$

Solving this, $\hat{p} = \frac{1}{n}\sum_{i=1}^{n} x_i$. The *estimator*, $\hat{p}$, should be written, $\frac{1}{n}\sum_{i=1}^{n} X_i$ as it is a random variable, but the *estimate* is $\frac{1}{n}\sum_{i=1}^{n} x_i$, where the $X_i$'s is replaced by the sample values, the $x_i$'s.

This is a very simple example with only one parameter. The principles are the same for distributions with more than one parameter.

The score statistic, $U(p)$, is the derivative of the log-likelihood, $U(p) = \frac{d\log(L)}{dp}$, which for the example is, $\frac{1}{p}\sum_{i=1}^{n} x_i - \frac{1}{1-p}\sum_{i=1}^{n}(1-x_i)$. In general, the expected value of the score statistic is zero, when the $X_i$'s truly have the particular parameter values. So for this example, if $X_i$ is Bern$(p)$, the mean is zero, but if $X_i$ is Bern$(p+0.1)$, the mean is not zero.

The information matrix, $I(p)$, is the expected value of the score statistic, $I(p) = \mathrm{E}[U(p)]$. For our example, this is a $1 \times 1$ matrix, i.e. a scalar, but will be a $k \times k$ matrix when there are $k$ parameters. It can be shown that $I(p)$ is also equal to $-\mathrm{E}\left[\frac{d^2\log(L)}{dp^2}\right]$. Hence for the example, using $\mathrm{E}[X_i] = p$,

$$I(p) = \mathrm{E}\left[\frac{1}{p^2}\sum_{i=1}^{n} X_i + \frac{1}{(1-p)^2}\sum_{i=1}^{n}(1-X_i)\right] = \frac{n}{p(1-p)},$$

after a little algebra.

For large sample sizes, the maximum likelihood estimator has an approximate normal distribution, with mean equal to the parameter vector and variance equal to the inverse of the information matrix, $I(p)^{-1}$. For the example, this is just the variance since there is only one parameter, and so $\hat{p} \sim N(p, p(1-p)/n)$.

There are other methods for estimating parameters, for example, by the method of moments.

Comparison of models to find the most acceptable one, can be carried out using *Akaike's information criterion* (AIC), which is $2k - 2\log(L(\hat{\boldsymbol{\theta}}))$, where there are $k$ parameters, $\boldsymbol{\theta}$, in the likelihood. A model with a lower AIC value is preferable to one with a higher AIC value. Alternatively, the *Bayesian information criterion* (BIC), $\log(n)k - 2\log(L(\hat{\boldsymbol{\theta}}))$, can be used.

Sometimes a restricted maximum likelihood (REML) approach is used that adjusts the likelihood to allow for nuisance parameters.

Confidence intervals for parameters are found using the distribution of the parameter estimator. For our example, we find a confidence interval $[p_l, p_u]$, that has confidence, $100(1-\alpha)\%$, as the values, $p_l$ and $p_u$, that satisfy, $\Pr[p_l \leq \hat{p} \leq p_u] = 1 - \alpha$. There is no unique solution, and so we share the probability, $\alpha$, between the two ends of the interval, i.e. $\alpha/2$ each. Now, $\hat{p} \sim N(p, p(1-p)/n)$, and so after a little manipulation of the normal distribution to the standard normal distribution, $N(0,1)$, the $100(1-\alpha)\%$ estimated confidence interval for $p$ is $[\hat{p} - z_{\alpha/2}\sqrt{\hat{p}(1-\hat{p})/n}, \ \hat{p} + z_{\alpha/2}\sqrt{\hat{p}(1-\hat{p})/n}]$.

## A.2.2 Hypothesis testing

Hypothesis testing is very widely used. The idea is to weight up the evidence, based on the data, to test whether a parameter takes a certain value, to test whether the population means for two groups are equal, to test whether the whole distributions of a variable for two groups are equal, not just the means, etc. For example, we want to compare the quality of life (QOL) measured by variable, $X$, for two groups, and see whether it is better for one group than the other. We use the sample means for the comparison, assumed normally distributed. We set up a null hypothesis, $H_0$, that the means are equal, and an alternative hypothesis, $H_1$, that the means are not equal, and hope to reject $H_0$ in favour of $H_1$. So the hypotheses are,

$$H_0 : \mu_1 = \mu_2 \text{ versus } H_1 : \mu_1 \neq \mu_2,$$

where $\mu_1$ and $\mu_2$ are the population means for the two groups, and we assume the variances are equal and have the known value, $\sigma^2$. Let the sample values be, $x_1, \ldots, x_{n_1}$ and $y_1, \ldots, y_{n_2}$ for the two groups. We need a test statistic to test the hypotheses, which is a function of the data and does not contain any unknown parameter. Now, $\bar{X} - \bar{Y} \sim N(\mu_1 - \mu_2, \frac{\sigma^2}{n_1} + \frac{\sigma^2}{n_2})$, and our test statistic based on this is,

$$Z = \frac{\bar{X} - \bar{Y}}{\sigma^2(n_1^{-1} + n_2^{-1})},$$

which has a normal distribution, $N((\mu_1 - \mu_2)/\{\sigma^2(n_1^{-1} + n_2^{-1})\}, 1)$. When the data are substituted into $Z$: we reject $H_0$, in favour of $H_1$, if the value of $Z$ is large; we do not reject $H_0$ if the value of $Z$ is small.

Now, it is possible to make mistakes, we may have a large value of $Z$ and reject $H_0$, when actually, $H_0$ is true – this is a Type I error. We may have a small value of $Z$ and do not reject $H_0$, when actually, $H_0$ is false ($H_1$ is true) – this is a Type II error. We set the probability of a Type I error as $\alpha$, called the significance level, so $\Pr(\text{reject } H_0 | H_0 \text{ true}) = \alpha$. The values of $Z$ where $H_0$ will be rejected is called the critical region, $R$, found as the values, $z_l$ and $z_u$, where $z_l$ and $z_u$ satisfy, $\Pr(Z_l < z_l | H_0 \text{ true}) + \Pr(Z_l > z_u | H_0 \text{ true}) = \alpha$. For the example, $Z \sim N(0, 1)$ when $H_0$ is true, and the critical region consists of all the values outside the interval, $[-z_{\alpha/2}, z_{\alpha/2}]$.

Once the significance level and critical region are set, we can calculate the probability of a Type II error, $\beta$, as $\Pr(\text{do not reject } H_0 | H_0 \text{ false})$, hence,

$$\beta = \Pr(-z_{\alpha/2} \leq Z \leq z_{\alpha/2}) = \Phi(z_{\alpha/2} - (\mu_1 - \mu_2)/\{\sigma^2(n_1^{-1} + n_2^{-1})\})$$
$$- \Phi(-z_{\alpha/2} - (\mu_1 - \mu_2)/\{\sigma^2(n_1^{-1} + n_2^{-1})\}),$$

where $\Phi$ is the standard normal cumulative distribution function. So $\beta$ depends on the difference in the population means.

Power is defined as the $\Pr(\text{reject } H_0 | H_0 \text{ false})$ and it is easily seen that power $= 1 - \beta$. Power is usually quoted rather than the Type II error, as it better conveys the probability of rejecting the null hypothesis when it is false. This is how the hypotheses were set up – we try to knock down the null hypothesis that the two treatment means are equal say, giving evidence that one treatment is better than another.

The *Wald test* for testing a parameter, $\theta$, is equal to a particular value, $\theta_0$, usually set as zero, has test statistic, $W = (\hat{\theta} - \theta_0)^2/\text{var}(\hat{\theta})$, where $\hat{\theta}$ is the maximum likelihood of $\theta$. For large samples, $W$ has a $\chi_1^2$ distribution.

The *score test* for testing whether a parameter, $\theta$, is equal to a particular value $\theta_0$, uses the score statistic. We have $U(\theta_0)/I(\theta_0)$ has a $\chi_1^2$ distribution when $H_0$ is true.

The *likelihood ratio test* can be a useful test when we have nested parameter sets. For example, suppose, for two normal distribution samples, we have the hypotheses,

$$H_0 : \mu_1 = \mu_2, \sigma_1^2 = \sigma_2^2, \quad H_1 : \mu_1 \neq \mu_2, \sigma_1^2 \neq \sigma_2^2.$$

The parameters are nested because the parameter values under $H_0$ are a subset of the parameter values under $H_1$. The likelihood ratio test statistic is, $\lambda_{LR} = -2\log(L_0/L_1)$, where $L_0$ and $L_1$ are the maximum values of the likelihood under $H_0$ and $H_1$ respectively. For large samples, $\lambda_{LR}$ has a $\chi^2$ distribution with degrees of freedom equal to the number of restrictions needed on the parameters of $H_1$ to get to the parameters of $H_0$. For the example, this is 2.

## A.3   Regression analysis

There are three types of regression models: parametric regression models, non-parametric regression models and semiparametric regression models.

- Parametric regression models use a completely specified probability distribution for the dependent variable, $y$, and will involve parameters (unknown constants). For example, $y_i = \beta_0 + \beta_1 x_i + \epsilon_i$, with $\beta_0, \beta_1$ parameters, $x$ an independent variable and $\epsilon \sim N(0, \sigma^2)$. Thus $y_i$ has a $N(\beta_0 + \beta_1 x_i, \sigma^2)$ distribution.

- Non-parametric regression models do not have a probability distribution specified for $y$, nor have any parameters. For example, $y_i = m(x_i) + \epsilon_i$, and $m(x_i)$ is a "smooth" function, estimated by some smoothing procedure such as a moving average.

- Semiparametric regression models have aspects of the distribution of $y$ specified and will have parameters. For example, the model above, $y_i = \beta_0 + \beta_1 x_i + \epsilon_i$, but this time $\epsilon_i$ is not given a distribution. The parameters, $\beta_0$ and $\beta_1$ are estimated by ordinary least squares (OLS).

Regression analysis is a ubiquitous data analysis technique at the heart of statistical theory and practice, where dependent (or response) variables are "regressed" on independent (or explanatory, or regressor) variables. When learning about regression, the order in which models are encountered is, *simple linear regression, multiple linear regression, general linear models, generalised linear models and non-linear regression.* The key-word "linear" refers to the regression model being linear in its coefficients for the explanatory variables. For example, both the following models are multiple linear regression models:

$$
\begin{aligned}
y &= \beta_0 + \beta_1 x_1 + \beta_2 x_2 + \epsilon \\
y &= \beta_0 + \beta_1 x + \beta_2 \log(x) + \epsilon,
\end{aligned}
$$

since the model equations are linear in $\beta_0$, $\beta_1$ and $\beta_2$. The "error" is $\epsilon$, which is assumed to be normally distributed, with mean 0 and variance $\sigma^2$.

The explanatory variables can be continuous, e.g. age, or indicator variable, taking values 0 or 1 (but sometimes $-1$ and $+1$ are used), e.g. treatment regime, chemotherapy or radiotherapy. When there are more than two categories for a categorical variable, several indicator variables have to be used, e.g. for three treatments use

control : $x_1 = 0, x_2 = 0$;    chemo : $x_1 = 1, x_2 = 0$;    radio : $x_1 = 0, x_2 = 1$.

In practice, you will not have to worry about choosing indicator variables for categorical variables, since statistical software packages sort this out for you.

For convenience, let expressions like $\beta_0 + \beta_1 x_1 + \beta_2 x_2$ be denoted by $\boldsymbol{\beta}^\mathsf{T} \mathbf{x}$ and called a *linear predictor*.

Another way of writing the model is

$$Y_i \sim \mathrm{N}(\mu_i, \sigma^2); \quad \mu_i = \boldsymbol{\beta}^\mathsf{T} \boldsymbol{x}_i, \quad (i = 1, \ldots, n),$$

where $n$ is the number of observations.

This leads to generalised linear models where the distribution of $Y$ can be any distribution belonging to the exponential family of distributions, and includes, the *Normal, Log-normal, Exponential, Weibull (fixed k), Gamma, Poisson, Binomial* (fixed $n$) distributions, plus others. A generalised linear model is formulated as

$$
\begin{aligned}
\text{Distribution} \quad &: \quad Y \text{ from exp. family, with mean } \mu \\
\text{Linear predictor} \quad &: \quad \eta = \boldsymbol{\beta}^\mathsf{T} \mathbf{x} \\
\text{Link function} \quad &: \quad \eta = g(\mu).
\end{aligned}
$$

For example, for $Y$ having an exponential distribution, the mean is $\lambda^{-1}$ and we choose $\eta = \beta_0 + \beta_1 x_1 + \beta_2 x_2$ and $g$ as $g(\mu) = 1/\mu$. This corresponds to $\lambda = \beta_0 + \beta_1 x_1 + \beta_2 x_2$.

The *deviance*, $D$, is a measure of fit of a generalised linear model. It is defined as, $D = -2\{l(\hat{\boldsymbol{\beta}}) - l(\hat{\boldsymbol{\beta}}_{sat})\}$, where $l(\hat{\boldsymbol{\beta}})$ is the log-likelihood for the fitted model and $l(\hat{\boldsymbol{\beta}}_{sat})$ is the log-likelihood for the saturated model. The saturated model has one parameter for every observation and gives a "perfect fit". The deviance can be used to compare two fitted models, $m_0$ and $m_1$, where model $m_0$ is nested in model $m_1$, i.e. the parameters of $m_0$ are a subset of those for $m_1$. The aim is to test the null hypothesis that the extra parameters in model $m_1$ are all zero. The test statistic is $-2(l_1 - l_0)$, where $l_0$ and $l_1$ are the log-likelihoods for the two models. Its distribution under $H_0$ is $\chi_k^2$, where $k$ is the number of extra parameters in $m_1$ compared to $m_0$.

## A.3.0.1  Choosing a subset of predictor variables

**Forward selection**
Forward selection fits each predictor variable in turn and selects the one that gives the best fitting model, based on p-values, AIC or other criterion. Then another predictor variable is added, the one that together with the first predictor variable selected, gives the best fitting model. If the fit is no better than for the first model, the second predictor variable is not added and the process stops. If the fit is better, a third variable is tried. This continues until no more predictor variables can be added.

**Backward selection**
This is like forward selection, except all the variables are placed in a model to start with and then variables are excluded, one at a time, and stops when the exclusion of another variable does not improve the model fit.

**Stepwise selection**

Stepwise selection is a mixture of both forward and backward selection. It starts with forward selection, but at any step, one of the included variables can be excluded from the model. It could start with all the variables selected and stepwise exclude/include variables.

**Best subsets regression**

As the name implies, a model is fitted for each possible subset of predictor variables and the best fitting model chosen. This becomes impossible when there are many predictor variables.

There are various ways to check if a models fits the data well, mostly using residual analyses, Section 8.3.1 briefly discusses some of these.

## A.4   Bayesian methods

Bayesian statistical methods are not regularly used in the analysis of data from cancer trials and other cancer studies. But, sometimes they can be useful, especially for studies of rare cancers where sample sizes are necessarily small. Bayesian methods place distributions on the parameters of statistical models, called *prior distributions*. These, combined with the data collected after the priors have been set, are turned into *posterior distributions* for the parameters. The key equation is,

$$p(\boldsymbol{\theta}|\mathbf{x}) = \text{Likelihood}(\mathbf{x}|\boldsymbol{\theta}) \times p(\boldsymbol{\theta}), \qquad (A.1)$$

where $\boldsymbol{\theta}$ is a vector of parameters, $\mathbf{x}$ is the data, $p(\boldsymbol{\theta}|\mathbf{x})$ is the posterior distribution and $p(\boldsymbol{\theta})$ is the prior distribution of $\boldsymbol{\theta}$, remembering the likelihood is the joint distribution of the sample values.

**Example** We use a normal distribution for the data, $X \sim N(\theta, 1)$ and another normal distribution for the prior $\theta \sim N(\mu_0, \tau^2)$. The likelihood and prior are,

$$p(\boldsymbol{x}|\theta) = \prod_{i=1}^{n} \frac{1}{\sqrt{2\pi}} \exp\{-(x_i - \theta)^2/2\}$$

$$p(\theta) = \frac{1}{\sqrt{2\pi}} \exp\{-(\theta - \mu_0)^2/2\tau^2\}.$$

Thus the posterior distribution for $\theta$ is,

$$p(\theta|\boldsymbol{x}) = C \times \exp\left\{-\sum_{i=1}^{n}(\theta - x_i)^2/2\right\} \times \exp\{-(\hat{\theta} - \mu_0)^2/2\tau^2\}$$

$$= C \times \exp\left\{-\sum_{i=1}^{n}\left((\theta - \bar{x}) - (\bar{x} - x_i)\right)^2/2\right\} \times \exp\{-(\theta - \mu_0)^2/2\tau^2\}$$

$$= C \times \exp\left\{-\frac{1}{2}\sum_{i=1}^{n}(\theta - \bar{x})^2 + \sum_{i=1}^{n}(\theta - \bar{x})(\bar{x} - x_i) - \frac{1}{2}\sum_{i=1}^{n}(\bar{x} - x_i)^2/2\right\}$$

$$\times \exp\{-(\theta - \mu_0)^2/2\tau^2\}$$

$$= C \times \exp\{-n(\theta - \bar{x})^2/2\} \times \exp\{-(\theta - \mu_0)^2/2\tau^2\},$$

where $C$ is a constant with respect to $\theta$ and noting that in the third line of the above algebra, the term $\sum_{i=1}^{n}(\theta - \bar{x})(\bar{x} - x_i)$ equals zero. After a little more algebra, we see the posterior distribution has the form of a normal distribution with mean $(n\tau^2\bar{x} + \mu_0)/(n\tau^2 + 1)$ and variance, $\tau^2/(n\tau^2 + 1)$.

We cannot form confidence intervals (CI) for our parameters, but we can form *confidence regions* (CR). A 95% CR for $\theta$ is defined as the interval $[\theta_L, \theta_U]$, where $\Pr(\theta_L \leq \theta \leq \theta_U) = 1 - \alpha$ for the posterior distribution of $\theta$. Like, CIs, CRs are not unique and so we usually choose the one with the smallest range, $\theta_U - \theta_L$.

## A.4.1   MCMC: Markov Chain Monte Carlo

A problem with Equation A.1 is that we can get stuck with almost impossible integrations to carry out in order to obtain the posterior distribution. Markov Chain Monte Carlo methods (MCMC) revolutionised Bayesian inference in the 1990's as a relatively easy way of simulating posterior distributions. Although the technique took off in the 1990's, its roots are much earlier, with a paper by Metropolis in 1953 and then another key paper by Hastings in 1970, leading to the Metropolis-Hastings algorithm.

Here, we give a short introduction to MCMC, indicating how it all works. First, we start with the *Acceptance-Rejection Method* of generating a sample from a distribution with pdf $f(x)$, where for some reason, simulating values from this distribution is difficult or impossible. Suppose we have another distribution with pdf, $g(x)$, where we can easily generate a random sample. Then, as if by magic, we can turn our random sample from $g(x)$ into one from $f(x)$ by using the acceptance-rejection algorithm. We will need to generate a larger sample from our $g(x)$ than for the required sample size for $f(x)$, sometimes much larger, but as it is easy to generate values from $g(x)$, that is not a problem.

We need a constant $c$ so that $\max_x\{f(x)/g(x)\} \leq c$ and we do not want $c$ to be too large. The acceptance-rejection algorithm is:

- Generate a value $y$ from distribution $g$.

- Accept this $y$ with probability $\frac{f(y)}{cg(y)}$, i.e. generate a random number, $u$, from a uniform distribution on $[0, 1]$, and accept this $y$ and put $x = y$, if $u \leq \frac{f(y)}{cg(y)}$, otherwise reject $y$.

- Continue until the required sample size is reached.

If $c$ is too large, there will be many rejected $y$'s. The resulting $x$'s are a random sample from distribution $f$.

We show how this works. We have, using Bayes theorem,

$$\Pr\left(Y \leq y \middle| U \leq \frac{f(Y)}{cg(Y)}\right) = \frac{\Pr\left(U \leq \frac{f(Y)}{cg(Y)} \middle| Y \leq y\right) \Pr(Y \leq y)}{\Pr\left(U \leq \frac{f(Y)}{cg(Y)}\right)}.$$

Now the three terms on the right hand side of this equation can be written as

$$\Pr\left(U \leq \frac{f(Y)}{cg(Y)} \middle| Y \leq y\right) = \frac{1}{G(y)} \int_{-\infty}^{y} \frac{f(w)}{cg(w)} g(w) \mathrm{d}w = \frac{F(y)}{cG(y)},$$

$$\Pr\left(U \leq \frac{f(Y)}{cg(Y)}\right) = \int_{-\infty}^{\infty} \frac{f(y)}{cg(y)} g(y) \mathrm{d}y = \frac{1}{c} \int_{-\infty}^{\infty} f(y) \mathrm{d}y = \frac{1}{c},$$

$$\Pr(Y \leq y) = G(y).$$

Substituting these back into the equation gives

$$\Pr\left(Y \leq y \middle| U \leq \frac{f(Y)}{cg(Y)}\right) = \frac{F(y)}{cG(y)} \times G(y) \div 1/c = F(y),$$

and hence the algorithm does generate random samples from distribution $F$. Note, as $\Pr\left(U \leq \frac{f(Y)}{cg(Y)}\right) = 1/c$, there will be on average $c$ draws from $g(y)$ for every $f(x)$ and so $c$ ought not to be too large,

The acceptance-rejection algorithm uses a random sample of independent random numbers chosen from $g(x)$. Improvements can be made to the algorithm by allowing the next value of $x_{i+1}$ to be generated, to be dependent on the current value, $x_i$ that has been generated. This is where *Markov chains* come in. Markov chains are a branch of *stochastic process theory*. A Markov chain moves from state to state in a series of steps. A simple example is a *random walk* on the integers, $0, 1, 2, \ldots N$ with steps of unit length, in either direction, taken at the time points, $t = 0, 1, 2, \ldots$, with a starting point given at $t = 0$. At each time point, a step is taken to the right with probability $p$ and to the left with probability $1 - p$. So a realisation of a walk starting at integer 10, might be $10, 11, 12, 11, 10, 9, 8, 7, 8, 7, 6, \ldots$ . If the walk hits a boundary at 0 or $N$, we will say that the walk reflects back to 1 or $N - 1$ respectively and then carries on. Bayesian MCMC uses a Markov chain when generating random numbers from posterior distributions – the next number to be generated depends on the current number that has been generated.

For MCMC, we need a $g(y)$ distribution from which to generate potential random values as for the acceptance-rejection algorithm, but now we call this a *proposal distribution*. We need to define a probability, $\alpha$, that we will accept/reject the "proposed" next value $y$, and here we describe a simple version of the *Metropolis-Hastings* algorithm, which is

$$\alpha(x^{(t)}, y) = \min \left\{ 1, \frac{p(y)}{p(x^{(t)})} \right\}, \tag{A.2}$$

where $p()$ is the posterior distribution for which we require a random sample, and $x^{(1)}, x^{(2)}, \ldots, x^{(t)}$ are the generated $x$ values, in sequence, up to current time $t$. If $y$ is accepted, $x^{(t+1)} = y$, otherwise $x^{(t+1)} = x^{(t)}$.

Returning to our prior and posterior distributions of the normal example above, we want a random sample from $p(\boldsymbol{\theta}|\mathbf{x})$. Let the sample values be $\theta^{(1)}, \ldots, \theta^{(t)}, \ldots, \theta^{(N)}$, noting we have dropped "$|\boldsymbol{x}$" from the notation. We will choose the proposer distribution to be a Cauchy distribution that has pdf $q(y) = 1/\{\pi(1 + y^2)\}$ and we will set $\mu = 5$ and $\tau^2 = 10$. The probability of acceptance/rejection of a proposed $y$ is from equation A.2, $\alpha(\theta^{(t)}) = \min\{1, p(y)/p(\theta^{(t-1)})\}$.

## TABLE A.1
MCMC results for a simple normal distribution example: $\bar{x} = 5.0$ in all cases

| Params and theoretical | N | thin | mean | sd | CR | $\rho_1$ | $\rho_2$ |
|---|---|---|---|---|---|---|---|
| $n = 20$, $\mu_0 = 7$, $\tau^2 = 100$ | 100 | 1 | 5.02 | 0.19 | [4.66, 5.37] | 0.69 | 0.52 |
| $\mu_1 = 5.00$, $\sigma_1 = 0.22$ | 100 | 5 | 5.00 | 0.26 | [4.45, 5.50] | 0.11 | −0.09 |
| CR = [4.56, 5.44] | 1000 | 1 | 4.99 | 0.21 | [4.58, 5.41] | 0.68 | 0.44 |
|  | 1000 | 10 | 5.00 | 0.21 | [4.46, 5.40] | 0.10 | 0.03 |
|  | 1000 | 100 | 4.99 | 0.22 | [4.55, 5.42] | 0.02 | −0.00 |
|  | 10K | 100 | 5.00 | 0.21 | [4.56, 5.42] | -0.01 | −0.01 |
| $n = 20$, $\mu_0 = 7$, $\tau^2 = 1$ | 1000 | 1 | 5.11 | 0.20 | [4.67, 5.52] | 0.72 | 0.53 |
| $\mu_1 = 5.09$, $\sigma_1 = 0.22$ | 1000 | 10 | 5.09 | 0.21 | [4.69, 5.49] | 0.03 | 0.02 |
| CR = [4.67, 5.52] | 1000 | 100 | 5.08 | 0.21 | [4.66, 5.51] | -0.03 | 0.03 |
|  | 10K | 100 | 5.09 | 0.21 | [4.68, 5.52] | -0.01 | −0.00 |
| $n = 20$, $\mu_0 = 7$, $\tau^2 = 0.1$ | 1000 | 1 | 5.63 | 0.18 | [5.34, 5.96] | 0.84 | 0.72 |
| $\mu_1 = 5.67$, $\sigma_1 = 0.18$ | 1000 | 10 | 5.63 | 0.18 | [5.26, 5.97] | 0.23 | 0.02 |
| CR = [5.31, 6.02] | 1000 | 100 | 5.62 | 0.18 | [5.28, 5.97] | -0.01 | −0.04 |
|  | 10K | 100 | 5.64 | 0.18 | [5.28, 5.98] | -0.00 | 0.01 |
| $n = 5$, $\mu_0 = 7$, $\tau^2 = 1$ | 1000 | 100 | 5.28 | 0.36 | [4.57, 6.01] | -0.04 | 0.02 |
| $n = 100$, $\mu_0 = 7$, $\tau^2 = 1$ | 1000 | 100 | 5.02 | 0.10 | [4.82, 5.22] | 0.03 | 0.03 |
| $n = 20$, $\mu_0 = 15$, $\tau^2 = 1$ | 1000 | 100 | 5.45 | 0.21 | [5.04, 5.86] | 0.07 | −0.01 |

It is clear that the sample values for the posterior distribution are correlated. To overcome this a *thinning process* can can be applied where we select
every $r$th value and reject the rest, which will reduce the correlation (sample values closest together will have the highest correlation). Also a *burn-in*
period is allowed where the first $N_0$ sample values are rejected, thus allowing the Markov chain to "settle down" to a *stationary state*. Lastly, several
"chains", i.e. several different samples of the posterior can be generated and
then combined.

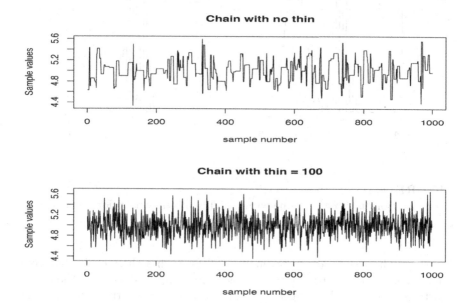

**FIGURE A.1**
Two MCMC chains, the first with no thinning, the second with a thin of 100.

Here, we do not need to generate the data values, $x_i$, we can just choose
a value of $n$ and a value of $\bar{x}$. We will put $\bar{x} = 5.0$ and use various values $n$,
$\mu_0$, $\tau^2$ and choose the posterior sample size, $N$, the burn in length and the
trim. For example, if $N = 1000$, the burn-in also set at 1000 and a thin of 5
is used, a chain of length 6000 is generated, but with the first value set at a
sensibly chosen starting value. The proposal distribution is Cauchy($\bar{x}$,1) and
note that the proposal distribution should be chosen carefully, otherwise the
MCMC simulation might not go as planned. Also, there are better ways of
choosing the proposal distribution, but these are not discussed here.

Table A.1 shows a series of runs of the program. The first six runs all
have $n = 20$, $\mu_0 = 7$ and $\tau^2 = 100$ as shown in the first column. Also shown
are the theoretical mean, standard deviation and confidence region as derived
from the theory. The first run had $N = 100$, a burn-in of 100 and no thinning

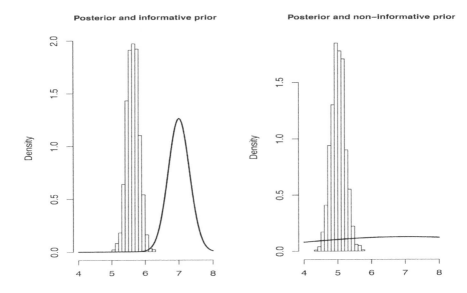

**FIGURE A.2**
Two posterior distributions; one with a highly informative prior and one with a non-informative prior.

(thin=1). The results are surprisingly good for such a short chain, but note the high correlation of adjacent sample values ($\rho_1 = 0.69$) and those two apart ($\rho_2 = 0.52$). Changing the thin to 5 in the second run shows $\rho_1$ and $\rho_2$ much reduced. Changing $N$ to 1000 and 10,000, with a burn in of 1000 (runs, 3–6) shows improved accuracy, but noting no thinning still gives highly correlated sample values, and a very large $N$ and thin, does not increase accuracy significantly.

Figure A.1 shows the chains for the third and fifth runs of the program, the first with no thinning and the second with a thin of 100. You can see a huge difference in the chains, with the first exhibiting the high correlation between sample values.

The first batch of runs in Table A.1 used a non-informative prior, $\tau^2 = 100$. The second batch of runs (7–10) use a more informative prior, $\tau^2 = 1$ and again we see that thinning is a good idea. Note the posterior mean has moved from 5.00 for the non-informative prior, slightly towards the mean of the prior distribution of 7. This happens more so in the third batch of runs (11–14). The last batch of three runs shows what happens when $n$ and $\mu_0$ change. As expected, the larger the $n$, the smaller the variance of the posterior distribution, the further the distance of the prior mean from the sample mean $\bar{x}$, the further the posterior mean moves away from the sample mean $\bar{x}$. This is illustrated in Figure A.2 which shows the histograms for samples for two posterior distributions together with plots of scaled pdf's of the prior distributions used, the first plot has prior, N(7, 0.01), and the second N(7, 100). The value of $\bar{x}$ is 5.0 in both cases. The second prior is so flat that only a very small part is plotted.

Our example has had just one parameter with a univariate prior distribution. The above can be generalised for multiple parameters, $p$ of them say, with a multivariate $p$-dimensional prior distribution. The proposal distribution will also need to be multivariate which might prove to be difficult to implement. Each step of the algorithm updates all $p$ parameters together, There is a special case of the Metropolis-Hastings algorithm, the single component Metropolis-Hastings, that updates one parameter at a time, but it can be inefficient. Gibbs sampling is related to the single component Metropolis-Hastings algorithm and is commonly used in MCMC, and is useful when there are many parameters. It is briefly discussed in the next section.

### A.4.1.1 Gibbs sampling

Let $p(\theta_1, \theta_2, \ldots, \theta_p)$ be the multivariate posterior distribution we are trying to sample from, there being $p$ parameters. Rather than sample from the $p$-dimensional prior distribution, Gibbs sampling samples, in turn, from each of the conditional distributions, $p(\theta_1|\theta_2, \theta_3, \ldots, \theta_p)$, $p(\theta_2|\theta_1, \theta_3, \ldots, \theta_p)$, $p(\theta_3, |\theta_1, \theta_2, \theta_4 \ldots, \theta_p)$, ..., $p(\theta_p|\theta_1, \theta_2, \ldots, \theta_{p-1})$. As soon as one parameter is updated, it is used immediately, not waiting for the complete set of updates to be produced in each round. So

Sample $\theta_1^{(n+1)}$ from $p(\theta_1|\theta_2^{(n)}, \theta_3^{(n)}, \ldots, \theta_p^{(n)})$

Sample $\theta_2^{(n+1)}$ from $p(\theta_2|\theta_1^{(n+1)}, \theta_3^{(n)}, \ldots, \theta_p^{(n)})$

Sample $\theta_3^{(n+1)}$ from $p(\theta_3|\theta_1^{(n+1)}, \theta_2^{(n+1)}, \theta_4^{(n)}, \ldots, \theta_p^{(n)})$

$\vdots$

Sample $\theta_p^{(n+1)}$ from $p(\theta_p|\theta_1^{(n+1)}, \theta_2^{(n+1)}, \ldots, \theta_{p-1}^{(n+1)})$.

Figure A.3 illustrates how Gibbs sampling manages to obtain a sample from a multivariate distribution. The plot shows a bivariate normal pdf for $(\theta_1, \theta_2)$ in the form of a contour plot. The highest point is 0.050 at $(\theta_1, \theta_2) = (5.0, 5.0)$. The starting value for the algorithm is chosen as $\theta_2^{(0)} = 2$; $\theta_1^{(0)}$ does not need to be set. We now sample $\theta_1^{(1)}$ from the conditional distribution of $p(\theta_1|\theta_2^{(0)})$. It can be shown that the conditional distribution of $\theta_1|\theta_2$ is a univariate normal distribution with a certain mean and variance, the values of which will not concern us here. If you imagine the bivariate normal distribution as a bell-shape and were to cut it vertically downwards along the $y = 2$ axis, the profile of the cut would be the normal pdf of the conditional distribution

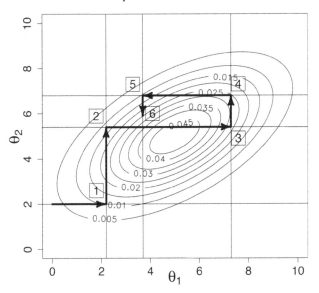

**FIGURE A.3**
Illustration of Gibbs sampling

of $\theta_1|\theta_2 = 2$. We generate a random number from this univariate normal distribution, say its value is 2.2, and so $\theta_1^{(1)} = 2.2$, point "1" in the plot. The vertical axis at $x = 2.2$ cuts through the bell-shape giving the pdf of $\theta_2|\theta_1 = 2.2$; we generate a random number from this distribution, say 5.4, and so $\theta_2^{(1)} = 5.4$, point "2" on the plot. Thus the first bivariate sample value is $(\theta_1^{(1)}, \theta_2^{(1)}) = (2.2, 5.4)$. The axis at $y = 5.4$ gives the pdf of $\theta_1|\theta_2 = 5.4$, etc., etc., giving the bivariate sample values sequentially. In this example, the next two samples are $(\theta_1^{(2)}, \theta_2^{(2)}) = (7.3, 6.8)$ and $(\theta_1^{(3)}, \theta_2^{(3)}) = (3.7, 5.9)$. You can see that as more and more rounds of the algorithm take place, the regions of high density of the pdf will have more sample values that the regions of low density and the sample will be representative of the bivariate distribution.

As we have seen, for Gibbs sampling we need the conditional distribution for each parameter. For our examples these were easy to find and to generate a random sample value from them. Finding the full algebraic form of the conditional probability distributions may not be possible, and so we have to revert to other methods. For example, for a particular parameter, $\theta_i$ say, we might use the Metropolis-Hastings algorithm. The conditional distribution of

$\theta_i$ given the rest of the parameters is given by

$$p(\theta_i | \theta_1, \ldots \theta_{i-1}, \theta_{i+1}, \ldots, \theta_p) = \frac{p(\theta_1, \ldots \theta_{i-1}, \theta_i, \theta_{i+1}, \ldots, \theta_p)}{p(\theta_1, \ldots \theta_{i-1}, \theta_{i+1}, \ldots, \theta_p)}, \qquad (A.3)$$

$$\propto p(\theta_1, \ldots \theta_{i-1}, \theta_i, \theta_{i+1}, \ldots, \theta_p), \qquad (A.4)$$

$$\propto \text{Likelihood}(\boldsymbol{x}|\boldsymbol{\theta}) \times p(\boldsymbol{\theta}). \qquad (A.5)$$

Equation A.3 is from the definition of conditional probability, Equation A.4 shows the conditional density is proportional to the joint density of all the parameters. Equation A.5 just shows our usual equation of posterior proportional to the likelihood times the prior.

Now to generate a random sample value, $\theta_i^{(n)}$ for $\theta_i$ conditioned on all the other parameters, we can substitute the current sample values for all the other parameters into the joint density function, $p(\theta_1, \theta_{i-1}, \theta_i, \theta_{i+1}, \ldots, \theta_p)$, and use this in the Metropolis-Hastings algorithm (or other algorithm).

## A.5 Multivariate data analysis

Multivariate data analysis is needed when data are collected on several variables. A key property of multivariate data is the correlational structure between the variables. We just outline three multivariate methods that can be used.

**Principal components analysis**
Let the variables under consideration be, $\mathbf{X} = (X_1, \ldots, X_p)$, with $n$ observations for analysis. Let $\mathbf{X}$ have covariance matrix, $\boldsymbol{\Sigma}$. The vector form of the variables is essential for multivariate analysis. Principal components analysis (PCA) finds new variables, $\mathbf{Y} = (Y_1, \ldots, Y_p)$ as linear combinations of the $X$'s,

$$Y_1 = a_{11}X_1 + a_{12}X_2, \ldots, a_{1p}X_p = \boldsymbol{a}_1^\mathsf{T}\mathbf{X}$$
$$Y_2 = a_{21}X_1 + a_{22}X_2, \ldots, a_{2p}X_p = \boldsymbol{a}_2^\mathsf{T}\mathbf{X}$$

$$\vdots \qquad \vdots \qquad \vdots \qquad \vdots \qquad \vdots \qquad \vdots$$

$$Y_p = a_{p1}X_1 + a_{p2}X_2, \ldots, a_{pp}X_p = \boldsymbol{a}_p^\mathsf{T}\mathbf{X},$$

where the $a_{ij}$'s are constants, placed in vectors, $\boldsymbol{a}_i$. These are to be found, such that the $Y$'s are uncorrelated and $\boldsymbol{a}_i^\mathsf{T}\boldsymbol{a}_i = 1$.

The solution is given by the eigenvectors of $\boldsymbol{\Sigma}$. These are found as the $p$ solutions to the equation, $\boldsymbol{\Sigma}\boldsymbol{\eta} = \lambda\boldsymbol{\eta}$. The $\lambda$'s are the eigenvalues, the $\boldsymbol{\eta}$'s the eigen vectors. The first principal component (PC) is given by the eigenvector with the largest eigenvalue, the second by the eigenvector with the second largest eigenvalue, etc. Thus $Y_1 = \boldsymbol{\eta}_1^\mathsf{T}\mathbf{X}$, $Y_2 = \boldsymbol{\eta}_2^\mathsf{T}\mathbf{X}$ , etc. The variance of

$Y_i$ is equal to $\lambda_i$, the largest for $Y_1$, the smallest for $Y_p$. It can be shown $\sum_{i=1}^{p} \text{var}(Y_i) = \sum_{i=1}^{p} \text{var}(X_i)$.

In practice, the sample covariance matrix, $S$, is used in place of the population covariance matrix, $\Sigma$. PCA is useful for dimension reduction, because the first few PCs may explain most of the variation in the data, i.e. $\sum_{i=1}^{p} \text{var}(X_i)$. The other useful property is that the PCs are uncorrelated. Each observation vector can have its PC values easily calculated. Plots of one PC against another are usually made. See Figure 10.12 for an example.

## Multidimensional scaling

First, we need a short discussion on *dissimilarity measure*. For two observational vectors, $x_i$ and $x_j$, we would like a measure of how dissimilar they are. We do this using a dissimilarity measure, $\delta_{ij}$. So for the two vectors, $x_1 = (2, 4, 7, 5)$ and $x_2 = (3, 3, 2, 5)$, we could define $\delta_{12} = |2 - 3| + |4 - 3| + |7 - 2| + |5 - 5| = 7$ . This measure is called the city block metric, $\delta_{ij} = \sum_{k=1}^{p} |x_{ik} - x_{jk}|$. For two vectors that are equal, $\delta_{ij} = 0$. The reason why we work with dissimilarity rather than similarity is that we require the measure to be like a distance – the more the vectors are different, the larger the distance and the dissimilarity. The closer the vectors, the smaller the distance and the dissimilarity. Another dissimilarity measure is Euclidean distance (as if you were using a ruler), $\delta_{ij} = \left\{ \sum_{k=1}^{p} (x_{ik} - x_{jk})^2 \right\}^{1/2}$. Dissimilarity measures can be defined for binary and categorical data, not just quantitative data.

Multidimensional scaling (MDS) uses the dissimilarities, $\delta_{ij}$, between all pairs of observation vectors and produces a map of points, each point representing one of the observations, and where the distances between the points, $d_{ij}$ match, as well as possible, the original dissimilarities, $\delta_{ij}$. The distance measure used is usually Euclidean. There are various ways of finding the map of points, using either a metric approach (metric MDS) or a non-metric approach (non-metric MDS). Non-metric MDS uses a goodness of fit measure called STRESS, with a rough guide to the fit: STRESS $\leq 5\%$ – excellent; $5\% < \text{STRESS} \leq 10\%$ – good; $10\% < \text{STRESS} \leq 20\%$ – fair; STRESS $> 20\%$ – poor. See Figure 10.4 for an example.

## Cluster analysis

Cluster analysis attempts to put observations into groups called clusters. It can also be used to cluster the variables. There are various cluster analysis methods, but we will only consider one, *hierarchical cluster analysis*. The observations are sequentially put into clusters based on the dissimilarities, $\delta_{ij}$ between the pairs of observations, using a chosen method, such as the *nearest neighbour method, furthest neighbour method* or *Ward's method*. For instance, the nearest neighbour method forms the first cluster of two observations with the two observations that have the smallest $\delta_{ij}$. The next step either forms a new cluster of two observations using the next smallest $\delta_{ij}$, or places another observation, say the $k$th, into the first cluster to form a cluster of three, depending on how "close" the $k$th observation is the the first cluster, compared to the closeness of the two observations that would form the second cluster pair. This continues sequentially, with new pairs forming a cluster, or existing pairs of clusters joining to form a larger cluster.

Clustering of variables needs a dissimilarity measure for variables, for example, $\delta_{ij} = (1 - r_{ij})/2$, where $r_{ij}$ is the correlation of the $i$th and $j$th variables.

A *dendrogram* is a plot of the hierarchical clustering plotted against the dissimilarity values at which clusters are formed or joined together. A *heatmap* is a plot of the data values, observations by variables, using a colour scale, and simultaneously displaying a cluster analyses of the observations and a cluster analysis of the variables. See Figure 10.10 for an example.

# Bibliography

[1]   ARMSTRONG, BG: Comparing standardized mortality ratios. In: *Ann. Epidemiol.* 5 (1995), Nr. 1, 60–64. doi.org/10.1016/1047-2797(94)00032-o

[2]   BRESLOW, NE ; DAY, NE: *Statistical Methods in Cancer Research Volume II: The Design and Analysis of Cohort Studies.* IARC Scientific Publications No. 32, Lyon, France, 1987

[3]   CHONGSUVIVATWONG, V: Case-control study on oesophageal cancer in southern Thailand. In: *Journal of Gastroenterology and Hepatology* 3 (1990), Nr. 5, 391–394. doi.org/10.1111/j.1440-1746.1990.tb01415.x

[4]   CHOUDHURI, S: *Bioinformatics for beginners: Genes, genomes, molecular evolution, databases and analytical tools.* Academic Press, Amsterdam, 2014

[5]   CHOW, SC ; J, Shao ; H, Wang: *Sample size calculations in clinical research, 2nd Ed.* Chapman and Hall/CRC Press, Boca Raton, FL, 2008

[6]   COLLETT, D: *Modelling survival data in medical research, 3rd Ed.* CRC Press, Boca Raton, FL, 2015

[7]   DATTA, D ; NETTLETON, D E.: *Statistical analysis of next generation sequencing data.* Springer International Publishing, Switzerland, 2014

[8]   DOLL, R ; HILL, AB: Smoking and carcinoma of the lung. In: *Br. Med. J.* 2 (1950), Nr. 4682, 739–748. doi.org/10.1136/bmj.2.4682.739

[9]   DOLL, R ; HILL, AB: The mortality of doctors in relation to their smoking habits. In: *Br. Med. J.* 1 (1954), Nr. 4877, 1451–1455. doi.org/10.1136/bmj.1.4877.1451

[10]  DOLL, R ; PETO, R: Mortality in relation to smoking: 20 years' observations on male British doctors. In: *Br. Med. J.* 2 (1976), Nr. 6051, 1525–1536. doi.org/10.1136/bmj.2.6051.1525

[11]  FALCON, S ; MORGAN, M ; ....: *An Introduction to Bioconductor's ExpressionSet Class.* Online available at https://www.bioconductor.org/packages/devel/bioc/vignettes/Biobase/ inst/doc/ExpressionSet Introduction.pdf, 2007

[12] FDA – NIH Biomarker Working Group: *BEST (Biomarkers, End-pointS, and other Tools) Resource.* Silver Spring (MD): Food and Drug Administration (US); Bethesda (MD): National Institutes of Health (US), 2016 www.ncbi.nlm.nih.gov/books/NBK326791/

[13] FLOM, P ; CASSELL, D: *Stopping stepwise: Why stepwise variable selection methods are bad, and what you should use.* Talk given at NESUG, SGF and WUSS. available at https://www.lexjansen.com/pnwsug/2008/DavidCassell-StoppingStepwise.pdf, 2020

[14] FOX, R ; BERHANE, S ; ....: Biomarker-based prognosis in hepatocellular carcinoma: validation and extension of the BALAD model. In: *British Journal of Cancer* 110 (2014), 2090–2098. doi.org/10.1038/bjc.2014.130

[15] GAMBLE, C ; KRISHAN, A ; ....: Guidelines for the content of statistical analysis plans in clinical trials. In: *JAMA* 318 (2017), Nr. 23, 2337–2343. doi.10.1001/jama.2017.18556

[16] GAO, Y ; ZHOU, X: Analysis of clinical features and prognostic factors of lung cancer patients: A population-based cohort study. In: *Clin. Respir. J.* 14 (2020), Nr. 8, 712–724. doi.org/10.1111/crj.13188

[17] GRAZIOLIA, S ; PACIARONIB, M ; ....: Cancer-associated ischemic stroke: A retrospective multicentre cohort study. In: *Thrombosis Research* 165 (2018), 33–37. doi.org/10.1016/j.thromres.2018.03.011

[18] GREEN, S ; SMITH, A ; ....: *Clinical trials in oncology, 3rd Ed.* CRC Press, Boca Raton, FL, 2012

[19] GREENHALF, W ; GHANEH, P ; ....: Pancreatic cancer hENT1 expression and survival from gemcitabine in patients from the ESPAC-3 Trial. In: *J. Natl. Cancer Inst.* 106 (2014), Nr. 1. doi.10.1093/jnci/djt347

[20] HACKSHAW, A: *A concise guide to clinical trials.* Wiley, Oxford, 2009

[21] HAHNE, F ; W, Huber ; ....: *Bioconductor case studies.* Springer, Switzerland, 2008

[22] HANSEN, AR ; GRAHAM, DM ; ....: Phase 1 Trial Design: Is 3 + 3 the Best? In: *Cancer Control* 21 (2014), Nr. 3, 200–208. doi.10.1177/107327481402100304

[23] IRIZARRY, RA ; HOBBS, B ; ....: Exploration, normalization, and summaries of high density oligonucleotide array probe level data. In: *Biostatistics* 4 (2003), Nr. 2, 249–264. doi.org/10.1093/biostatistics/4.2.249

[24] JACKSON, CH: flexsurv: A platform for parametric survival modeling in R. In: *Journal of Statistical Software* 70 (2016), Nr. 8, 1–33. doi.org/10.18637/jss.v070.i08

[25] JADAD, AR ; MOORE, RA ; ....: Assessing the quality of reports of randomized clinical trials: is blinding necessary? In: *Control Clin. Trials* 17 (1996), Nr. 1, 1–12. doi.org/10.1016/0197-2456(95)00134-4

[26] JENNISON, C ; TURNBULL, BW: *Group sequential methods with applications to cliniucal trials.* Chapman and Hall/CRC Press, Boca Raton, FL, 2000

[27] JOHNSON, PJ ; PIRRIE, SJ ; ....: The detection of hepatocellular carcinoma using a prospectively developed and validated model based on serological biomarkers. In: *Cancer Epidemiol Biomarkers Prev.* 23 (2014), Nr. 1, 144–153. doi.org/10.1158/1055-9965.EPI-13-0870

[28] JONES, RP ; PSARELLI, EE ; ....: Patterns of recurrence after resection of pancreatic ductal adenocarcinoma: A secondary analysis of the ESPAC-4 randomized adjuvant chemotherapy trial. In: *JAMA Surg.* 154 (2019), Nr. 11, 1038–1048. doi.10.1001/jamasurg.2019.3337

[29] JUNG, SH: *Randomised phase II cancer clinical trials.* Chapman and Hall/CRC Press, Boca Raton, FL, 2013

[30] KCHOUK, M ; GIBRAT, JF ; ....: Generations of sequencing technologies: from first to next generation. In: *Biol. Med. (Aligarh)* 9 (2017), Nr. 3. doi.org/10.4172/0974-8369.1000395

[31] KIDANE, B ; COUGHLIN, S ; ....: Preoperative chemotherapy for resectable thoracic esophageal cancer. In: *Cochrane Database of Systematic Reviews* (2015), Nr. Issue 5, Art. No. CD001556. doi.org/10.1002/14651858.CD001556.pub3

[32] KLAUS, B ; REISENAUER, S: *An end to end workflow for differential gene expression using Affymetrix microarrays. Version 2.* Online available at https:///doi.org/10.12688/f1000research.8967.2, 2018

[33] KONTOPANTELIS, E: A comparison of one-stage vs two-stage individual patient data meta-analysis methods: A simulation study. In: *Res. Syn. Meth.* 9 (2018), Nr. 3, 417–430. doi.org/10.1002/jrsm.1303

[34] LARISCY, JY ; HUMMER, A ; ....: Lung cancer mortality among never-smokers in the United States: estimating smoking-attributable mortality with nationally representative data. In: *Annals of Epidemiology* 45 (2020), 5–11. doi.org/10.1016/j.annepidem.2020.03.008

[35] LAWSON, AB ; BANERJEE, S ; (EDS), ....: *Handbook of Spatial Epidemiology.* Chapman and Hall/CRC Press, Boca Raton, FL, 2020

[36] LESK, AM: *Introduction to bioinformatics, 5th Ed.* Oxford University Press, Oxford, 2019

[37] LIAO, WC ; CHIEN, KL ; ....: Adjuvant treatments for resected pancreatic adenocarcinoma: a systematic review and network meta-analysis. In: *Lancet Oncol.* 14 (2013), Nr. 11, 1095–1103. doi.org/10.1016/S1470-2045(13)70388-7

[38] MACHIN, D ; FAYERS, PM ; ....: *Randomised clinical trials: Design, practice and reporting, 2nd Ed.* Wiley, Oxford, 2021

[39] MACLEAN, D: *R bioinformatics cookbook: Use R and Bioconductor to perform RNAseq, genomics, data visualization, and bioinformatic analysis.* Packt Publishing, Birmingham, 2019

[40] MARTINELLI, P ; SANTA PAU, E Carrillo-de ; ....: GATA6 regulates EMT and tumour dissemination, and is a marker of response to adjuvant chemotherapy in pancreatic cancer. In: *Gut* 66 (2017), 1665–1676. doi.10.1136/gutjnl-2015-311256

[41] MOCELLIN, S ; GOODWIN, A ; ....: Risk-reducing medications for primary breast cancer: a network meta-analysis. In: *Cochrane Database of Systematic Reviews* (2019), Nr. Issue 4, Art. No. CD012191. doi.org/10.1002/14651858.CD012191.pub2

[42] MOORE, DF: *Applied survival analysis using R.* Springer International Publishing, Switzerland, 2016

[43] MURRAY, JS: Multiple imputation: A review of practical and theoretical findings. In: *Statistical Science* 33 (2018), Nr. 2, 142–159. doi.org/10.1214/18-STS644

[44] NEOPTOLEMOS, JP ; PALMER, DH ; ....: Comparison of adjuvant gemcitabine and capecitabine with gemcitabine monotherapy in patients with resected pancreatic cancer (ESPAC-4): a multicentre, open-label, randomised, phase 3 trial. In: *The Lancet* 389 (2017), Nr. 10073, 1011–1024. doi.org/10.1016/S0140-6736(16)32409-6

[45] O'KELLY, M ; RATITCH, B: *Clinical trials with missing data: A guide for practitioners.* Wiley, Chichester, 2014

[46] O'QUIGLEY, J: *Survival analysis: Proportional and non-proportional hazards regression.* Springer Nature Switzerland AG, 2021

[47] ORTUTAY, C ; ORTUTAY, Z: *Molecular data analysis Using R.* Wiley, Hoboken, NJ, 2017

[48] PARMAR, A ; CHAVES-PORRAS, J ; ....: Adjuvant treatments for resected pancreatic adenocarcinoma: a systematic review and network meta-analysis. In: *Crit. Rev. Oncol. Hematol.* 145 (2020), Nr. 102817. doi.org/10.1016/j.critrevonc.2019.102817

[49] Pos, Z ; Spivey, TL ; ....: Longitudinal study of recurrent metastatic melanoma cell lines underscores the individuality of cancer biology. In: *J. Invest. Dermatol.* 134 (2014), Nr. 5, 1389–1396. doi.org/10.1038/jid.2013.495

[50] Ready, NE ; Pang, HH ; ....: Chemotherapy with or without maintenance Sunitinib for untreated extensive-stage smal-cell lung cancer: A randomized, double-blind, placebo-controlled Phase II study – CALGB 30504 (Alliance). In: *J. Clin. Oncol.* 33 (2015), Nr. 15, 1160-1665. doi.org/10.1200/JCO.2014.57.3105

[51] Riley, RD ; Lambert, PC ; ....: Meta-analysis of individual participant data:rationale, conduct, and reporting. In: *BMJ* 340 (2010), Nr. c221. doi.org/10.1136/bmj.c221

[52] Royston, P ; Parmar, KBM: Flexible parametric proportional-hazards and proportional-odds models for censored survival data, with application to prognostic modelling and estimation of treatment effects. In: *Statistics in Medicine* 21 (2002), Nr. 15, 2175–2197. doi.org/10.1002/sim.1203

[53] Royston, P ; Parmar, MKB: Restricted mean survival time: an alternative to the hazard ratio for the design and analysis of randomized trials with a time-to-event outcome. In: *BMC Med. Res. Methodol.* 13 (2013), Nr. 152. doi.org/10.1186/1471-2288-13-152

[54] Salanti, S: Indirect and mixed-treatment comparison, network, or multipletreatments meta-analysis: many names, many benefits, many concerns for the next generation evidence synthesis tool. In: *Res. Syn. Meth.* 3 (2012), Nr. 2, 80–97. doi.org/10.1002/jrsm.1037

[55] Sauerbrei, W ; Taube, SE ; ....: Reporting recommendations for tumor marker prognostic studies (REMARK): An abridged explanation and elaboration. In: *J. Natl. Cancer Inst.* 110 (2018), Nr. 8, 803–811. doi.org/10.1093/jnci/djy088

[56] Schöffski, P ; Adkins, D ; ....: An open-label, phase 2 study evaluating the efficacy and safety of the anti-IGF-1R antibody cixutumumab in patients with previously treated advanced or metastatic soft-tissue sarcoma or Ewing family of tumours. In: *Eur. J. Cancer* 49 (2013), Nr. 15, 3219–28. doi.10.1016/j.ejca.2013.06.010

[57] Schr oder, MD ; Hugosson, J: Screening and prostate cancer mortality in a randomized European study. In: *N. Engl. J. Med.* 360 (2009), 1320–1328. doi.org/10.1056/NEJMoa0810084

[58] Schwarzer, G ; Carpenter, JR ; ....: *Meta-analysis with R (Use R!)*. Springer, Switzerland, 2015

[59] Silcocks, P: An easy approach to the Robins-Breslow-Greenland variance estimator. In: *Epidemiologic Perspectives & Innovations* 2 (2005), Nr. 9. doi.org/10.1186/1742-5573-2-9

[60] SMITH, G: Step away from stepwise. In: *J Big Data* 5 (2018), Nr. 32. doi.org/10.1186/s40537-018-0143-6

[61] STERNE, JA ; WHITE, IR ; ....: Multiple imputation for missing data in epidemiological and clinical research: potential and pitfalls. In: *BMJ* 338 (2009), Nr. b239. doi.org/10.1136/bmj.b2393

[62] STERNE, JAC ; SUTTON, AJ ; ....: Recommendations for examining and interpreting funnel plot asymmetry in meta-analyses of randomised controlled trials. In: *BMJ* 342 (2011), Nr. d4002. doi.org/10.1136/bmj.d4002

[63] SUTTON, AJ ; ABRAMS, KR: Bayesian methods in meta-analysis and evidence synthesis. In: *Stat. Methods Med. Res.* 10 (2001), Nr. 4, 277–303. doi.org/10.1177/096228020101000404

[64] TAJIMA, H ; MAKINO, I ; ....: A phase I study of preoperative (neoadjuvant) chemotherapy with gemcitabine plus nab-paclitaxel for resectable pancreatic cancer. In: *Mol. Clin. Oncol.* 14 (Epub 2021), Nr. 2:26. doi.10.3892/mco.2020.2188

[65] TAKAHASHI, K ; ISHIBASHI, E ; ....: A phase II basket trial of combination therapy with trastuzumab and pertuzumab in patients with solid cancers harbouring human epidermal growth factor receptor 2 amplification (JUPITER trial). In: *Medicine* 99 (2020), Nr. 32, p e21457. doi.org/10.1097/MD.0000000000021457

[66] THERNEAU, T ; CROWSON, C ; ....: *Using time dependent covariates and time dependent coefficients in the Cox model.* The Comprehensive R Archive Network: https://cran.r-project.org/web/packages/survival/vignettes/timedep.pdf (23/08/2021), 2020

[67] TOYODA, H ; KUMADA, T ; ....: Staging hepatocellular carcinoma by a novel scoring system (BALAD score) based on serum markers. In: *Clin. Gastroenterol. Hepatol.* 4 (2006), 1528—1536. doi.org/10.1016/j.cgh.2006.09.021

[68] TUNYNS, AJ ; PEQUIGNOT, G ; ....: Esophageal cancer in Ille-et-Vilaine in relation to levels of alcohol and tobacco consumption. Risks are multiplying. In: *Bull. Cancer* 64 (1977), Nr. 1, S. 45–60

[69] VAN BURREN, S: *Flexible imputation of missing data, 2nd Ed.* Chapman and Hall/CRC Press, Boca Raton, FL, 2018

[70] VAN BURREN, S ; GROOTHUIS-OUDSHOORN, K: mice: Multivariate imputation by chained equations in R. In: *Journal of Statistical Software* 45 (2011), Nr. 3, 1–67. doi.org/10.18637/jss.v045.i03

[71] VICKERS, AJ ; CRONIN, AM: A simple decision analytic solution to the comparison of two binary diagnostic tests. In: *Stat. Med.* 32 (2018), Nr. 11, 1865–1876. `doi.org/10.1002/sim.5601`

[72] VIECHTBAUER, W: Conducting meta-analyses in R with the metafor package. In: *Journal of Statistical Software* 36 (2010), Nr. 3, 1–48. `doi.org/10.18637/jss.v036.i03`

[73] WASSMER, G ; BRANNATH, W: *Group sequential and confirmatory adaptive designs in clinical trials.* Springer, Switzerland, 2016

[74] WHITEHEAD, A: *Meta-analysis of cintrolled clinical trials.* Wiley, Chichester, 2002

[75] WU, Z ; IRIZARRY, RA ; ....: A model-based background adjustment for oligonucleotide expression arrays. In: *J. American Statistical Association* 99 (2011), Nr. 468, 909–917. `doi.org/10.1198/016214504000000683`

[76] ZHOU, XH ; OBUCHOWSKI, NA ; ....: *Statistical methods in diagnostic medicine, 2nd Ed.* Wiley, Hoboken, New Jersey, 2011

[77] ZLOTTA, AR ; EGAWA, S ; ....: Prevalence of prostate cancer on autopsy: cross-sectional study on unscreened Caucasian and Asian men. In: *J. Natl. Cancer Inst.* 105 (2013), Nr. 14, 1050–1058. `doi.org/10.1093/jnci/djt151`

# *Index*

Printed in the United States
by Baker & Taylor Publisher Services